The Chemical Elements

" The Building Blocks of Molecules "

Edited by Paul F. Kisak

Contents

Chapter 1

Chemical element

A **chemical element** (or **element**) is a chemical substance consisting of atoms having the same number of protons in their atomic nuclei (i.e. the same atomic number, Z).[1] There are 118 elements that have been identified, of which the first 98 occur naturally on Earth with the remaining 20 being Synthetic elements. There are 80 elements that have at least one stable isotope and 38 that have exclusively radioactive isotopes, which decay over time into other elements. Iron is the most abundant element (by mass) making up the Earth, while oxygen is the most common element in the crust of the earth.[2]

Chemical elements constitute approximately 15% of the matter in the universe: the remainder is dark matter, the composition of it is unknown, but it is not composed of chemical elements.[3] The two lightest elements, hydrogen and helium were mostly formed in the Big Bang and are the most common elements in the universe. The next three elements (lithium, beryllium and boron) were formed mostly by cosmic ray spallation, and are thus more rare than those that follow. Formation of elements with from six to twenty six protons occurred and continues to occur in main sequence stars via stellar nucleosynthesis. The high abundance of oxygen, silicon, and iron on Earth reflects their common production in such stars. Elements with greater than twenty six protons are formed by supernova nucleosynthesis in supernovae, which, when they explode, blast these elements far into space as planetary nebulae, where they may become incorporated into planets when they are formed.[4]

When different elements are chemically combined, with the atoms held together by chemical bonds, they form chemical compounds. Only a minority of elements are found uncombined as relatively pure minerals. Among the more common of such "native elements" are copper, silver, gold, carbon (as coal, graphite, or diamonds), and sulfur. All but a few of the most inert elements, such as noble gases and noble metals, are usually found on Earth in chemically combined form, as chemical compounds. While about 32 of the chemical elements occur on Earth in native uncombined forms, most of these occur as mixtures. For example, atmospheric air is primarily a mixture of nitrogen, oxygen, and argon, and native solid elements occur in alloys, such as that of iron and nickel.

The history of the discovery and use of the elements began with primitive human societies that found native elements like carbon, sulfur, copper and gold. Later civilizations extracted elemental copper, tin, lead and iron from their ores by smelting, using charcoal. Alchemists and chemists subsequently identified many more, with almost all of the naturally-occurring elements becoming known by 1900.

The properties of the chemical elements are summarized on the periodic table, which organizes the elements by increasing atomic number into rows ("periods") in which the columns ("groups") share recurring ("periodic") physical and chemical properties. Save for unstable radioactive elements with short half-lives, all of the elements are available industrially, most of them in high degrees of purity.

1.1 Description

The lightest chemical elements are hydrogen and helium, both created by Big Bang nucleosynthesis during the first 20 minutes of the universe[5] in a ratio of around 3:1 by mass (or 12:1 by number of atoms).[6][7] Almost all other elements found in nature were made by various natural methods of nucleosynthesis.[8] On Earth, small amounts of new atoms are naturally produced in nucleogenic reactions, or in cosmogenic processes, such as cosmic ray spallation. New atoms are also naturally produced on Earth as radiogenic daughter isotopes of ongoing radioactive decay processes such as alpha decay, beta decay, spontaneous fission, cluster decay, and other rarer modes of decay.

Of the 98 naturally occurring elements, those with atomic numbers 1 through 82 each have at least one stable isotope, (except for technetium, element 43 and promethium, element 61, which have no stable isotopes). Isotopes considered stable are those for which no radioactive decay has yet been observed. Elements with atomic numbers 83 through

1

98 are unstable to the point that radioactive decay of all isotopes can be detected. Some of these elements, notably bismuth (atomic number 83), thorium (atomic number 90), uranium (atomic number 92) and plutonium (atomic number 94), have one or more isotopes with half-lives long enough to survive as remnants of the explosive stellar nucleosynthesis that produced the heavy elements before the formation of our solar system. For example, at over 1.9×10^{19} years, over a billion times longer than the current estimated age of the universe, bismuth-209 (atomic number 83) has the longest known alpha decay half-life of any naturally occurring element.[9][10] The very heaviest elements (those beyond californium, atomic number 98) undergo radioactive decay with half-lives so short that they do not occur in nature and must be synthesized.

As of 2010, there are 118 known elements (in this context, "known" means observed well enough, even from just a few decay products, to have been differentiated from other elements).[11][12] Of these 118 elements, 98 occur naturally on Earth.[13] Ten of these occur in extreme trace quantities: technetium, atomic number 43; promethium, number 61; astatine, number 85; francium, number 87; neptunium, number 93; plutonium, number 94; americium, number 95; curium, number 96; berkelium, number 97; and californium, number 98. These 98 elements have been detected in the universe at large, in the spectra of stars and also supernovae, where short-lived radioactive elements are newly being made. The first 98 elements have been detected directly on Earth as primordial nuclides present from the formation of the solar system, or as naturally-occurring fission or transmutation products of uranium and thorium.

The remaining 20 heavier elements, not found today either on Earth or in astronomical spectra, have been produced artificially: these are all radioactive, with very short half-lives; if any atoms of these elements were present at the formation of Earth, they are extremely likely, to the point of certainty, to have already decayed, and if present in novae, have been in quantities too small to have been noted. Technetium was the first purportedly non-naturally occurring element synthesized, in 1937, although trace amounts of technetium have since been found in nature (and also the element may have been discovered naturally in 1925).[14] This pattern of artificial production and later natural discovery has been repeated with several other radioactive naturally-occurring rare elements.[15]

Lists of the elements are available by name, by symbol, by atomic number, by density, by melting point, and by boiling point as well as ionization energies of the elements. The nuclides of stable and radioactive elements are also available as a list of nuclides, sorted by length of half-life for those that are unstable. One of the most convenient, and certainly the most traditional presentation of the elements, is in the form of the periodic table, which groups together elements with similar chemical properties (and usually also similar electronic structures).

1.1.1 Atomic number

Main article: atomic number

The atomic number of an element is equal to the number of protons in each atom, and defines the element.[16] For example, all carbon atoms contain 6 protons in their atomic nucleus; so the atomic number of carbon is 6.[17] Carbon atoms may have different numbers of neutrons; atoms of the same element having different numbers of neutrons are known as isotopes of the element.[18]

The number of protons in the atomic nucleus also determines its electric charge, which in turn determines the number of electrons of the atom in its non-ionized state. The electrons are placed into atomic orbitals that determine the atom's various chemical properties. The number of neutrons in a nucleus usually has very little effect on an element's chemical properties (except in the case of hydrogen and deuterium). Thus, all carbon isotopes have nearly identical chemical properties because they all have six protons and six electrons, even though carbon atoms may, for example, have 6 or 8 neutrons. That is why the atomic number, rather than mass number or atomic weight, is considered the identifying characteristic of a chemical element.

The symbol for atomic number is Z.

1.1.2 Isotopes

Main articles: Isotope, Stable isotope and List of nuclides

Isotopes are atoms of the same element (that is, with the same number of protons in their atomic nucleus), but having *different* numbers of neutrons. Most (66 of 94) naturally occurring elements have more than one stable isotope. Thus, for example, there are three main isotopes of carbon. All carbon atoms have 6 protons in the nucleus, but they can have either 6, 7, or 8 neutrons. Since the mass numbers of these are 12, 13 and 14 respectively, the three isotopes of carbon are known as carbon-12, carbon-13, and carbon-14, often abbreviated to ^{12}C, ^{13}C, and ^{14}C. Carbon in everyday life and in chemistry is a mixture of ^{12}C (about 98.9%), ^{13}C (about 1.1%) and about 1 atom per trillion of ^{14}C.

Except in the case of the isotopes of hydrogen (which differ greatly from each other in relative mass—enough to cause chemical effects), the isotopes of a given element are chemically nearly indistinguishable.

All of the elements have some isotopes that are radioactive

(radioisotopes), although not all of these radioisotopes occur naturally. The radioisotopes typically decay into other elements upon radiating an alpha or beta particle. If an element has isotopes that are not radioactive, these are termed "stable" isotopes. All of the known stable isotopes occur naturally (see primordial isotope). The many radioisotopes that are not found in nature have been characterized after being artificially made. Certain elements have no stable isotopes and are composed *only* of radioactive isotopes: specifically the elements without any stable isotopes are technetium (atomic number 43), promethium (atomic number 61), and all observed elements with atomic numbers greater than 82.

Of the 80 elements with at least one stable isotope, 26 have only one single stable isotope. The mean number of stable isotopes for the 80 stable elements is 3.1 stable isotopes per element. The largest number of stable isotopes that occur for a single element is 10 (for tin, element 50).

1.1.3 Isotopic mass and atomic mass

Main articles: atomic mass and relative atomic mass

The mass number of an element, A, is the number of nucleons (protons and neutrons) in the atomic nucleus. Different isotopes of a given element are distinguished by their mass numbers, which are conventionally written as a superscript on the left hand side of the atomic symbol (e.g., ^{238}U). The mass number is always a simple whole number and has units of "nucleons." An example of a referral to a mass number is "magnesium-24," which is an atom with 24 nucleons (12 protons and 12 neutrons).

Whereas the mass number simply counts the total number of neutrons and protons and is thus a natural (or whole) number, the atomic mass of a single atom is a real number for the mass of a particular isotope of the element, the unit being **u**. In general, when expressed in **u** it differs in value slightly from the mass number for a given nuclide (or isotope) since the mass of the protons and neutrons is not exactly 1 **u**, since the electrons contribute a lesser share to the atomic mass as neutron number exceeds proton number, and (finally) because of the nuclear binding energy. For example, the atomic mass of chlorine-35 to five significant digits is 34.969 **u** and that of chlorine-37 is 36.966 **u**. However, the atomic mass in **u** of each isotope is quite close to its simple mass number (always within 1%). The only isotope whose atomic mass is exactly a natural number is ^{12}C, which by definition has a mass of exactly 12, because **u** is defined as 1/12 of the mass of a free neutral carbon-12 atom in the ground state.

The relative atomic mass (historically and commonly also called "atomic weight") of an element is the *average* of the atomic masses of all the chemical element's isotopes as found in a particular environment, weighted by isotopic abundance, relative to the atomic mass unit (**u**). This number may be a fraction that is *not* close to a whole number, due to the averaging process. For example, the relative atomic mass of chlorine is 35.453 **u**, which differs greatly from a whole number due to being made of an average of 76% chlorine-35 and 24% chlorine-37. Whenever a relative atomic mass value differs by more than 1% from a whole number, it is due to this averaging effect resulting from significant amounts of more than one isotope being naturally present in the sample of the element in question.

1.1.4 Chemically pure and isotopically pure

Chemists and nuclear scientists have different definitions of a *pure element*. In chemistry, a pure element means a substance whose atoms all (or in practice almost all) have the same atomic number, or number of protons. Nuclear scientists, however, define a pure element as one that consists of only one stable isotope.[19]

For example, a copper wire is 99.99% chemically pure if 99.99% of its atoms are copper, with 29 protons each. However it is not isotopically pure since ordinary copper consists of two stable isotopes, 69% ^{63}Cu and 31% ^{65}Cu, with different numbers of neutrons.

1.1.5 Allotropes

Main article: Allotropy

Atoms of chemically pure elements may bond to each other chemically in more than one way, allowing the pure element to exist in multiple structures (spatial arrangements of atoms), known as allotropes, which differ in their properties. For example, carbon can be found as diamond, which has a tetrahedral structure around each carbon atom; graphite, which has layers of carbon atoms with a hexagonal structure stacked on top of each other; graphene, which is a single layer of graphite that is very strong; fullerenes, which have nearly spherical shapes; and carbon nanotubes, which are tubes with a hexagonal structure (even these may differ from each other in electrical properties). The ability of an element to exist in one of many structural forms is known as 'allotropy'.

The standard state, also known as reference state, of an element is defined as its thermodynamically most stable state at 1 bar at a given temperature (typically at 298.15 K). In thermochemistry, an element is defined to have an enthalpy of formation of zero in its standard state. For example, the reference state for carbon is graphite, because the structure of graphite is more stable than that of the other allotropes.

1.1.6 Properties

Several kinds of descriptive categorizations can be applied broadly to the elements, including consideration of their general physical and chemical properties, their states of matter under familiar conditions, their melting and boiling points, their densities, their crystal structures as solids, and their origins.

General properties

Several terms are commonly used to characterize the general physical and chemical properties of the chemical elements. A first distinction is between metals, which readily conduct electricity, nonmetals, which do not, and a small group, (the *metalloids*), having intermediate properties and often behaving as semiconductors.

A more refined classification is often shown in colored presentations of the periodic table. This system restricts the terms "metal" and "nonmetal" to only certain of the more broadly defined metals and nonmetals, adding additional terms for certain sets of the more broadly viewed metals and nonmetals. The version of this classification used in the periodic tables presented here includes: actinides, alkali metals, alkaline earth metals, halogens, lanthanides, transition metals, post-transition metals; metalloids, noble gases, polyatomic nonmetals, diatomic nonmetals, and transition metals. In this system, the alkali metals, alkaline earth metals, and transition metals, as well as the lanthanides and the actinides, are special groups of the metals viewed in a broader sense. Similarly, the polyatomic nonmetals, diatomic nonmetals and the noble gases are nonmetals viewed in the broader sense. In some presentations, the halogens are not distinguished, with astatine identified as a metalloid and the others identified as nonmetals.

States of matter

Another commonly used basic distinction among the elements is their state of matter (phase), whether solid, liquid, or gas, at a selected standard temperature and pressure (STP). Most of the elements are solids at conventional temperatures and atmospheric pressure, while several are gases. Only bromine and mercury are liquids at 0 degrees Celsius (32 degrees Fahrenheit) and normal atmospheric pressure; caesium and gallium are solids at that temperature, but melt at $28.4 \,^\circ\text{C}$ ($83.2 \,^\circ\text{F}$) and $29.8 \,^\circ\text{C}$ ($85.6 \,^\circ\text{F}$), respectively.

Melting and boiling points

Melting and boiling points, typically expressed in degrees Celsius at a pressure of one atmosphere, are commonly used in characterizing the various elements. While known for most elements, either or both of these measurements is still undetermined for some of the radioactive elements available in only tiny quantities. Since helium remains a liquid even at absolute zero at atmospheric pressure, it has only a boiling point, and not a melting point, in conventional presentations.

Densities

Main article: Densities of the elements (data page)

The density at a selected standard temperature and pressure (STP) is frequently used in characterizing the elements. Density is often expressed in grams per cubic centimeter (g/cm^3). Since several elements are gases at commonly encountered temperatures, their densities are usually stated for their gaseous forms; when liquefied or solidified, the gaseous elements have densities similar to those of the other elements.

When an element has allotropes with different densities, one representative allotrope is typically selected in summary presentations, while densities for each allotrope can be stated where more detail is provided. For example, the three familiar allotropes of carbon (amorphous carbon, graphite, and diamond) have densities of 1.8–2.1, 2.267, and 3.515 g/cm^3, respectively.

Crystal structures

The elements studied to date as solid samples have eight kinds of crystal structures: cubic, body-centered cubic, face-centered cubic, hexagonal, monoclinic, orthorhombic, rhombohedral, and tetragonal. For some of the synthetically produced transuranic elements, available samples have been too small to determine crystal structures.

Occurrence and origin on Earth

Chemical elements may also be categorized by their origin on Earth, with the first 98 considered naturally occurring, while those with atomic numbers beyond 98 have only been produced artificially as the synthetic products of man-made nuclear reactions.

Of the 98 naturally occurring elements, 84 are considered primordial and either stable or weakly radioactive. The remaining 14 naturally occurring elements possess half lives too short for them to have been present at the beginning of the Solar System, and are therefore considered transient elements. Of these 14 transient elements, 7 (polonium, astatine, radon, francium, radium, actinium, and protactinium) are relatively common decay products of thorium, uranium, and plutonium. The remaining 7 transient elements (technetium, promethium, neptunium, americium, curium, berkelium, and californium) occur only rarely, as products of rare nuclear reaction processes involving uranium or other heavy elements.

Elements with atomic numbers 1 through 40 are all stable, while those with atomic numbers 41 through 82 (except technetium and promethium) are metastable. The half-lives of these metastable "theoretical radionuclides" are so long (at least 100 million times longer than the estimated age of the universe) that their radioactive decay has yet to be detected by experiment. Elements with atomic numbers 83 through 98 are unstable to the point that their radioactive decay can be detected. Some of these elements, notably thorium (atomic number 90) and uranium (atomic number 92), have one or more isotopes with half-lives long enough to survive as remnants of the explosive stellar nucleosynthesis that produced the heavy elements before the formation of our solar system. For example, at over 1.9×10^{19} years, over a billion times longer than the current estimated age of the universe, bismuth-209 (atomic number 83) has the longest known alpha decay half-life of any naturally occurring element.[9][10] The very heaviest elements (those beyond californium, atomic number 98) undergo radioactive decay with short half-lives and do not occur in nature.

1.1.7 The periodic table

Main article: Periodic table

The properties of the chemical elements are often summarized using the periodic table, which powerfully and elegantly organizes the elements by increasing atomic number into rows ("periods") in which the columns ("groups") share recurring ("periodic") physical and chemical properties. The current standard table contains 118 confirmed elements as of 10 April 2010.

Although earlier precursors to this presentation exist, its invention is generally credited to the Russian chemist Dmitri Mendeleev in 1869, who intended the table to illustrate recurring trends in the properties of the elements. The layout of the table has been refined and extended over time as new elements have been discovered and new theoretical models have been developed to explain chemical behavior.

Use of the periodic table is now ubiquitous within the academic discipline of chemistry, providing an extremely useful framework to classify, systematize and compare all the many different forms of chemical behavior. The table has also found wide application in physics, geology, biology, materials science, engineering, agriculture, medicine, nutrition, environmental health, and astronomy. Its principles are especially important in chemical engineering.

1.2 Nomenclature and symbols

The various chemical elements are formally identified by their unique atomic numbers, by their accepted names, and by their symbols.

1.2.1 Atomic numbers

The known elements have atomic numbers from 1 through 118, conventionally presented as Arabic numerals. Since the elements can be uniquely sequenced by atomic number, conventionally from lowest to highest (as in a periodic table), sets of elements are sometimes specified by such notation as "through", "beyond", or "from ... through", as in "through iron", "beyond uranium", or "from lanthanum through lutetium". The terms "light" and "heavy" are sometimes also used informally to indicate relative atomic numbers (not densities!), as in "lighter than carbon" or "heavier than lead", although technically the weight or mass of atoms of an element (their atomic weights or atomic masses) do not always increase monotonically with their atomic numbers.

1.2.2 Element names

Main article: List of chemical element name etymologies

The naming of various substances now known as elements precedes the atomic theory of matter, as names were given locally by various cultures to various minerals, metals, compounds, alloys, mixtures, and other materials, although at the time it was not known which chemicals were elements and which compounds. As they were identified as elements, the existing names for anciently-known elements (e.g., gold, mercury, iron) were kept in most countries. National differences emerged over the names of elements either for convenience, linguistic niceties, or nationalism. For a few illustrative examples: German speakers use "Wasserstoff" (water substance) for "hydrogen", "Sauerstoff" (acid substance) for "oxygen" and "Stickstoff" (smothering substance) for "nitrogen", while English and some romance languages use "sodium" for "natrium" and "potassium" for "kalium", and the French, Italians, Greeks, Portuguese and Poles prefer "azote/azot/azoto" (from roots meaning "no life") for "nitrogen".

For purposes of international communication and trade, the official names of the chemical elements both ancient and more recently recognized are decided by the International Union of Pure and Applied Chemistry (IUPAC), which has decided on a sort of international English language, drawing on traditional English names even when an element's chemical symbol is based on a Latin or other traditional word, for example adopting "gold" rather than "aurum" as the name for the 79th element (Au). IUPAC prefers the British spellings "aluminium" and "caesium" over the U.S. spellings "aluminum" and "cesium", and the U.S. "sulfur" over the British "sulphur". However, elements that are practical to sell in bulk in many countries often still have locally used national names, and countries whose national language does not use the Latin alphabet are likely to use the IUPAC element names.

According to IUPAC, chemical elements are not proper nouns in English; consequently, the full name of an element is not routinely capitalized in English, even if derived from a proper noun, as in californium and einsteinium. Isotope names of chemical elements are also uncapitalized if written out, *e.g.,* carbon-12 or uranium-235. Chemical element *symbols* (such as Cf for californium and Es for einsteinium), are always capitalized (see below).

In the second half of the twentieth century, physics laboratories became able to produce nuclei of chemical elements with half-lives too short for an appreciable amount of them to exist at any time. These are also named by IUPAC, which generally adopts the name chosen by the discoverer. This practice can lead to the controversial question of which research group actually discovered an element, a question that has delayed naming of elements with atomic number of 104 and higher for a considerable time. (See element naming controversy).

Precursors of such controversies involved the nationalistic namings of elements in the late 19th century. For example, *lutetium* was named in reference to Paris, France. The Germans were reluctant to relinquish naming rights to the French, often calling it *cassiopeium*. Similarly, the British discoverer of *niobium* originally named it *columbium*, in reference to the New World. It was used extensively as such by American publications prior to international standardization.

1.2.3   Chemical symbols

For listings of current chemical symbols, symbols not currently used, and other symbols that may look like chemical symbols, see Symbol (chemistry).

Specific chemical elements

Before chemistry became a science, alchemists had designed arcane symbols for both metals and common compounds. These were however used as abbreviations in diagrams or procedures; there was no concept of atoms combining to form molecules. With his advances in the atomic theory of matter, John Dalton devised his own simpler symbols, based on circles, to depict molecules.

The current system of chemical notation was invented by Berzelius. In this typographical system, chemical symbols are not mere abbreviations—though each consists of letters of the Latin alphabet. They are intended as universal symbols for people of all languages and alphabets.

The first of these symbols were intended to be fully universal. Since Latin was the common language of science at that time, they were abbreviations based on the Latin names of metals. Cu comes from Cuprum, Fe comes from Ferrum, Ag from Argentum. The symbols were not followed by a period (full stop) as with abbreviations. Later chemical elements were also assigned unique chemical symbols, based on the name of the element, but not necessarily in English. For example, sodium has the chemical symbol 'Na' after the Latin *natrium*. The same applies to "W" (wolfram) for tungsten, "Fe" (ferrum) for iron, "Hg" (hydrargyrum) for mercury, "Sn" (stannum) for tin, "K" (kalium) for potassium, "Au" (aurum) for gold, "Ag" (argentum) for silver, "Pb" (plumbum) for lead, "Cu" (cuprum) for copper, and "Sb" (stibium) for antimony.

Chemical symbols are understood internationally when element names might require translation. There have sometimes been differences in the past. For example, Germans in the past have used "J" (for the alternate name Jod) for iodine, but now use "I" and "Iod."

The first letter of a chemical symbol is always capitalized, as in the preceding examples, and the subsequent letters, if any, are always lower case (small letters). Thus, the symbols for californium or einsteinium are Cf and Es.

General chemical symbols

There are also symbols in chemical equations for groups of chemical elements, for example in comparative formulas. These are often a single capital letter, and the letters are reserved and not used for names of specific elements. For example, an "X" indicates a variable group (usually a halogen) in a class of compounds, while "R" is a radical, meaning a compound structure such as a hydrocarbon chain. The letter "Q" is reserved for "heat" in a chemical reaction. "Y" is also often used as a general chemical symbol, although it is also the symbol of yttrium. "**Z**" is also frequently used as a general variable group. "**E**" is used in organic chemistry to denote an electron-withdrawing group. "**L**" is used to represent a general ligand in inorganic and organometallic chemistry. "**M**" is also often used in place of a general metal.

At least two additional, two-letter generic chemical symbols are also in informal usage, "**Ln**" for any lanthanide element and "**An**" for any actinide element. "**Rg**" was formerly used for any rare gas element, but the group of rare gases has now been renamed noble gases and the symbol "**Rg**" has now been assigned to the element roentgenium.

Isotope symbols

Isotopes are distinguished by the atomic mass number (total protons and neutrons) for a particular isotope of an element, with this number combined with the pertinent element's symbol. IUPAC prefers that isotope symbols be written in superscript notation when practical, for example ^{12}C and ^{235}U. However, other notations, such as carbon-12 and uranium-235, or C-12 and U-235, are also used.

As a special case, the three naturally occurring isotopes of the element hydrogen are often specified as **H** for ^{1}H (protium), **D** for ^{2}H (deuterium), and **T** for ^{3}H (tritium). This convention is easier to use in chemical equations, replacing the need to write out the mass number for each atom. For example, the formula for heavy water may be written D_2O instead of ^2H_2O.

1.3 Origin of the elements

Only about 4% of the total mass of the universe is made of atoms or ions, and thus represented by chemical elements. This fraction is about 15% of the total matter, with the remainder of the matter (85%) being dark matter. The nature of dark matter is unknown, but it is not composed of atoms of chemical elements because it contains no protons, neutrons, or electrons. (The remaining non-matter part of the mass of the universe is composed of the even more mysterious dark energy).

The universe's 98 naturally occurring chemical elements are thought to have been produced by at least four cosmic processes. Most of the hydrogen and helium in the universe was produced primordially in the first few minutes of the Big Bang. Three recurrently occurring later processes are thought to have produced the remaining elements. Stellar nucleosynthesis, an ongoing process, produces all elements from carbon through iron in atomic number, but little lithium, beryllium, or boron. Elements heavier in atomic

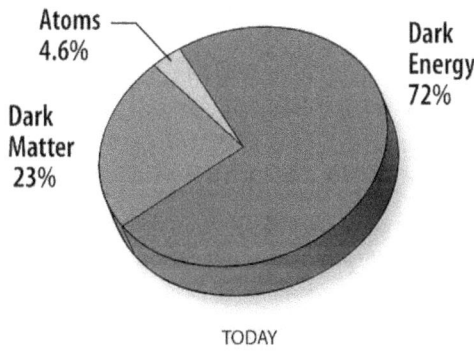

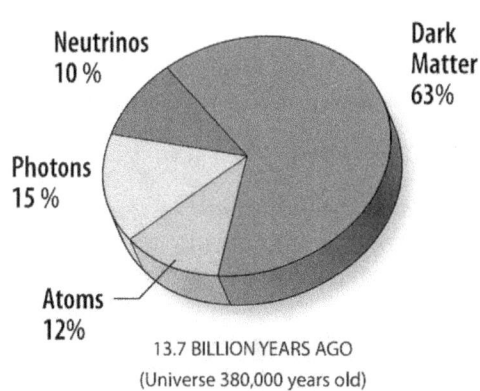

Estimated distribution of dark matter and dark energy in the universe. Only the fraction of the mass and energy in the universe labeled "atoms" is composed of chemical elements.

number than iron, as heavy as uranium and plutonium, are produced by explosive nucleosynthesis in supernovas and other cataclysmic cosmic events. Cosmic ray spallation (fragmentation) of carbon, nitrogen, and oxygen is important to the production of lithium, beryllium and boron.

During the early phases of the Big Bang, nucleosynthesis of hydrogen nuclei resulted in the production of hydrogen-1 (protium, ^{1}H) and helium-4 (^{4}He), as well as a smaller amount of deuterium (^{2}H) and very minuscule amounts (on the order of 10^{-10}) of lithium and beryllium. Even smaller amounts of boron may have been produced in the Big Bang, since it has been observed in some very old stars, while carbon has not.[20] It is generally agreed that no heavier elements than boron were produced in the Big Bang. As a result, the primordial abundance of atoms (or ions) consisted of roughly 75% ^{1}H, 25% ^{4}He, and 0.01% deuterium, with only tiny traces of lithium, beryllium, and perhaps boron.[21] Subsequent enrichment of galactic halos occurred due to stellar nucleosynthesis and supernova nucleosynthesis.[22] However, the element abundance in intergalactic space can still closely resemble primordial conditions, unless it has been enriched by some means.

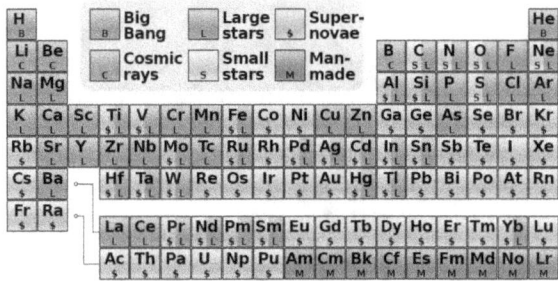

Periodic table showing the cosmogenic origin of each element in the Big Bang, or in large or small stars. Small stars can produce certain elements up to sulfur, by the alpha process. Supernovae are needed to produce "heavy" elements (those beyond iron and nickel) rapidly by neutron buildup, in the r-process. Certain large stars slowly produce other elements heavier than iron, in the s-process; these may then be blown into space in the off-gassing of planetary nebulae

On Earth (and elsewhere), trace amounts of various elements continue to be produced from other elements as products of natural transmutation processes. These include some produced by cosmic rays or other nuclear reactions (see cosmogenic and nucleogenic nuclides), and others produced as decay products of long-lived primordial nuclides.[23] For example, trace (but detectable) amounts of carbon-14 (^{14}C) are continually produced in the atmosphere by cosmic rays impacting nitrogen atoms, and argon-40 (^{40}Ar) is continually produced by the decay of primordially occurring but unstable potassium-40 (^{40}K). Also, three primordially occurring but radioactive actinides, thorium, uranium, and plutonium, decay through a series of recurrently produced but unstable radioactive elements such as radium and radon, which are transiently present in any sample of these metals or their ores or compounds. Seven other radioactive elements, technetium, promethium, neptunium, americium, curium, berkelium, and californium, occur only incidentally in natural materials, produced as individual atoms by natural fission of the nuclei of various heavy elements or in other rare nuclear processes.

Human technology has produced various additional elements beyond these first 98, with those through atomic number 118 now known.

1.4 Abundance

Main article: Abundance of the chemical elements

The following graph (note log scale) shows the abundance of elements in our solar system. The table shows the twelve most common elements in our galaxy (estimated spectro-scopically), as measured in parts per million, by mass.[24] Nearby galaxies that have evolved along similar lines have a corresponding enrichment of elements heavier than hydrogen and helium. The more distant galaxies are being viewed as they appeared in the past, so their abundances of elements appear closer to the primordial mixture. As physical laws and processes appear common throughout the visible universe, however, scientist expect that these galaxies evolved elements in similar abundance.

The abundance of elements in the Solar System is in keeping with their origin from nucleosynthesis in the Big Bang and a number of progenitor supernova stars. Very abundant hydrogen and helium are products of the Big Bang, but the next three elements are rare since they had little time to form in the Big Bang and are not made in stars (they are, however, produced in small quantities by the breakup of heavier elements in interstellar dust, as a result of impact by cosmic rays). Beginning with carbon, elements are produced in stars by buildup from alpha particles (helium nuclei), resulting in an alternatingly larger abundance of elements with even atomic numbers (these are also more stable). In general, such elements up to iron are made in large stars in the process of becoming supernovas. Iron-56 is particularly common, since it is the most stable element that can easily be made from alpha particles (being a product of decay of radioactive nickel-56, ultimately made from 14 helium nuclei). Elements heavier than iron are made in energy-absorbing processes in large stars, and their abundance in the universe (and on Earth) generally decreases with their atomic number.

The abundance of the chemical elements on **Earth** varies from air to crust to ocean, and in various types of life. The abundance of elements in Earth's crust differs from that in the Solar system (as seen in the Sun and heavy planets like Jupiter) mainly in selective loss of the very lightest elements (hydrogen and helium) and also volatile neon, carbon (as hydrocarbons), nitrogen and sulfur, as a result of solar heating in the early formation of the solar system. Oxygen, the most abundant Earth element by mass, is retained on Earth by combination with silicon. Aluminum at 8% by mass is more common in the Earth's crust than in the universe and solar system, but the composition of the far more bulky mantle, which has magnesium and iron in place of aluminum (which occurs there only at 2% of mass) more closely mirrors the elemental composition of the solar system, save for the noted loss of volatile elements to space, and loss of iron which has migrated to the Earth's core.

The composition of the human body, by contrast, more closely follows the composition of seawater—save that the human body has additional stores of carbon and nitrogen necessary to form the proteins and nucleic acids, together with phosphorus in the nucleic acids and energy transfer molecule adenosine triphosphate (ATP) that occurs in the cells of all living organisms. Certain kinds of organisms require particular additional elements, for example the magnesium in chlorophyll in green plants, the calcium in mollusc shells, or the iron in the hemoglobin in vertebrate animals' red blood cells.

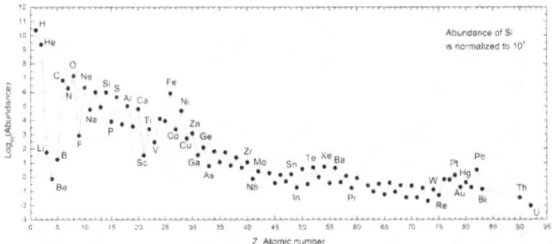

Abundances of the chemical elements in the Solar system. Hydrogen and helium are most common, from the Big Bang. The next three elements (Li, Be, B) are rare because they are poorly synthesized in the Big Bang and also in stars. The two general trends in the remaining stellar-produced elements are: (1) an alternation of abundance in elements as they have even or odd atomic numbers (the Oddo-Harkins rule), and (2) a general decrease in abundance as elements become heavier. Iron is especially common because it represents the minimum energy nuclide that can be made by fusion of helium in supernovae.

1.5 History

ОПЫТЪ СИСТЕМЫ ЭЛЕМЕНТОВЪ,
ОСНОВАННОЙ НА ИХЪ АТОМНОМЪ ВѢСЪ И ХИМИЧЕСКОМЪ СХОДСТВѢ.

		Ti=50	Zr=90	?=180.
		V=51	Nb=94	Ta=182.
		Cr=52	Mo=96	W=186.
		Mn=55	Rh=104,4	Pt=197,1.
		Fe=56	Ru=104,4	Ir=198.
		Ni=Co=59	Pd=106,6	Os=199.
H=1		Cu=63,4	Ag=108	Hg=200.
	Be= 9,4 Mg=24	Zn=65,2	Cd=112	
	B=11 Al=27,3	?=68	Ur=116	Au=197?
	C=12 Si=28	?=70	Sn=118	
	N=14 P=31	As=75	Sb=122	Bi=210?
	O=16 S=32	Se=79,4	Te=128?	
	F=19 Cl=35,5	Br=80	I=127	
Li=7 Na=23	K=39	Rb=85,4	Cs=133	Tl=204.
	Ca=40	Sr=87,6	Ba=137	Pb=207.
	?=45	Ce=92		
	?Er=56	La=94		
	?Yt=60	Di=95		
	?In=75,6	Th=118?		

Д. Менделѣевъ

Mendeleev's 1869 periodic table: An experiment on a system of elements. Based on their atomic weights and chemical similarities.

1.5.1 Evolving definitions

The concept of an "element" as an undivisible substance has developed through three major historical phases: Classical definitions (such as those of the ancient Greeks), chemical definitions, and atomic definitions.

Classical definitions

Ancient philosophy posited a set of classical elements to explain observed patterns in nature. These *elements* originally referred to *earth*, *water*, *air* and *fire* rather than the chemical elements of modern science.

The term 'elements' (*stoicheia*) was first used by the Greek philosopher Plato in about 360 BCE in his dialogue Timaeus, which includes a discussion of the composition of inorganic and organic bodies and is a speculative treatise on chemistry. Plato believed the elements introduced a century earlier by Empedocles were composed of small polyhedral forms: tetrahedron (fire), octahedron (air), icosahedron (water), and cube (earth).[25][26]

Aristotle, c. 350 BCE, also used the term *stoicheia* and added a fifth element called aether, which formed the heavens. Aristotle defined an element as:

> Element – one of those bodies into which other bodies can decompose, and that itself is not capable of being divided into other.[27]

Chemical definitions

In 1661, Robert Boyle proposed his theory of corpuscularism which favoured the analysis of matter as constituted by irreducible units of matter (atoms) and, choosing to side with neither Aristotle's view of the four elements nor Paracelsus' view of three fundamental elements, left open the question of the number of elements.[28] The first modern list of chemical elements was given in Antoine Lavoisier's 1789 *Elements of Chemistry*, which contained thirty-three elements, including light and caloric.[29] By 1818, Jöns Jakob Berzelius had determined atomic weights for forty-five of the forty-nine then-accepted elements. Dmitri Mendeleev had sixty-six elements in his periodic table of 1869.

From Boyle until the early 20th century, an element was defined as a pure substance that could not be decomposed into any simpler substance.[28] Put another way, a chemical element cannot be transformed into other chemical elements by chemical processes. Elements during this time were generally distinguished by their atomic weights, a property measurable with fair accuracy by available analytical techniques.

Dmitri Mendeleev

Henry Moseley

Atomic definitions

The 1913 discovery by English physicist Henry Moseley that the nuclear charge is the physical basis for an atom's atomic number, further refined when the nature of protons and neutrons became appreciated, eventually led to the current definition of an element based on atomic number (number of protons per atomic nucleus). The use of atomic numbers, rather than atomic weights, to distinguish elements has greater predictive value (since these numbers are integers), and also resolves some ambiguities in the chemistry-based view due to varying properties of isotopes and allotropes within the same element. Currently, IUPAC defines an element to exist if it has isotopes with a lifetime longer than the 10^{-14} seconds it takes the nucleus to form an electronic cloud.[30]

By 1914, seventy-two elements were known, all naturally occurring.[31] The remaining naturally occurring elements were discovered or isolated in subsequent decades, and various additional elements have also been produced synthetically, with much of that work pioneered by Glenn T. Seaborg. In 1955, element 101 was discovered and named mendelevium in honor of D.I. Mendeleev, the first to arrange the elements in a periodic manner. Most recently, the

synthesis of element 118 was reported in October 2006, and the synthesis of element 117 was reported in April 2010.[32]

1.5.2   Discovery and recognition of various elements

See also: Timeline of chemical elements discoveries

Ten materials familiar to various prehistoric cultures are now known to be chemical elements: Carbon, copper, gold, iron, lead, mercury, silver, sulfur, tin, and zinc. Three additional materials now accepted as elements, arsenic, antimony, and bismuth, were recognized as distinct substances prior to 1500 AD. Phosphorus, cobalt, and platinum were isolated before 1750.

Most of the remaining naturally occurring chemical elements were identified and characterized by 1900, including:

- Such now-familiar industrial materials as aluminium, silicon, nickel, chromium, magnesium, and tungsten

- Reactive metals such as lithium, sodium, potassium,

and calcium

- The halogens fluorine, chlorine, bromine, and iodine

- Gases such as hydrogen, oxygen, nitrogen, helium, argon, and neon

- Most of the rare-earth elements, including cerium, lanthanum, gadolinium, and neodymium.

- The more common radioactive elements, including uranium, thorium, radium, and radon

Elements isolated or produced since 1900 include:

- The three remaining undiscovered regularly occurring stable natural elements: hafnium, lutetium, and rhenium

- Plutonium, which was first produced synthetically in 1940 by Glenn T. Seaborg, but is now also known from a few long-persisting natural occurrences

- The seven incidentally occurring natural elements (americium, berkelium, californium, curium, neptunium, promethium, and technetium), which were all first produced synthetically but later discovered in trace amounts in certain geological samples

- Three scarce decay products of uranium or thorium, (astatine, francium, and protactinium), and

- Various synthetic transuranic elements, beginning with einsteinium, fermium, mendelevium, nobelium, and lawrencium

1.5.3   Recently discovered elements

The first transuranium element (element with atomic number greater than 92) discovered was neptunium in 1940. Since 1999 claims for the discovery of new elements have been considered by the IUPAC/IUPAP Joint Working Party. As of May 2012, only the elements up to 112, copernicium, as well as element 114 Flerovium and element 116 Livermorium have been confirmed as discovered by IUPAC, while claims have been made for synthesis of elements 113, 115, 117[33] and 118. The discovery of element 112 was acknowledged in 2009, and the name 'copernicium' and the atomic symbol 'Cn' were suggested for it.[34] The name and symbol were officially endorsed by IUPAC on 19 February 2010.[35] The heaviest element that is believed to have been synthesized to date is element 118, ununoctium, on 9 October 2006, by the Flerov Laboratory of Nuclear Reactions in Dubna, Russia.[12][36] Element 117 was the latest element claimed to be discovered, in 2009.[33]

IUPAC officially recognized flerovium and livermorium, elements 114 and 116, in June 2011 and approved their names in May 2012.[37]

1.6   List of the 118 known chemical elements

The following sortable table includes the 118 known chemical elements, with the names linking to the *Wikipedia* articles on each.

- **Atomic number**, **name**, and **symbol** all serve independently as unique identifiers.

- **Names** are those accepted by IUPAC; provisional names for recently produced elements not yet formally named are in parentheses.

- **Group, period**, and **block** refer to an element's position in the periodic table. Group numbers here show the currently accepted numbering; for older alternate numberings, see Group (periodic table).

- **State of matter** (*solid, liquid,* or *gas*) applies at standard temperature and pressure conditions (STP).

- **Occurrence** distinguishes naturally occurring elements, categorized as either *primordial* or *transient* (from decay), and additional *synthetic* elements that have been produced technologically, but are not known to occur naturally.

- **Description** summarizes an element's properties using the broad categories commonly presented in periodic tables: Actinide, alkali metal, alkaline earth metal, halogen, lanthanide, metal, metalloid, noble gas, non-metal, and transition metal.

1.7   See also

- Discovery of the chemical elements

- Element collecting

- Fictional element

- Goldschmidt classification

- Island of stability

- List of elements by name

- List of the elements' densities

- List of nuclides

- Periodic Systems of Small Molecules

- Prices of elements and their compounds

- Symbol (chemical element)#Symbols not currently used

- Systematic element name

- Table of nuclides

- The Mystery of Matter: Search for the Elements (PBS film)

1.8 References

[1] IUPAC (ed.). "chemical element". *http://iupac.org". doi:10.1351/goldbook.C01022.*

[2] Los Alamos National Laboratory (2011). "Periodic Table of Elements: Oxygen". Los Alamos, New Mexico: Los Alamos National Security, LLC. Retrieved 7 May 2011.

[3] Oerter, Robert (2006). *The Theory of Almost Everything: The Standard Model, the Unsung Triumph of Modern Physics.* Penguin. p. 223. ISBN 978-0-452-28786-0.

[4] E. M. Burbidge, G. R. Burbidge, W. A. Fowler, F. Hoyle (1957). "Synthesis of the Elements in Stars". *Reviews of Modern Physics* **29** (4): 547–650. Bibcode:1957RvMP...29..547B. doi:10.1103/RevModPhys.29.547.

[5] See the timeline on p.10 in Oganessian, Yu. Ts.; Utyonkov, V.; Lobanov, Yu.; Abdullin, F.; Polyakov, A.; Sagaidak, R.; Shirokovsky, I.; Tsyganov, Yu. et al. (2006). "Evidence for Dark Matter" (PDF). *Physical Review C* **74** (4): 044602. Bibcode:2006PhRvC..74d4602O. doi:10.1103/PhysRevC.74.044602.

[6] lbl.gov (2005). "The Universe Adventure Hydrogen and Helium". *Lawrence Berkeley National Laboratory U.S. Department of Energy.*

[7] astro.soton.ac.uk (January 3, 2001). "Formation of the light elements". *University of Southampton.*

[8] foothill.edu (October 18, 2006). "How Stars Make Energy and New Elements" (PDF). *Foothill College.*

[9] Dumé, B (23 April 2003). "Bismuth breaks half-life record for alpha decay". *Physicsworld.com* (Bristol, England: Institute of Physics). Retrieved 14 July 2015.

[10] de Marcillac, P; Coron, N; Dambier, G; Leblanc, J; Moalic, J-P (2003). "Experimental detection of alpha-particles from the radioactive decay of natural bismuth". *Nature* **422** (6934): 876–8. Bibcode:2003Natur.422..876D. doi:10.1038/nature01541. PMID 12712201.

[11] Sanderson, K (17 October 2006). "Heaviest element made – again". Nature News. doi:10.1038/news061016-4.

[12] Schewe, P; Stein, B (17 October 200). "Elements 116 and 118 Are Discovered". *Physics News Update.* American Institute of Physics. Retrieved 19 October 2006. Check date values in: |date= (help)

[13] scienceline.ucsb.edu/. "How many elements are there in the known universe?". *University of California, Santa Cruz.*

[14] *United States Environmental Protection Agency.* "Technetium-99". epa.gov. Retrieved 26 February 2013.

[15] *Harvard–Smithsonian Center for Astrophysics.* "ORIGIN OF HEAVY ELEMENTS". cfa.harvard.edu. Retrieved 26 February 2013.

[16] "ATOMIC NUMBER AND MASS NUMBERS". ndted.org. Retrieved 17 February 2013.

[17] periodic.lanl.gov. "PERIODIC TABLE OF ELEMENTS: LANL Carbon". *Los Alamos National Laboratory.*

[18] Katsuya Yamada. "Atomic mass, isotopes, and mass number." (PDF). *Los Angeles Pierce College.*

[19] "Pure element". European Nuclear Society.

[20] Wilford, JN (14 January 1992). "Hubble Observations Bring Some Surprises". *New York Times.*

[21] Wright, EL (12 September 2004). "Big Bang Nucleosynthesis". UCLA, Division of Astronomy. Retrieved 22 February 2007.

[22] Wallerstein, George; Iben, Icko; Parker, Peter; Boesgaard, Ann; Hale, Gerald; Champagne, Arthur; Barnes, Charles; Käppeler, Franz et al. (1999). "Synthesis of the elements in stars: forty years of progress" (PDF). *Reviews of Modern Physics* **69** (4): 995–1084. Bibcode:1997RvMP...69..995W. doi:10.1103/RevModPhys.69.995.

[23] Earnshaw, A; Greenwood, N (1997). *Chemistry of the Elements* (2nd ed.). Butterworth-Heinemann.

[24] Croswell, K (1996). *Alchemy of the Heavens.* Anchor. ISBN 0-385-47214-5.

[25] Plato (2008) [c. 360 BC]. *Timaeus.* Forgotten Books. p. 45. ISBN 978-1-60620-018-6.

[26] Hillar, M (2004). "The Problem of the Soul in Aristotle's De anima". NASA/WMAP. Retrieved 10 August 2006.

[27] Partington, JR (1937). *A Short History of Chemistry.* New York: Dover Publications. ISBN 0-486-65977-1.

[28] Boyle, R (1661). *The Sceptical Chymist.* London. ISBN 0-922802-90-4.

[29] Lavoisier, AL (1790). *Elements of chemistry translated by Robert Kerr*. Edinburgh. pp. 175–6. ISBN 978-0-415-17914-0.

[30] Transactinide-2. www.kernchemie.de

[31] Carey, GW (1914). *The Chemistry of Human Life*. Los Angeles. ISBN 0-7661-2840-7.

[32] Glanz, J (6 April 2010). "Scientists Discover Heavy New Element". *New York Times*.

[33] Greiner, W. "Recommendations" (PDF). *31st meeting, PAC for Nuclear Physics*. Joint Institute for Nuclear Research.

[34] "IUPAC Announces Start of the Name Approval Process for the Element of Atomic Number 112" (PDF). IUPAC. 20 July 2009. Retrieved 27 August 2009.

[35] "IUPAC (International Union of Pure and Applied Chemistry): Element 112 is Named Copernicium". IUPAC. 20 February 2010.

[36] Oganessian, Yu. Ts.; Utyonkov, V.; Lobanov, Yu.; Abdullin, F.; Polyakov, A.; Sagaidak, R.; Shirokovsky, I.; Tsyganov, Yu. et al. (2006). "Evidence for Dark Matter" (PDF). *Physical Review C* **74** (4): 044602. Bibcode:2006PhRvC..74d4602O. doi:10.1103/PhysRevC.74.044602.

[37] "Two ultra-heavy elements added to the periodic table". 6 June 2011.

1.9   Further reading

- Ball, P (2004). *The Elements: A Very Short Introduction*. Oxford University Press. ISBN 0-19-284099-1.

- Emsley, J (2003). *Nature's Building Blocks: An A-Z Guide to the Elements*. Oxford University Press. ISBN 0-19-850340-7.

- Gray, T (2009). *The Elements: A Visual Exploration of Every Known Atom in the Universe*. Black Dog & Leventhal Publishers Inc. ISBN 1-57912-814-9.

- Scerri, ER (2007). *The Periodic Table, Its Story and Its Significance*. Oxford University Press.

- Strathern, P (2000). *Mendeleyev's Dream: The Quest for the Elements*. Hamish Hamilton Ltd. ISBN 0-241-14065-X.

- Kean, Sam (2011). *The Disappearing Spoon: And Other True Tales of Madness, Love, and the History of the World from the Periodic Table of the Elements*. Back Bay Books.

- Compiled by A. D. McNaught and A. Wilkinson. (1997). Blackwell Scientific Publications, Oxford, ed. *Compendium of Chemical Terminology, 2nd ed. (the "Gold Book")*. doi:10.1351/goldbook. ISBN 0-9678550-9-8.

    XML on-line corrected version: created by M. Nic, J. Jirat, B. Kosata; updates compiled by A. Jenkins.

1.10   External links

- Videos for each element by the University of Nottingham

Chapter 2

Chemical substance

"Chemical" redirects here. For other uses, see Chemical (disambiguation).

A **chemical substance** is a form of matter that has con-

Steam and liquid water are two different forms of the same chemical substance, water.

stant chemical composition and characteristic properties.[1] It cannot be separated into components by physical separation methods, i.e., without breaking chemical bonds. Chemical substances can be chemical elements, chemical compounds, ions or alloys.

Chemical substances are often called 'pure' to set them apart from mixtures. A common example of a chemical substance is pure water; it has the same properties and the same ratio of hydrogen to oxygen whether it is isolated from a river or made in a laboratory. Other chemical substances commonly encountered in pure form are diamond (carbon), gold, table salt (sodium chloride) and refined sugar (sucrose). However, in practice, no substance is entirely pure, and chemical purity is specified according to the intended use of the chemical.

Chemical substances exist as solids, liquids, gases or plasma, and may change between these phases of matter with changes in temperature or pressure. Chemical reactions convert one chemical substance into another.

Forms of energy, such as light and heat, are not considered to be matter, and thus they are not "substances" in this regard.

2.1 Definition

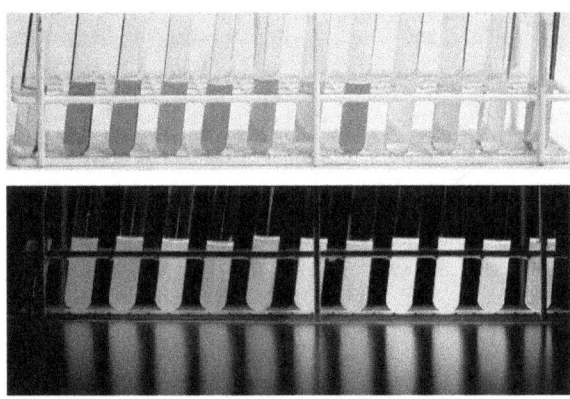

Colors of a single chemical (Nile red) in different solvents, under visible and UV light, showing how the chemical interacts dynamically with its solvent environment.

A chemical substance (also called a pure substance) may well be defined as "any material with a definite chemical composition" in an introductory general chemistry textbook.[2] According to this definition a chemical substance can either be a pure chemical element or a pure chemical compound. But, there are exceptions to this definition; a pure substance can also be defined as a form of matter that has both definite composition and distinct properties.[3] The chemical substance index published by CAS also includes several alloys of uncertain composition.[4] Non-stoichiometric compounds are a special case (in inorganic chemistry) that violates the law of constant composition, and for them, it is sometimes difficult to draw the line between a mixture and a compound, as in the case of palladium hydride. Broader definitions of chemicals or chemical substances can be found, for example: "the term 'chemical substance' means any organic or inorganic substance of a particular molecular identity, including – (i) any combination of such substances occurring in whole or in part as a result of a chemical reaction or occurring in nature"[5]

In geology, substances of uniform composition are called minerals, while physical mixtures (aggregates) of several minerals (different substances) are defined as rocks. Many minerals, however, mutually dissolve into solid solutions, such that a single rock is a uniform substance despite being a mixture in stoichiometric terms. Feldspars are a common example: anorthoclase is an alkali aluminium silicate, where the alkali metal is interchangeably either sodium or potassium.

2.2 History

The concept of a "chemical substance" became firmly established in the late eighteenth century after work by the chemist Joseph Proust on the composition of some pure chemical compounds such as basic copper carbonate.[6] He deduced that, "All samples of a compound have the same composition; that is, all samples have the same proportions, by mass, of the elements present in the compound." This is now known as the law of constant composition.[7] Later with the advancement of methods for chemical synthesis particularly in the realm of organic chemistry; the discovery of many more chemical elements and new techniques in the realm of analytical chemistry used for isolation and purification of elements and compounds from chemicals that led to the establishment of modern chemistry, the concept was defined as is found in most chemistry textbooks. However, there are some controversies regarding this definition mainly because the large number of chemical substances reported in chemistry literature need to be indexed.

Isomerism caused much consternation to early researchers, since isomers have exactly the same composition, but differ in configuration (arrangement) of the atoms. For example, there was much speculation for the chemical identity of benzene, until the correct structure was described by Friedrich August Kekulé. Likewise, the idea of stereoisomerism - that atoms have rigid three-dimensional structure and can thus form isomers that differ only in their three-dimensional arrangement - was another crucial step in understanding the concept of distinct chemical substances. For example, tartaric acid has three distinct isomers, a pair of diastereomers with one diastereomer forming two enantiomers.

2.3 Chemical elements

Native sulfur crystals. Sulfur occurs naturally as elemental sulfur, in sulfide and sulfate minerals and in hydrogen sulfide.

Main article: Chemical element
See also: List of elements

An element is a chemical substance that is made up of a particular kind of atoms and hence cannot be broken down or transformed by a chemical reaction into a different element, though it can be transmutated into another element through a nuclear reaction. This is so, because all of the atoms in a sample of an element have the same number of protons, though they may be different isotopes, with differing numbers of neutrons.

As of 2012, there are 118 known elements, about 80 of which are stable – that is, they do not change by radioactive decay into other elements. Some elements can occur as more than a single chemical substance (allotropes). For instance, oxygen exists as both diatomic oxygen (O_2) and ozone (O_3). The majority of elements are classified as metals. These are elements with a characteristic lustre such as iron, copper, and gold. Metals typically conduct electricity and heat well, and they are malleable and ductile.[8] Around a dozen elements,[9] such as carbon, nitrogen, and oxygen, are classified as non-metals. Non-metals lack the metallic properties described above, they also have a high electronegativity and a tendency to form negative ions. Certain elements such as silicon sometimes resemble metals and sometimes resemble non-metals, and are known as metalloids.

Potassium ferricyanide is a compound of potassium, iron, carbon and nitrogen; although it contains cyanide anions, it does not release them and is nontoxic.

2.4   Chemical compounds

Main article: Chemical compound
See also: List of organic compounds and List of inorganic compounds

A pure chemical compound is a chemical substance that is composed of a particular set of molecules or ions. Two or more elements combined into one substance through a chemical reaction form a chemical compound. All compounds are substances, but not all substances are compounds.

A chemical compound can be either atoms bonded together in molecules or crystals in which atoms, molecules or ions form a crystalline lattice. Compounds based primarily on carbon and hydrogen atoms are called organic compounds, and all others are called inorganic compounds. Compounds containing bonds between carbon and a metal are called organometallic compounds.

Compounds in which components share electrons are known as covalent compounds. Compounds consisting of oppositely charged ions are known as ionic compounds, or salts.

In organic chemistry, there can be more than one chemical compound with the same composition and molecular weight. Generally, these are called isomers. Isomers usually have substantially different chemical properties, may be isolated and do not spontaneously convert to each other. A common example is glucose vs. fructose. The former is an aldehyde, the latter is a ketone. Their interconversion requires either enzymatic or acid-base catalysis. However, there are also tautomers, where isomerization occurs spontaneously, such that a pure substance cannot be isolated into its tautomers. A common example is glucose, which has open-chain and ring forms. One cannot manufacture pure

open-chain glucose because glucose spontaneously cyclizes to the hemiacetal form. Materials may also comprise other entities such as polymers. These may be inorganic or organic and sometimes a combination of inorganic and organic.

2.5   Substances versus mixtures

Cranberry glass, while it looks homogeneous, is a mixture *consisting of glass and gold colloidal particles of ca. 40 nm diameter, which give it a red color.*

Main article: Mixture

All matter consists of various elements and chemical compounds, but these are often intimately mixed together. Mixtures contain more than one chemical substance, and they do not have a fixed composition. In principle, they can be separated into the component substances by purely mechanical processes. Butter, soil and wood are common examples of mixtures.

Grey iron metal and yellow sulfur are both chemical elements, and they can be mixed together in any ratio to form a yellow-grey mixture. No chemical process occurs, and the material can be identified as a mixture by the fact that the sulfur and the iron can be separated by a mechanical process, such as using a magnet to attract the iron away from the sulfur.

In contrast, if iron and sulfur are heated together in a certain ratio (1 atom of iron for each atom of sulfur, or by weight, 56 grams (1 mol) of iron to 32 grams (1 mol) of sulfur), a chemical reaction takes place and a new substance is formed, the compound iron(II) sulfide, with chemical formula FeS. The resulting compound has all the properties of a chemical substance and is not a mixture. Iron(II) sulfide has its own distinct properties such as melting point and solubility, and the two elements cannot be separated using normal mechanical processes; a magnet will be unable to recover the iron, since there is no metallic iron present in the compound.

2.6 Chemicals versus chemical substances

While the term *chemical substance* is a precise technical term that is synonymous with "chemical" for professional chemists, the meaning of the word *chemical* varies for non-chemists within the English speaking world or those using English. For industries, government and society in general in some countries,[10] the word *chemical* includes a wider class of substances that contain many mixtures of such chemical substances, often finding application in many vocations.[11] In countries that require a list of ingredients in products, the "chemicals" listed would be equated with "chemical substances".[12]

Within the chemical industry, manufactured "chemicals" are chemical substances, which can be classified by production volume into bulk chemicals, fine chemicals and chemicals found in research only:

- Bulk chemicals are produced in very large quantities, usually with highly optimized continuous processes and to a relatively low price.

- Fine chemicals are produced at a high cost in small quantities for special low-volume applications such as biocides, pharmaceuticals and speciality chemicals for technical applications.

- Research chemicals are produced individually for research, such as when searching for synthetic routes or screening substances for pharmaceutical activity. In effect, their price per gram is very high, although they are not sold.

The cause of the difference in production volume is the complexity of the molecular structure of the chemical. Bulk chemicals are usually much less complex. While fine chemicals may be more complex, many of them are simple enough to be sold as "building blocks" in the synthesis of more complex molecules targeted for single use, as named above. The *production* of a chemical includes not only its synthesis but also its purification to eliminate by-products and impurities involved in the synthesis. The last step in production should be the analysis of batch lots of chemicals in order to identify and quantify the percentages of impurities for the buyer of the chemicals. The required purity and analysis depends on the application, but higher tolerance of impurities is usually expected in the production of bulk chemicals. Thus, the user of the chemical in the US might choose between the bulk or "technical grade" with higher amounts of impurities or a much purer "pharmaceutical grade" (labeled "USP", United States Pharmacopeia).

2.7 Naming and indexing

Every chemical substance has one or more systematic names, usually named according to the IUPAC rules for naming. An alternative system is used by the Chemical Abstracts Service (CAS).

Many compounds are also known by their more common, simpler names, many of which predate the systematic name. For example, the long-known sugar glucose is now systematically named 6-(hydroxymethyl)oxane-2,3,4,5-tetrol. Natural products and pharmaceuticals are also given simpler names, for example the mild pain-killer Naproxen is the more common name for the chemical compound (S)−6-methoxy-α-methyl-2-naphthaleneacetic acid.

Chemists frequently refer to chemical compounds using chemical formulae or molecular structure of the compound. There has been a phenomenal growth in the number of chemical compounds being synthesized (or isolated), and then reported in the scientific literature by professional chemists around the world.[13] An enormous number of chemical compounds are possible through the chemical combination of the known chemical elements. As of May 2011, about sixty million chemical compounds are known.[14] The names of many of these compounds are often nontrivial and hence not very easy to remember or cite accurately. Also it is difficult to keep the track of them in the literature. Several international organizations like IUPAC and CAS have initiated steps to make such tasks easier. CAS provides the abstracting services of the chemical literature, and provides a numerical identifier, known as CAS registry number to each chemical substance that has been reported in the chemical literature (such as chemistry journals and patents). This information is compiled as a database and is popularly known as the Chemical substances index. Other computer-friendly systems that have been developed for substance information, are: SMILES and the International Chemical Identifier or InChI.

2.8 Isolation, purification, characterization, and identification

Often a pure substance needs to be isolated from a mixture, for example from a natural source (where a sample often contains numerous chemical substances) or after a chemical reaction (which often give mixtures of chemical substances).

2.9 See also

- Chemical safety signs
- IUPAC nomenclature
- Prices of elements and their compounds

2.10 Notes and references

[1] IUPAC, *Compendium of Chemical Terminology*, 2nd ed. (the "Gold Book") (1997). Online corrected version: (2006–) "Chemical Substance".

[2] Hill, J. W.; Petrucci, R. H.; McCreary, T. W.; Perry, S. S. *General Chemistry*, 4th ed., p5, Pearson Prentice Hall, Upper Saddle River, New Jersey, 2005

[3] "Pure Substance – DiracDelta Science & Engineering Encyclopedia". Diracdelta.co.uk. Retrieved 2013-06-06.

[4] Appendix IV: Chemical Substance Index Names

[5] "What is the TSCA Chemical Substance Inventory?". US Environmental Protection Agency. Retrieved 2009-10-19.

[6] Hill, J. W.; Petrucci, R. H.; McCreary, T. W.; Perry, S. S. *General Chemistry*, 4th ed., p37, Pearson Prentice Hall, Upper Saddle River, New Jersey, 2005.

[7] Law of Definite Proportions

[8] Hill, J. W.; Petrucci, R. H.; McCreary, T. W.; Perry, S. S. *General Chemistry*, 4th ed., pp 45–46, Pearson Prentice Hall, Upper Saddle River, New Jersey, 2005.

[9] The boundary between metalloids and non-metals is imprecise, as explained in the previous reference.

[10] "What is a chemical". Nicnas.gov.au. 2005-06-01. Retrieved 2013-06-06.

[11] "BfR – Chemicals". Bfr.bund.de. 1980-09-18. Retrieved 2013-06-06.

[12] There is only one definition for "chemical", that of a substance, in the US Unabridged Edition of the Random House Dictionary of the English Language, New York, 1966.

[13] Joachim Schummer. "Coping with the Growth of Chemical Knowledge: Challenges for Chemistry Documentation, Education, and Working Chemists". Rz.uni-karlsruhe.de. Retrieved 2013-06-06.

[14] "Chemical Abstracts substance count". Cas.org. Retrieved 2013-06-06.

Inorganic polymers, N. H. Ray, Academic, John Wiley and Sons Inc.New York, 1979, 173 pp.

2.11 External links

- eChemPortal substance and property search
- Chemical Reactions

Inorganic polymers, N. H. Ray, Academic, John Wiley and Sons Inc.New York, 1979, 173 pp.

Chapter 3

Synthetic element

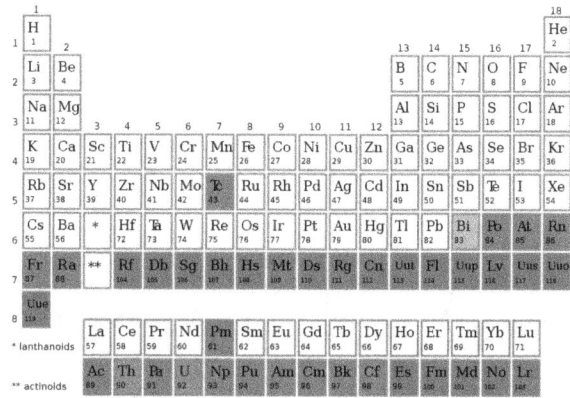

Synthetic elements
Rare radioactive natural elements; often produced artificially

In chemistry, a **synthetic element** is a chemical element that does not occur naturally on Earth, and can only be created artificially. So far, 20 synthetic elements have been created (those with atomic numbers 99–118). All are unstable, decaying with half-lives ranging from a year to a few hundred microseconds.

Nine other elements were first created artificially and thus considered synthetic, but later discovered to exist naturally (in trace quantities) as well; among them plutonium—first synthesized in 1940—the one best known to laypeople, because of its use in atomic bombs and nuclear reactors.

3.1   Properties

Synthetic elements are radioactive and decay rapidly into lighter elements—possessing half-lives so short, relative to the age of the Earth (which formed approximately 4.6 billion years ago), that any atoms of these elements that may have existed when the Earth formed have long since decayed. Atoms of synthetic elements only occur on Earth as the product of atomic bombs or experiments that involve nuclear reactors or particle accelerators, via nuclear fusion or neutron absorption.

Atomic mass for natural life is based on weighted average abundance of natural isotopes that occur in the Earth's crust and atmosphere. For *synthetic* elements, the isotope depends on the means of synthesis, so the concept of natural isotope abundance has no meaning. Therefore, for synthetic elements the total nucleus count (protons plus neutrons) of the most stable isotope, i.e. the isotope with the longest half-life—is listed in brackets as the atomic mass.

Not all radioactive elements are synthetic. For instance, uranium and thorium have no stable isotopes but occur naturally in the Earth's crust and atmosphere. Unstable elements such as polonium, radium, and radon—which form through the decay of uranium and thorium—are also found in nature, despite their short half-lives. Plutonium is an outlier: Its half-life, depending on the isotope, can be as long as 376,000 years. (The *principal* plutonium isotope in use has a half-life of 24,100 years.)

3.2   History

The first element discovered through synthesis was technetium—its discovery being definitely confirmed in 1936. This discovery filled a gap in the periodic table, and the fact that no stable isotopes of technetium exist explains its natural absence on Earth (and the gap). With the longest-lived isotope of technetium, Tc-98, having a 4.2-million-year half-life, no technetium remains from the formation of the Earth. Only minute traces of technetium occur naturally in the Earth's crust—as a spontaneous fission product of uranium-238 or by neutron capture in molybdenum ores—but technetium is present naturally in red giant stars.

The first discovered synthetic elements were einsteinium and fermium in 1952, by a team of scientists led by Albert Ghiorso in 1952 while studying the radioactive debris from the detonation of the first hydrogen bomb. The isotopes discovered were einsteinium-253, with a half-life of 20.5 days, and fermium-255, with a half-life of about 20 hours.

The discoveries of mendelevium, lawrencium, and nobelium followed. During the height of the Cold War, the Soviet Union and United States independently discovered rutherfordium and dubnium. The naming and credit for discovery of those elements remained unresolved for many years but eventually shared credit was recognized by IUPAC/IUPAP in 1992. In 1997, IUPAC decided to give dubnium its current name honoring the city of Dubna where the Russian team made their discoveries since American-chosen names had already been used for many existing synthetic elements, while the name *rutherfordium* (chosen by the American team) was accepted for element 104.

No element with an atomic number greater than 98 has any use outside of scientific research, as they have extremely short half-lives.

3.3 List of synthetic elements

The following elements do not occur naturally on Earth. All are transuranium elements and have atomic numbers of 99 and higher.

3.3.1 Other elements usually produced through synthesis

All elements with atomic numbers 1 through 98 are naturally occurring at least in trace quantities, but the following elements are usually produced through synthesis. Except for francium, they were all discovered through synthesis before being found in nature.

3.4 References

- http://www.britannica.com/EBchecked/topic/ 181416/einsteinium-Es

- http://www.britannica.com/EBchecked/topic/ 374759/mendelevium-Md

- http://encyclopedia2.thefreedictionary.com/ synthetic+elements

- http://education.jlab.org/itselemental/ele100.html

Chapter 4

Isotope

This article is about the atomic variants of chemical elements. For other uses, see Isotope (disambiguation).

Isotopes are variants of a particular chemical element

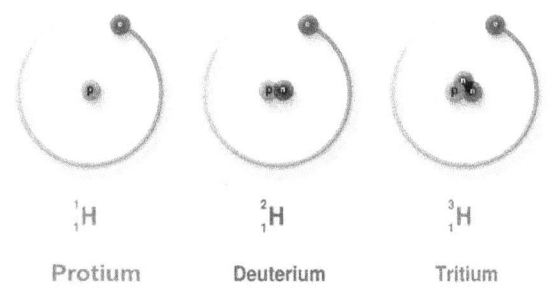

The three naturally-occurring isotopes of hydrogen. The fact that each isotope has one proton makes them all variants of hydrogen: the identity of the isotope is given by the number of neutrons. From left to right, the isotopes are protium (^1H) with zero neutrons, deuterium (^2H) with one neutron, and tritium (^3H) with two neutrons.

which differ in neutron number, although all isotopes of a given element have the same number of protons in each atom. The term isotope is formed from the Greek roots isos (ἴσος "equal") and topos (τόπος "place"), meaning "the same place"; thus, the meaning behind the name it is that different isotopes of a single element occupy the same position on the periodic table. The number of protons within the atom's nucleus is called atomic number and is equal to the number of electrons in the neutral (non-ionized) atom. Each atomic number identifies a specific element, but not the isotope; an atom of a given element may have a wide range in its number of neutrons. The number of nucleons (both protons and neutrons) in the nucleus is the atom's mass number, and each isotope of a given element has a different mass number.

For example, carbon-12, carbon-13 and carbon-14 are three isotopes of the element carbon with mass numbers 12, 13 and 14 respectively. The atomic number of carbon is 6, which means that every carbon atom has 6 protons, so that the neutron numbers of these isotopes are 6, 7 and 8

respectively.

4.1 Isotope vs. nuclide

Nuclide refers to a nucleus rather than to an atom. Identical nuclei belong to one nuclide, for example each nucleus of the carbon-13 nuclide is composed of 6 protons and 7 neutrons. The *nuclide* concept (referring to individual nuclear species) emphasizes nuclear properties over chemical properties, while the *isotope* concept (grouping all atoms of each element) emphasizes chemical over nuclear. The neutron number has large effects on nuclear properties, but its effect on chemical properties is negligible for most elements. Even in the case of the very lightest elements where the ratio of neutron number to atomic number varies the most between isotopes it usually has only a small effect, although it does matter in some circumstances (for hydrogen, the lightest element, the isotope effect is large enough to strongly affect biology). Since *isotope* is the older term, it is better known than *nuclide*, and is still sometimes used in contexts where *nuclide* might be more appropriate, such as nuclear technology and nuclear medicine.

4.2 Notation

An isotope and/or nuclide is specified by the name of the particular element (this indicates the atomic number) followed by a hyphen and the mass number (e.g. helium-3, helium-4, carbon-12, carbon-14, uranium-235 and uranium-239).[1] When a chemical symbol is used, e.g., "C" for carbon, standard notation (now known as "AZE notation" because A is the mass number, Z the atomic number, and E for element) is to indicate the mass number (number of nucleons) with a superscript at the upper left of the chemical symbol and to indicate the atomic number with a subscript at the lower left (e.g. 3
2He, 4
2He, 12

6C, 14
6C, 235
92U, and 239
92U).[2] Since the atomic number is given by the element symbol, it is common to state only the mass number in the superscript and leave out the atomic number subscript (e.g. 3He, 4He, 12C, 14C, 235U, and 239U). The letter *m* is sometimes appended after the mass number to indicate a nuclear isomer, a metastable or energetically-excited nuclear state (as opposed to the lowest-energy ground state), for example 180m
73Ta (tantalum-180m).

4.3 Radioactive, primordial, and stable isotopes

Some isotopes are radioactive, and are therefore described as radioisotopes or radionuclides, while others have never been observed to undergo radioactive decay and are described as stable isotopes or stable nuclides. For example, 14C is a radioactive form of carbon while 12C and 13C are stable isotopes. There are about 339 naturally occurring nuclides on Earth,[3] of which 288 are primordial nuclides, meaning that they have existed since the solar system's formation.

Primordial nuclides include 34 nuclides with very long half-lives (over 80 million years) and 254 that are formally considered as "stable nuclides",[3] since they have not been observed to decay. In most cases, for obvious reasons, if an element has stable isotopes, those isotopes predominate in the elemental abundance found on Earth and in the solar system. However, in the cases of three elements (tellurium, indium, and rhenium) the most abundant isotope found in nature is actually one (or two) extremely long lived radioisotope(s) of the element, despite these elements having one or more stable isotopes.

Theory predicts that many apparently "stable" isotopes/nuclides are radioactive, with extremely long half-lives (discounting the possibility of proton decay, which would make all nuclides ultimately unstable). Of the 254 nuclides never observed to decay, only 90 of these (all from the first 40 elements) are theoretically stable to all known forms of decay. Element 41 (niobium) is theoretically unstable via spontaneous fission, but this has never been detected. Many other stable nuclides are in theory energetically susceptible to other known forms of decay, such as alpha decay or double beta decay, but no decay products have yet been observed, and so these isotopes are described as "observationally stable". The predicted half-lives for these nuclides often greatly exceed the estimated age of the universe, and in fact there are

also 27 known radionuclides (see primordial nuclide) with half-lives longer than the age of the universe.

Adding in the radioactive nuclides that have been created artificially, there are more than 3100 currently known nuclides.[4] These include 905 nuclides that are either stable or have half-lives longer than 60 minutes. See list of nuclides for details.

4.4 History

4.4.1 Radioactive isotopes

The existence of isotopes was first suggested in 1913 by the radiochemist Frederick Soddy, based on studies of radioactive decay chains that indicated about 40 different species described as *radioelements* (i.e. radioactive elements) between uranium and lead, although the periodic table only allowed for 11 elements from uranium to lead.[5][6]

Several attempts to separate these new radioelements chemically had failed.[7] For example, Soddy had shown in 1910 that mesothorium (later shown to be ^{228}Ra), radium (^{226}Ra, the longest-lived isotope), and thorium X (^{224}Ra) are impossible to separate.[8] Attempts to place the radioelements in the periodic table led Soddy and Kazimierz Fajans independently to propose their radioactive displacement law in 1913, to the effect that alpha decay produced an element two places to the left in the periodic table, while beta decay emission produced an element one place to the right.[9] Soddy recognized that emission of an alpha particle followed by two beta particles led to the formation of an element chemically identical to the initial element but with a mass four units lighter and with different radioactive properties.

Soddy proposed that several types of atoms (differing in radioactive properties) could occupy the same place in the table. For example, the alpha-decay of uranium-235 forms thorium-231, while the beta decay of actinium-230 forms thorium-230.[7] The term "isotope", Greek for "at the same place", was suggested to Soddy by Margaret Todd, a Scottish physician and family friend, during a conversation in which he explained his ideas to her.[8][10][11][12][13][14]

In 1914 T. W. Richards found variations between the atomic weight of lead from different mineral sources, attributable to variations in isotopic composition due to different radioactive origins.[7][15]

4.4.2 Stable isotopes

The first evidence for multiple isotopes of a stable (non-radioactive) element was found by J. J. Thomson in 1913

4.5 Variation in properties between isotopes

See also: Neutron-proton ratio

4.5.1 Chemical and molecular properties

A neutral atom has the same number of electrons as protons. Thus different isotopes of a given element all have the same number of electrons and share a similar electronic structure. Because the chemical behavior of an atom is largely determined by its electronic structure, different isotopes exhibit nearly identical chemical behavior. The main exception to this is the kinetic isotope effect: due to their larger masses, heavier isotopes tend to react somewhat more slowly than lighter isotopes of the same element. This is most pronounced by far for protium (1H), deuterium (2H), and tritium (3H), because deuterium has twice the mass of protium and tritium has three times the mass of protium. These mass differences also affect the behavior of their respective chemical bonds, by changing the center of gravity (reduced mass) of the atomic systems. However, for heavier elements the relative mass difference between isotopes is much less, so that the mass-difference effects on chemistry are usually negligible. (Heavy elements also have relatively more neutrons than lighter elements, so the ratio of the nuclear mass to the collective electronic mass is slightly greater.)

Similarly, two molecules that differ only in the isotopes of their atoms (isotopologues) have identical electronic structure, and therefore almost indistinguishable physical and chemical properties (again with deuterium and tritium being the primary exceptions). The **vibrational modes** of a molecule are determined by its shape and by the masses of its constituent atoms; so different isotopologues have different sets of vibrational modes. Since vibrational modes allow a molecule to absorb photons of corresponding energies, isotopologues have different optical properties in the infrared range.

4.5.2 Nuclear properties and stability

See also: Stable isotope, List of nuclides and List of elements by stability of isotopes

Atomic nuclei consist of protons and neutrons bound together by the residual strong force. Because protons are positively charged, they repel each other. Neutrons, which are electrically neutral, stabilize the nucleus in two ways. Their copresence pushes protons slightly apart, reducing the

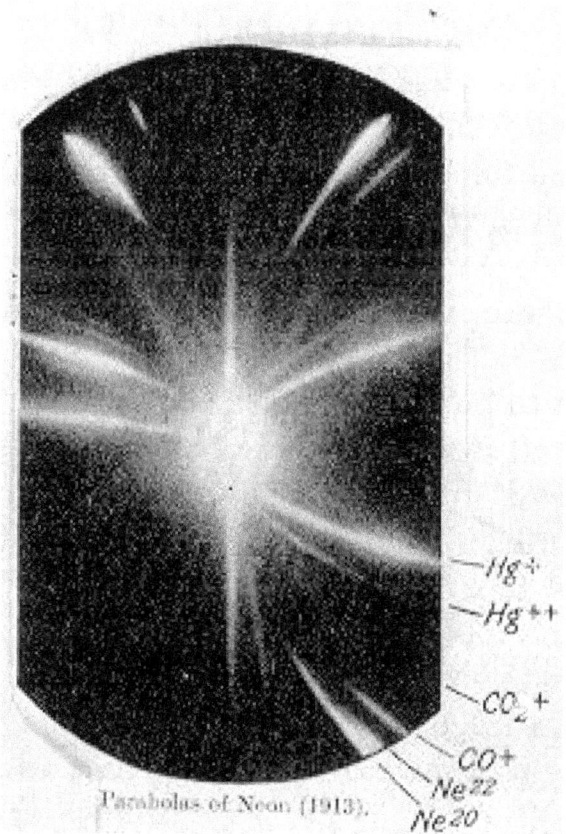

In the bottom right corner of J. J. Thomson's photographic plate are the separate impact marks for the two isotopes of neon: neon-20 and neon-22.

as part of his exploration into the composition of canal rays (positive ions).[16][17] Thomson channeled streams of neon ions through a magnetic and an electric field and measured their deflection by placing a photographic plate in their path. Each stream created a glowing patch on the plate at the point it struck. Thomson observed two separate patches of light on the photographic plate (see image), which suggested two different parabolas of deflection. Thomson eventually concluded that some of the atoms in the neon gas were of higher mass than the rest.

F. W. Aston subsequently discovered multiple stable isotopes for numerous elements using a mass spectrograph. In 1919 Aston studied neon with sufficient resolution to show that the two isotopic masses are very close to the integers 20 and 22, and that neither is equal to the known molar mass (20.2) of neon gas. This is an example of Aston's whole number rule for isotopic masses, which states that large deviations of elemental molar masses from integers are primarily due to the fact that the element is a mixture of isotopes. Aston similarly showed that the molar mass of chlorine (35.45) is a weighted average of the almost integral masses for the two isotopes ^{35}Cl and ^{37}Cl.[18]

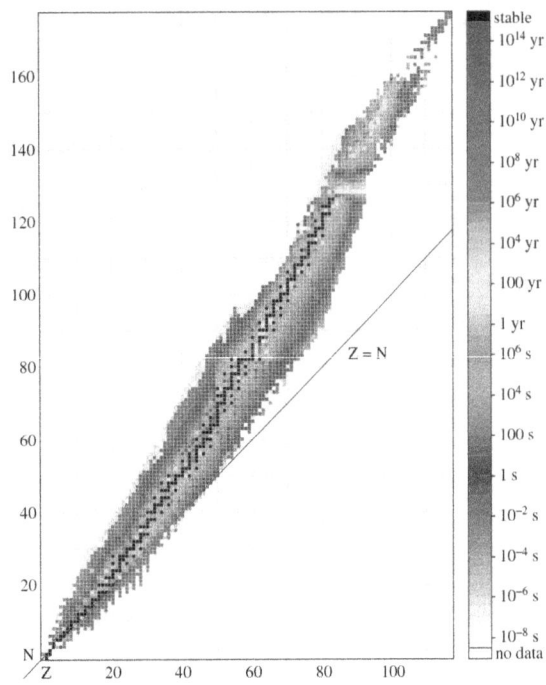

Isotope half-lives. The plot for stable isotopes diverges from the line Z = N as the element number Z becomes larger

electrostatic repulsion between the protons, and they exert the attractive nuclear force on each other and on protons. For this reason, one or more neutrons are necessary for two or more protons to bind into a nucleus. As the number of protons increases, so does the ratio of neutrons to protons necessary to ensure a stable nucleus (see graph at right). For example, although the neutron:proton ratio of ^3_2He is 1:2, the neutron:proton ratio of $^{238}_{92}U$ is greater than 3:2. A number of lighter elements have stable nuclides with the ratio 1:1 ($Z = N$). The nuclide $^{40}_{20}Ca$ (calcium-40) is observationally the heaviest stable nuclide with the same number of neutrons and protons; (theoretically, the heaviest stable one is sulfur-32). All stable nuclides heavier than calcium-40 contain more neutrons than protons.

4.5.3 Numbers of isotopes per element

Of the 81 elements with a stable isotope, the largest number of stable isotopes observed for any element is ten (for the element tin). No element has nine stable isotopes. Xenon is the only element with eight stable isotopes. Four elements have seven stable isotopes, eight have six stable isotopes, ten have five stable isotopes, nine have four stable isotopes, five have three stable isotopes, 16 have two stable isotopes (counting $^{180m}_{73}Ta$ as stable), and 26 elements have only a single sta-

ble isotope (of these, 19 are so-called mononuclidic elements, having a single primordial stable isotope that dominates and fixes the atomic weight of the natural element to high precision; 3 radioactive **mononuclidic** elements occur as well).[19] In total, there are 254 nuclides that have not been observed to decay. For the 80 elements that have one or more stable isotopes, the average number of stable isotopes is 254/80 = 3.2 isotopes per element.

4.5.4   Even and odd nucleon numbers

Main article: Even and odd atomic nuclei

The proton:neutron ratio is not the only factor affecting nuclear stability. It depends also on evenness or oddness of its atomic number Z, neutron number N and, consequently, of their sum, the mass number A. Oddness of both Z and N tends to lower the nuclear binding energy, making odd nuclei, generally, less stable. This remarkable difference of nuclear binding energy between neighbouring nuclei, especially of odd-A isobars, has important consequences: unstable isotopes with a nonoptimal number of neutrons or protons decay by beta decay (including positron decay), electron capture or other exotic means, such as spontaneous fission and cluster decay.

The majority of stable nuclides are even-proton-even-neutron, where all numbers Z, N, and A are even. The odd-A stable nuclides are divided (roughly evenly) into odd-proton-even-neutron, and even-proton-odd-neutron nuclides. Odd-proton-odd-neutron nuclei are the least common.

Even atomic number

The 148 even-proton, even-neutron (EE) nuclides comprise ~ 58% of all stable nuclides and all have spin 0 because of pairing. There are also 22 primordial long-lived even-even nuclides. As a result, each of the 41 even-numbered elements from 2 to 82 has at least one stable isotope, and most of these elements have *several* primordial isotopes. Half of these even-numbered elements have six or more stable isotopes. The extreme stability of helium-4 due to a double pairing of 2 protons and 2 neutrons prevents *any* nuclides containing five or eight nucleons from existing for long enough to serve as platforms for the buildup of heavier elements via nuclear fusion in stars (see triple alpha process).

These 53 stable nuclides have an even number of protons and an odd number of neutrons. They are a minority in comparison to the even-even isotopes, which are about 3 times as numerous. Among the 41 even-Z elements that have a stable nuclide, only three elements (argon, cerium, and lead) have no even-odd stable nuclides. One element (tin) has three. There are 24 elements that have one even-odd nuclide and 13 that have two odd-even nuclides. Of 35 primordial radionuclides there exist four even-odd nuclides (see table at right), including the fissile 235

92U. Because of their odd neutron numbers, the even-odd nuclides tend to have large neutron capture cross sections, due to the energy that results from neutron-pairing effects. These stable even-proton odd-neutron nuclides tend to be uncommon by abundance in nature, generally because, to form and enter into primordial abundance, they must have escaped capturing neutrons to form yet other stable even-even isotopes, during both the s-process and r-process of neutron capture, during nucleosynthesis in stars. For this reason, only 195

78Pt and 9

4Be are the most naturally abundant isotopes of their element.

Odd atomic number

48 stable odd-proton-even-neutron nuclides, stabilized by their even numbers of paired neutrons, form most of the stable isotopes of the odd-numbered elements; the very few odd-odd nuclides comprise the others. There are 41 odd-numbered elements with $Z = 1$ through 81, of which 39 have stable isotopes (the elements technetium (

43Tc) and promethium (

61Pm) have no stable isotopes). Of these 39 odd Z elements, 30 elements (including hydrogen-1 where 0 neutrons is even) have one stable odd-even isotope, and nine elements: chlorine (

17Cl), potassium (

19K), copper (

29Cu), gallium (

31Ga), bromine (

35Br), silver (

47Ag), antimony (

51Sb), iridium (

77Ir), and thallium (

81Tl), have two odd-even stable isotopes each. This makes a total $30 + 2(9) = 48$ stable odd-even isotopes.

There are also five primordial long-lived radioactive odd-even isotopes, 87

37Rb, 115

49In, 187

75Re, 151

63Eu, and 209

83Bi. The last two were only recently found to decay, with half-lives greater than 10^{18} years.

Only five stable nuclides contain both an odd number of protons *and* an odd number of neutrons. The first four "odd-odd" nuclides occur in low mass nuclides, for which changing a proton to a neutron or vice versa would lead to a very lopsided proton-neutron ratio (2

1H, 6

3Li, 10

5B, and 14

7N; spins 1, 1, 3, 1). The only other entirely "stable" odd-odd nuclide is 180m

73Ta (spin 9), the only primordial nuclear isomer, which has not yet been observed to decay despite experimental attempts.[20] Hence, all observationally stable odd-odd nuclides have nonzero integer spin. This is because the single unpaired neutron and unpaired proton have a larger nuclear force attraction to each other if their spins are aligned (producing a total spin of at least 1 unit), instead of anti-aligned. See deuterium for the simplest case of this nuclear behavior.

Many odd-odd radionuclides (like tantalum-180) with comparatively short half lives are known. Usually, they beta-decay to their nearby even-even isobars that have paired protons and paired neutrons. Of the nine primordial odd-odd nuclides (five stable and four radioactive with long half lives), only 14

7N is the most common isotope of a common element. This is the case because it is a part of the CNO cycle. The nuclides 6

3Li and 10

5B are minority isotopes of elements that are themselves rare compared to other light elements, while the other six isotopes make up only a tiny percentage of the natural abundance of their elements. For example, 180m

73Ta is thought to be the rarest of the 254 stable isotopes.

Odd neutron number

Actinides with odd neutron number are generally fissile (with thermal neutrons), while those with even neutron number are generally not, though they are fissionable with fast neutrons. Only 195

78Pt, 9

4Be and 14

7N have odd neutron number and are the most naturally abundant isotope of their element.

4.6   Occurrence in nature

See also: Abundance of the chemical elements

Elements are composed of one or more naturally occurring isotopes. The unstable (radioactive) isotopes are either primordial or postprimordial. Primordial isotopes were a product of stellar nucleosynthesis or another type of nucleosynthesis such as cosmic ray spallation, and have persisted down to the present because their rate of decay is so slow (e.g., uranium-238 and potassium-40). Postprimordial isotopes were created by cosmic ray bombardment as cosmogenic nuclides (e.g., tritium, carbon-14), or by the decay of a radioactive primordial isotope to a radioactive radiogenic nuclide daughter (e.g., uranium to radium). A few isotopes are naturally synthesized as nucleogenic nuclides, by some other natural nuclear reaction, such as when neutrons from natural nuclear fission are absorbed by another atom.

As discussed above, only 80 elements have any stable isotopes, and 26 of these have only one stable isotope. Thus, about two-thirds of stable elements occur naturally on Earth in multiple stable isotopes, with the largest number of stable isotopes for an element being ten, for tin (
50Sn). There are about 94 elements found naturally on Earth (up to plutonium inclusive), though some are detected only in very tiny amounts, such as plutonium-244. Scientists estimate that the elements that occur naturally on Earth (some only as radioisotopes) occur as 339 isotopes (nuclides) in total.[21] Only 254 of these naturally occurring isotopes are stable in the sense of never having been observed to decay as of the present time. An additional 35 primordial nuclides (to a total of 289 primordial nuclides), are radioactive with known half-lives, but have half-lives longer than 80 million years, allowing them to exist from the beginning of the solar system. See list of nuclides for details.

All the known stable isotopes occur naturally on Earth; the other naturally occurring-isotopes are radioactive but occur on Earth due to their relatively long half-lives, or else due to other means of ongoing natural production. These include the afore-mentioned cosmogenic nuclides, the nucleogenic nuclides, and any radiogenic radioisotopes formed by ongoing decay of a primordial radioactive isotope, such as radon and radium from uranium.

An additional ~3000 radioactive isotopes not found in nature have been created in nuclear reactors and in particle accelerators. Many short-lived isotopes not found naturally on Earth have also been observed by spectroscopic analysis, being naturally created in stars or supernovae. An example is aluminium-26, which is not naturally found on Earth, but is found in abundance on an astronomical scale.

The tabulated atomic masses of elements are averages that account for the presence of multiple isotopes with different masses. Before the discovery of isotopes, empirically determined noninteger values of atomic mass confounded scientists. For example, a sample of chlorine contains 75.8% chlorine-35 and 24.2% chlorine-37, giving an average atomic mass of 35.5 atomic mass units.

According to generally accepted cosmology theory, only isotopes of hydrogen and helium, traces of some isotopes of lithium and beryllium, and perhaps some boron, were created at the Big Bang, while all other isotopes were synthesized later, in stars and supernovae, and in interactions between energetic particles such as cosmic rays, and previously produced isotopes. (See nucleosynthesis for details of the various processes thought responsible for isotope production.) The respective abundances of isotopes on Earth result from the quantities formed by these processes, their spread through the galaxy, and the rates of decay for isotopes that are unstable. After the initial coalescence of the solar system, isotopes were redistributed according to mass, and the isotopic composition of elements varies slightly from planet to planet. This sometimes makes it possible to trace the origin of meteorites.

4.7 Atomic mass of isotopes

The atomic mass (m_r) of an isotope is determined mainly by its mass number (i.e. number of nucleons in its nucleus). Small corrections are due to the binding energy of the nucleus (see mass defect), the slight difference in mass between proton and neutron, and the mass of the electrons associated with the atom, the latter because the electron:nucleon ratio differs among isotopes.

The mass number is a dimensionless quantity. The atomic mass, on the other hand, is measured using the atomic mass unit based on the mass of the carbon-12 atom. It is denoted with symbols "u" (for unified atomic mass unit) or "Da" (for dalton).

The atomic masses of naturally occurring isotopes of an element determine the atomic mass of the element. When the element contains N isotopes, the expression below is applied for the average atomic mass $\overline{m}_a$:

$$\overline{m}_a = m_1 x_1 + m_2 x_2 + ... + m_N x_N$$

where m_1, m_2, ..., mN are the atomic masses of each individual isotope, and x_1, ..., xN are the relative abundances of these isotopes.

4.8 Applications of isotopes

4.8.1 Purification of isotopes

Main article: isotope separation

Several applications exist that capitalize on properties of the various isotopes of a given element. Isotope separation is a significant technological challenge, particularly with heavy elements such as uranium or plutonium. Lighter elements such as lithium, carbon, nitrogen, and oxygen are commonly separated by gas diffusion of their compounds such as CO and NO. The separation of hydrogen and deuterium is unusual since it is based on chemical rather than physical properties, for example in the Girdler sulfide process. Uranium isotopes have been separated in bulk by gas diffusion, gas centrifugation, laser ionization separation, and (in the Manhattan Project) by a type of production mass spectrometry.

4.8.2 Use of chemical and biological properties

Main articles: isotope geochemistry, cosmochemistry and paleoclimatology

- Isotope analysis is the determination of isotopic signature, the relative abundances of isotopes of a given element in a particular sample. For biogenic substances in particular, significant variations of isotopes of C, N and O can occur. Analysis of such variations has a wide range of applications, such as the detection of adulteration in food products[22] or the geographic origins of products using isoscapes. The identification of certain meteorites as having originated on Mars is based in part upon the isotopic signature of trace gases contained in them.[23]

- Isotopic substitution can be used to determine the mechanism of a chemical reaction via the kinetic isotope effect.

- Another common application is isotopic labeling, the use of unusual isotopes as tracers or markers in chemical reactions. Normally, atoms of a given element are indistinguishable from each other. However, by using isotopes of different masses, even different nonradioactive stable isotopes can be distinguished by mass spectrometry or infrared spectroscopy. For example, in 'stable isotope labeling with amino acids in cell culture (SILAC)' stable isotopes are used to quantify proteins. If radioactive isotopes are used, they can be detected by the radiation they emit (this is called *radioisotopic labeling*).

- Isotopes are commonly used to determine the concentration of various elements or substances using the isotope dilution method, whereby known amounts of isotopically-substituted compounds are mixed with the samples and the isotopic signatures of the resulting mixtures are determined with mass spectrometry.

4.8.3 Use of nuclear properties

- A technique similar to radioisotopic labeling is radiometric dating: using the known half-life of an unstable element, one can calculate the amount of time that has elapsed since a known concentration of isotope existed. The most widely known example is radiocarbon dating used to determine the age of carbonaceous materials.

- Several forms of spectroscopy rely on the unique nuclear properties of specific isotopes, both radioactive and stable. For example, nuclear magnetic resonance (NMR) spectroscopy can be used only for isotopes with a nonzero nuclear spin. The most common isotopes used with NMR spectroscopy are ^{1}H, ^{2}D, ^{15}N, ^{13}C, and ^{31}P.

- Mössbauer spectroscopy also relies on the nuclear transitions of specific isotopes, such as ^{57}Fe.

- Radionuclides also have important uses. Nuclear power and nuclear weapons development require relatively large quantities of specific isotopes. Nuclear medicine and radiation oncology utilize radioisotopes respectively for medical diagnosis and treatment.

4.9 See also

- Abundance of the chemical elements

- Atom

- Table of nuclides

- Table of nuclides (complete)

- List of isotopes

- List of isotopes by half-life

- List of elements by stability of isotopes

- Isotones

- Isobars

- Radionuclide (or radioisotope)

- Nuclear medicine (includes medical isotopes)

- Isotopomer

- List of particles

- Geotraces

- Isotope dilution

4.10   Notes

- Isotopes are nuclides having the same number of protons; compare:

  - Isotones are nuclides having the same number of neutrons. $N = A - Z$

  - Isobars are nuclides having the same mass number, i.e. sum of protons plus neutrons. A

  - Nuclear isomers are different excited states of the same type of nucleus. A transition from one isomer to another is accompanied by emission or absorption of a gamma ray, or the process of internal conversion. Isomers are by definition both isotopic and isobaric. (Not to be confused with chemical isomers.)

  - Isodiaphers are nuclides having the same neutron excess, i.e. number of neutrons minus number of protons. $D = N - Z$

- Bainbridge mass spectrometer

4.11   References

[1] IUPAC (Connelly, N. G.; Damhus, T.; Hartshorn, R. M.; and Hutton, A. T.), *Nomenclature of Inorganic Chemistry – IUPAC Recommendations 2005*, The Royal Society of Chemistry, 2005; IUPAC (McCleverty, J. A.; and Connelly, N. G.), *Nomenclature of Inorganic Chemistry II. Recommendations 2000*, The Royal Society of Chemistry, 2001; IUPAC (Leigh, G. J.), *Nomenclature of Inorganic Chemistry (recommendations 1990)*, Blackwell Science, 1990; IUPAC, *Nomenclature of Inorganic Chemistry, Second Edition*, 1970; probably in the 1958 first edition as well

[2] This notation seems to have been introduced in the second half of the 1930s. Before that, various notations were used, such as Ne(22) for neon-22 (1934), Ne^{22} for neon-22 (1935), or even Pb$_{210}$ for lead-210 (1933).

[3] "Radioactives Missing From The Earth".

[4] "NuDat 2 Description".

[5] Choppin, G.; Liljenzin, J. O. and Rydberg, J. (1995) *Radiochemistry and Nuclear Chemistry* (2nd ed.) Butterworth-Heinemann, pp. 3–5

[6] Others had also suggested the possibility of isotopes; e.g.,

  - Strömholm, Daniel and Svedberg, Theodor (1909) "Untersuchungen über die Chemie der radioactiven Grundstoffe II." (Investigations into the chemistry of the radioactive elements, part 2), *Zeitschrift für anorganischen Chemie*, **63**: 197–206; see especially page 206.

  - Alexander Thomas Cameron, *Radiochemistry* (London, England: J. M. Dent & Sons, 1910), p. 141. (Cameron also anticipated the displacement law.)

[7] Scerri, Eric R. (2007) *The Periodic Table* Oxford University Press, pp. 176–179 ISBN 0195305736

[8] Nagel, Miriam C. (1982). "Frederick Soddy: From Alchemy to Isotopes". *Journal of Chemical Education* **59** (9): 739–740. Bibcode:1982JChEd..59..739N. doi:10.1021/ed059p739.

[9] See:

  - Kasimir Fajans (1913) "Über eine Beziehung zwischen der Art einer radioaktiven Umwandlung und dem elektrochemischen Verhalten der betreffenden Radioelemente" (On a relation between the type of radioactive transformation and the electrochemical behavior of the relevant radioactive elements), *Physikalische Zeitschrift*, **14**: 131–136.

  - Soddy announced his "displacement law" in: Soddy, Frederick (1913). "The Radio-Elements and the Periodic Law". *Nature* **91** (2264): 57. Bibcode:1913Natur..91...57S. doi:10.1038/091057a0..

  - Soddy elaborated his displacement law in: Soddy, Frederick (1913) "Radioactivity," *Chemical Society Annual Report*, **10**: 262–288.

  - Alexander Smith Russell (1888–1972) also published a displacement law: Russell, Alexander S. (1913) "The periodic system and the radio-elements," *Chemical News and Journal of Industrial Science*, **107**: 49–52.

[10] Soddy first used the word "isotope" in: Soddy, Frederick (1913). "Intra-atomic charge". *Nature* **92** (2301): 399–400. Bibcode:1913Natur..92..399S. doi:10.1038/092399c0.

[11] Fleck, Alexander (1957). "Frederick Soddy". *Biographical Memoirs of Fellows of the Royal Society* **3**: 203–216. doi:10.1098/rsbm.1957.0014. p. 208: Up to 1913 we used the phrase 'radio elements chemically non-separable' and at that time the word isotope was suggested in a drawing-room discussion with Dr. Margaret Todd in the home of Soddy's father-in-law, Sir George Beilby.

[12] Budzikiewicz H and Grigsby RD (2006). "Mass spectrometry and isotopes: a century of research and discussion". *Mass spectrometry reviews* **25** (1): 146–57. doi:10.1002/mas.20061. PMID 16134128.

[13] Scerri, Eric R. (2007) *The Periodic Table*, Oxford University Press, ISBN 0195305736, Ch. 6, note 44 (p. 312) citing Alexander Fleck, described as a former student of Soddy's.

[14] In his 1893 book, William T. Preyer also used the word "isotope" to denote similarities among elements. From p. 9 of William T. Preyer, *Das genetische System der chemischen Elemente* [The genetic system of the chemical elements] (Berlin, Germany: R. Friedländer & Sohn, 1893):

"Die ersteren habe ich der Kürze wegen isotope Elemente genannt, weil sie in jedem der sieben Stämmme der gleichen Ort, nämlich dieselbe Stuffe, einnehmen." (For the sake of brevity, I have named the former "isotopic" elements, because they occupy the same place in each of the seven families [i.e., columns of the periodic table], namely the same step [i.e., row of the periodic table].)

[15] The origins of the conceptions of isotopes Frederick Soddy, Nobel prize lecture

[16] Thomson, J. J. (1912). "XIX. Further experiments on positive rays". *Philosophical Magazine Series 6* **24** (140): 209. doi:10.1080/14786440808637325.

[17] Thomson, J. J. (1910). "LXXXIII. Rays of positive electricity". *Philosophical Magazine Series 6* **20** (118): 752. doi:10.1080/14786441008636962.

[18] Mass spectra and isotopes Francis W. Aston, Nobel prize lecture 1922

[19] **Sonzogni**, Alejandro (2008). "Interactive Chart of Nuclides". National Nuclear Data Center: Brookhaven National Laboratory. Retrieved 2013-05-03.

[20] Hult, Mikael; Wieslander, J. S.; Marissens, Gerd; Gasparro, Joël; Wätjen, Uwe; Misiaszek, Marcin (2009). "Search for the radioactivity of 180mTa using an underground HPGe sandwich spectrometer". *Applied Radiation and Isotopes* **67** (5): 918–21. doi:10.1016/j.apradiso.2009.01.057. PMID 19246206.

[21] "Radioactives Missing From The Earth". Don-lindsay-archive.org. Retrieved 2012-06-16.

[22] Jamin, Eric; Guérin, Régis; Rétif, Mélinda; Lees, Michèle; Martin, Gérard J. (2003). "Improved Detection of Added Water in Orange Juice by Simultaneous Determination of the Oxygen-18/Oxygen-16 Isotope Ratios of Water and Ethanol Derived from Sugars". *J. Agric. Food Chem.* **51** (18): 5202. doi:10.1021/jf030167m.

[23] A. H. Treiman, J. D. Gleason and D. D. Bogard (2000). "The SNC meteorites are from Mars". *Planet. Space Sci.* **48** (12–14): 1213. Bibcode:2000P&SS...48.1213T. doi:10.1016/S0032-0633(00)00105-7.

4.12 External links

- The Nuclear Science web portal Nucleonica

- The Karlsruhe Nuclide Chart

- National Nuclear Data Center Portal to large repository of free data and analysis programs from NNDC

- National Isotope Development Center Coordination and management of the production, availability, and distribution of isotopes, and reference information for the isotope community

- Isotope Development & Production for Research and Applications (IDPRA) U.S. Department of Energy program for isotope production and production research and development

- International Atomic Energy Agency Homepage of International Atomic Energy Agency (IAEA), an Agency of the United Nations (UN)

- Atomic Weights and Isotopic Compositions for All Elements Static table, from NIST (National Institute of Standards and Technology)

- Atomgewichte, Zerfallsenergien und Halbwertszeiten aller Isotope

- Exploring the Table of the Isotopes at the LBNL

- Current isotope research and information isotope.info

- Emergency Preparedness and Response: Radioactive Isotopes by the CDC (Centers for Disease Control and Prevention)

- Chart of Nuclides Interactive Chart of Nuclides (National Nuclear Data Center)

- Interactive Chart of the nuclides, isotopes and Periodic Table

- The LIVEChart of Nuclides – IAEA with isotope data.

- Annotated bibliography for isotopes from the Alsos Digital Library for Nuclear Issues

Chapter 5

Periodic table

This article is about the table used in chemistry. For other uses, see Periodic table (disambiguation).

Standard form of the periodic table (color legend below)

The **periodic table** is a tabular arrangement of the chemical elements, ordered by their atomic number (number of protons in the nucleus), electron configurations, and recurring chemical properties. The table also shows four rectangular blocks: s-, p- d- and f-block. In general, within one row (period) the elements are metals on the lefthand side, and non-metals on the righthand side.

The rows of the table are called periods; the columns are called groups. Six groups (columns) have names as well as numbers: for example, group 17 elements are the halogens; and group 18, the noble gases. The periodic table can be used to derive relationships between the properties of the elements, and predict the properties of new elements yet to be discovered or synthesized. The periodic table provides a useful framework for analyzing chemical behavior, and is widely used in chemistry and other sciences.

Although precursors exist, Dmitri Mendeleev is generally credited with the publication, in 1869, of the first widely recognized periodic table. He developed his table to illustrate periodic trends in the properties of the then-known elements. Mendeleev also predicted some properties of then-unknown elements that would be expected to fill gaps in this table. Most of his predictions were proved correct when the elements in question were subsequently discov-

ered. Mendeleev's periodic table has since been expanded and refined with the discovery or synthesis of further new elements and the development of new theoretical models to explain chemical behavior.

All elements from atomic numbers 1 (hydrogen) to 118 (ununoctium) have been discovered or reportedly synthesized, with elements 113, 115, 117, and 118 having yet to be confirmed. The first 98 elements exist naturally, although some are found only in trace amounts and were synthesized in laboratories before being found in nature.[n 1] Elements with atomic numbers from 99 to 118 have only been synthesized in laboratories. It has been shown that einsteinium and fermium once occurred in nature but currently do not.[1] Synthesis of elements having higher atomic numbers is being pursued. Numerous synthetic radionuclides of naturally occurring elements have also been produced in laboratories.

5.1 Overview

For large cell versions, see Periodic table (large cells).

Some presentations include an element zero (i.e. a substance composed purely of neutrons), although this is uncommon. See, for example. Philip Stewart's Chemical Galaxy. Each chemical element has a unique atomic number representing the number of protons in its nucleus. Most elements have differing numbers of neutrons among different atoms, with these variants being referred to as isotopes. For example, carbon has three naturally occurring isotopes: all of its atoms have six protons and most have six neutrons as well, but about one per cent have seven neutrons, and a very small fraction have eight neutrons. Isotopes are never separated in the periodic table; they are always grouped together under a single element. Elements with no stable isotopes have the atomic masses of their most stable isotopes, where such masses are shown, listed in parentheses.[2]

In the standard periodic table, the elements are listed in order of increasing atomic number (the number of protons in the nucleus of an atom). A new row (*period*) is started when a new electron shell has its first electron. Columns (*groups*) are determined by the electron configuration of the atom; elements with the same number of electrons in a particular subshell fall into the same columns (e.g. oxygen and selenium are in the same column because they both have four electrons in the outermost p-subshell). Elements with similar chemical properties generally fall into the same group in the periodic table, although in the f-block, and to some respect in the d-block, the elements in the same period tend to have similar properties, as well. Thus, it is relatively easy to predict the chemical properties of an element if one knows the properties of the elements around it.[3]

As of 2014, the periodic table has 114 confirmed elements, comprising elements 1 (hydrogen) to 112 (copernicium), 114 (flerovium) and 116 (livermorium). Elements 113, 115, 117 and 118 have reportedly been synthesised in laboratories, but none of these claims have been officially confirmed by the International Union of Pure and Applied Chemistry (IUPAC), nor are they named. As such these elements are currently identified by their atomic number (e.g., "element 113"), or by their provisional systematic name ("ununtrium", symbol "Uut").[4]

A total of 98 elements occur naturally; the remaining 16 elements, from einsteinium to copernicium, and flerovium and livermorium, occur only when synthesised in laboratories. Of the 98 elements that occur naturally, 84 are primordial. The other 14 naturally occurring elements occur only in decay chains of primordial elements.[1] No element heavier than einsteinium (element 99) has ever been observed in macroscopic quantities in its pure form.[5]

5.1.1 Layout variants

In the most common graphic presentation of the periodic table, the main table has 18 columns and the lanthanides and the actinides are shown as two additional rows below the main body of the table,[6] with two placeholders shown in the main table, between barium and hafnium, and radium and rutherfordium, respectively. These placeholders can be asterisk-like markers, or a contracted range description of elements ("57–71"). This convention is entirely a matter of formatting practicality. The same table structure can be shown in a 32-column format, with the lanthanides and actinides in the main table's row 6 and 7.

However, based on the chemical and physical properties of elements, many alternative table *structures* have been constructed.

5.2 Grouping methods

5.2.1 Groups

Main article: Group (periodic table)

A *group* or *family* is a vertical column in the periodic table. Groups usually have more significant periodic trends than periods and blocks, explained below. Modern quantum mechanical theories of atomic structure explain group trends by proposing that elements within the same group generally have the same electron configurations in their valence shell.[7] Consequently, elements in the same group tend to have a shared chemistry and exhibit a clear trend in properties with increasing atomic number.[8] However, in some parts of the periodic table, such as the d-block and the f-block, horizontal similarities can be as important as, or more pronounced than, vertical similarities.[9][10][11]

Under an international naming convention, the groups are numbered numerically from 1 to 18 from the leftmost column (the alkali metals) to the rightmost column (the noble gases).[12] Previously, they were known by roman numerals. In America, the roman numerals were followed by either an "A" if the group was in the s- or p-block, or a "B" if the group was in the d-block. The roman numerals used correspond to the last digit of today's naming convention (e.g. the group 4 elements were group IVB, and the group 14 elements was group IVA). In Europe, the lettering was similar, except that "A" was used if the group was before group 10, and "B" was used for groups including and after group 10. In addition, groups 8, 9 and 10 used to be treated as one triple-sized group, known collectively in both notations as group VIII. In 1988, the new IUPAC naming system was put into use, and the old group names were deprecated.[13]

Some of these groups have been given trivial (unsystematic) names, as seen in the table below, although some are rarely used. Groups 3–10 have no trivial names and are referred to simply by their group numbers or by the name of the first member of their group (such as 'the scandium group' for Group 3), since they display fewer similarities and/or vertical trends.[12]

Elements in the same group tend to show patterns in atomic radius, ionization energy, and electronegativity. From top to bottom in a group, the atomic radii of the elements increase. Since there are more filled energy levels, valence electrons are found farther from the nucleus. From the top, each successive element has a lower ionization energy because it is easier to remove an electron since the atoms are less tightly bound. Similarly, a group has a top to bottom decrease in electronegativity due to an increasing distance between valence electrons and the nucleus.[14] There are exceptions to these trends, however, an example of which occurs in group 11 where electronegativity increases farther down the group.[15]

5.2.2 Periods

Main article: Period (periodic table)

A *period* is a horizontal row in the periodic table. Although groups generally have more significant periodic trends, there are regions where horizontal trends are more significant than vertical group trends, such as the f-block, where the lanthanides and actinides form two substantial horizontal series of elements.[16]

Elements in the same period show trends in atomic radius, ionization energy, electron affinity, and electronegativity. Moving left to right across a period, atomic radius usually decreases. This occurs because each successive element has an added proton and electron, which causes the electron to be drawn closer to the nucleus.[17] This decrease in atomic radius also causes the ionization energy to increase when moving from left to right across a period. The more tightly bound an element is, the more energy is required to remove an electron. Electronegativity increases in the same manner as ionization energy because of the pull exerted on the electrons by the nucleus.[14] Electron affinity also shows a slight trend across a period. Metals (left side of a period) generally have a lower electron affinity than nonmetals (right side of a period), with the exception of the noble gases.[18]

5.2.3 Blocks

Main article: Block (periodic table)
Specific regions of the periodic table can be referred to

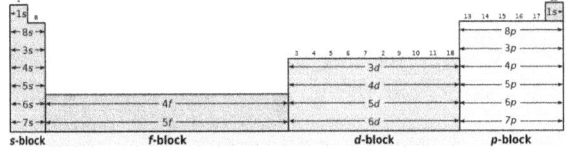

Left to right: s-, f-, d-, p-block in the periodic table

as *blocks* in recognition of the sequence in which the electron shells of the elements are filled. Each block is named according to the subshell in which the "last" electron notionally resides.[19][n 2] The s-block comprises the first two groups (alkali metals and alkaline earth metals) as well as hydrogen and helium. The p-block comprises the last six groups, which are groups 13 to 18 in IUPAC (3A to 8A in American) and contains, among other elements, all of the metalloids. The d-block comprises groups 3 to 12 (or 3B to 2B in American group numbering) and contains all of the transition metals. The f-block, often offset below the rest of the periodic table, has no group numbers and comprises lanthanides and actinides.[20]

5.2.4 Metals, metalloids and nonmetals

Metals, metalloids, nonmetals, and elements with unknown chemical properties in the periodic table. Sources disagree on the classification of some of these elements.

According to their shared physical and chemical properties, the elements can be classified into the major categories of metals, metalloids and nonmetals. Metals are generally shiny, highly conducting solids that form alloys with one another and salt-like ionic compounds with nonmetals (other than the noble gases). The majority of nonmetals are coloured or colourless insulating gases; nonmetals that form compounds with other nonmetals feature covalent bonding. In between metals and nonmetals are metalloids, which have intermediate or mixed properties.[21]

Metal and nonmetals can be further classified into subcategories that show a gradation from metallic to nonmetallic properties, when going left to right in the rows. The metals are subdivided into the highly reactive alkali metals, through the less reactive alkaline earth metals, lanthanides and actinides, via the archetypal transition metals, and ending in the physically and chemically weak posttransition metals. The nonmetals are simply subdivided into the polyatomic nonmetals, which, being nearest to the metalloids, show some incipient metallic character; the diatomic nonmetals, which are essentially nonmetallic; and the monatomic noble gases, which are nonmetallic and almost completely inert. Specialized groupings such as the refractory metals and the noble metals, which are subsets (in this example) of the transition metals, are also known[22] and occasionally denoted.[23]

Placing the elements into categories and subcategories based on shared properties is imperfect. There is a spectrum of properties within each category and it is not hard to find overlaps at the boundaries, as is the case with most classification schemes.[24] Beryllium, for example, is classified as an alkaline earth metal although its amphoteric chemistry and tendency to mostly form covalent compounds are both attributes of a chemically weak or post transition metal. Radon is classified as a nonmetal and a noble gas yet has some cationic chemistry that is more characteristic of a metal. Other classification schemes are possible such as the division of the elements into mineralogical occurrence categories, or crystalline structures. Categorising the elements in this fashion dates back to at least 1869 when Hinrichs[25] wrote that simple boundary lines could be drawn on the periodic table to show elements having like properties, such as the metals and the nonmetals, or the gaseous elements.

5.3 Periodic trends

Main article: Periodic trends

5.3.1 Electron configuration

Main article: Electronic configuration
The electron configuration or organisation of electrons or-

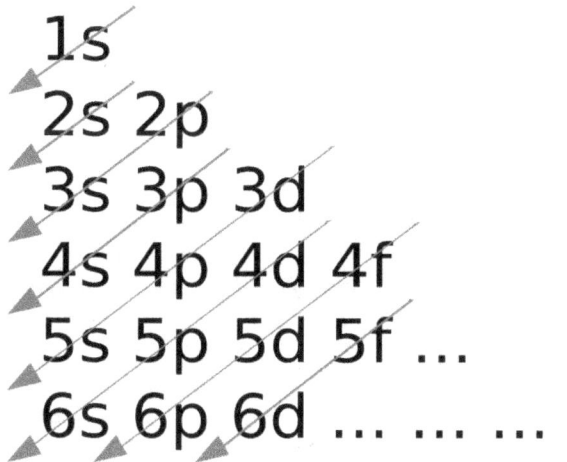

Approximate order in which shells and subshells are arranged by increasing energy according to the Madelung rule

biting neutral atoms shows a recurring pattern or periodicity. The electrons occupy a series of electron shells (numbered shell 1, shell 2, and so on). Each shell consists of one or more subshells (named s, p, d, f and g). As atomic number increases, electrons progressively fill these shells and subshells more or less according to the Madelung rule or energy ordering rule, as shown in the diagram. The electron configuration for neon, for example, is $1s^2\ 2s^2\ 2p^6$. With an atomic number of ten, neon has two electrons in the first shell, and eight electrons in the second shell—two in the s subshell and six in the p subshell. In periodic table terms, the first time an electron occupies a new shell corresponds to the start of each new period, these positions being occupied by hydrogen and the alkali metals.[26][27]

Since the properties of an element are mostly determined by its electron configuration, the properties of the elements likewise show recurring patterns or periodic behaviour, some examples of which are shown in the diagrams below for atomic radii, ionization energy and electron affinity. It is this periodicity of properties, manifestations of which were

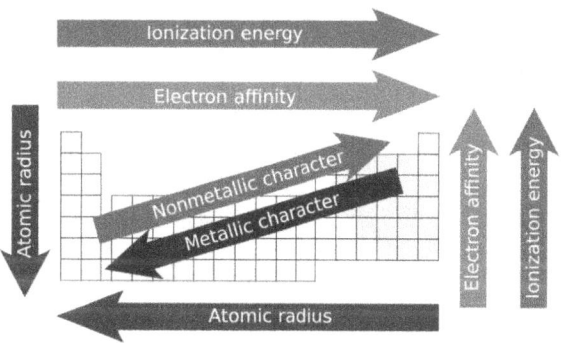

Periodic table trends (arrows direct an increase)

noticed well before the underlying theory was developed, that led to the establishment of the periodic law (the properties of the elements recur at varying intervals) and the formulation of the first periodic tables.[26][27]

5.3.2 Atomic radii

Main article: Atomic radius
Atomic radii vary in a predictable and explainable manner

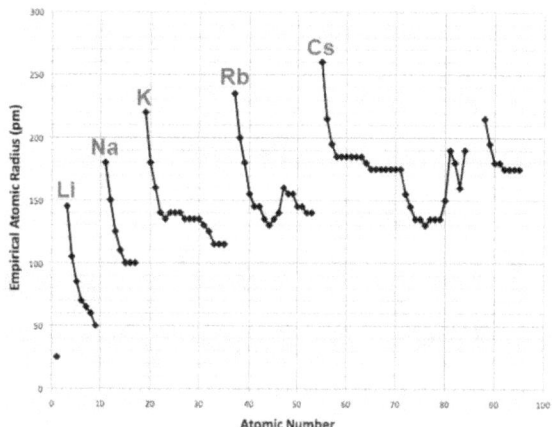

Atomic number plotted against atomic radius[n 3]

across the periodic table. For instance, the radii generally decrease along each period of the table, from the alkali metals to the noble gases; and increase down each group. The radius increases sharply between the noble gas at the end of each period and the alkali metal at the beginning of the next period. These trends of the atomic radii (and of various other chemical and physical properties of the elements) can be explained by the electron shell theory of the atom; they provided important evidence for the development and confirmation of quantum theory.[28]

The electrons in the 4f-subshell, which is progressively filled from cerium (element 58) to ytterbium (element 70), are not particularly effective at shielding the increasing nuclear charge from the sub-shells further out. The elements immediately following the lanthanides have atomic radii that are smaller than would be expected and that are almost identical to the atomic radii of the elements immediately above them.[29] Hence hafnium has virtually the same atomic radius (and chemistry) as zirconium, and tantalum has an atomic radius similar to niobium, and so forth. This is known as the lanthanide contraction. The effect of the lanthanide contraction is noticeable up to platinum (element 78), after which it is masked by a relativistic effect known as the inert pair effect.[30] The d-block contraction, which is a similar effect between the d-block and p-block, is less pronounced than the lanthanide contraction but arises from a similar cause.[29]

5.3.3 Ionization energy

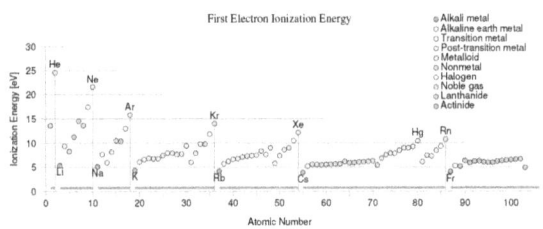

Ionization energy: each period begins at a minimum for the alkali metals, and ends at a maximum for the noble gases

Main article: Ionization energy

The first ionization energy is the energy it takes to remove one electron from an atom, the second ionization energy is the energy it takes to remove a second electron from the atom, and so on. For a given atom, successive ionization energies increase with the degree of ionization. For magnesium as an example, the first ionization energy is 738 kJ/mol and the second is 1450 kJ/mol. Electrons in the closer orbitals experience greater forces of electrostatic attraction; thus, their removal requires increasingly more energy. Ionization energy becomes greater up and to the right of the periodic table.[30]

Large jumps in the successive molar ionization energies occur when removing an electron from a noble gas (complete electron shell) configuration. For magnesium again, the first two molar ionization energies of magnesium given above correspond to removing the two 3s electrons, and the third ionization energy is a much larger 7730 kJ/mol, for the removal of a 2p electron from the very stable neon-like configuration of Mg^{2+}. Similar jumps occur in the ionization energies of other third-row atoms.[30]

5.3.4 Electronegativity

Main article: Electronegativity
Electronegativity is the tendency of an atom to attract

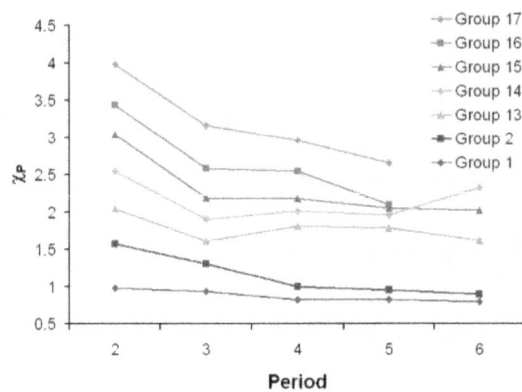

Graph showing increasing electronegativity with growing number of selected groups

electrons.[31] An atom's electronegativity is affected by both its atomic number and the distance between the valence electrons and the nucleus. The higher its electronegativity, the more an element attracts electrons. It was first proposed by Linus Pauling in 1932.[32] In general, electronegativity increases on passing from left to right along a period, and decreases on descending a group. Hence, fluorine is the most electronegative of the elements,[n 4] while caesium is the least, at least of those elements for which substantial data is available.[15]

There are some exceptions to this general rule. Gallium and germanium have higher electronegativities than aluminium and silicon respectively because of the d-block contraction. Elements of the fourth period immediately after the first row of the transition metals have unusually small atomic radii because the 3d-electrons are not effective at shielding the increased nuclear charge, and smaller atomic size correlates with higher electronegativity.[15] The anomalously high electronegativity of lead, particularly when compared to thallium and bismuth, appears to be an artifact of data selection (and data availability)—methods of calculation other than the Pauling method show the normal periodic trends for these elements.[33]

5.3.5 Electron affinity

Main article: Electron affinity
The electron affinity of an atom is the amount of energy released when an electron is added to a neutral atom to form a negative ion. Although electron affinity varies greatly, some patterns emerge. Generally, nonmetals have more positive

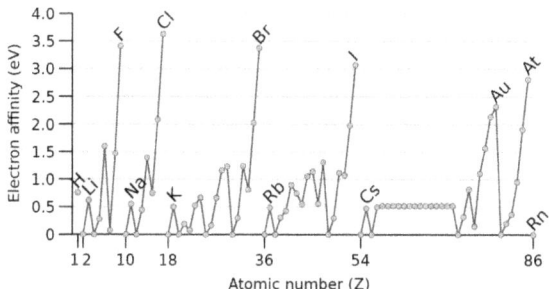

Dependence of electron affinity on atomic number.[34] Values generally increase across each period, culminating with the halogens before decreasing precipitously with the noble gases. Examples of localized peaks seen in hydrogen, the alkali metals and the group 11 elements are caused by a tendency to complete the s-shell (with the 6s shell of gold being further stabilized by relativistic effects and the presence of a filled 4f sub shell). Examples of localized troughs seen in the alkaline earth metals, and nitrogen, phosphorus, manganese and rhenium are caused by filled s-shells, or half-filled p- or d-shells.[35]

electron affinity values than metals. Chlorine most strongly attracts an extra electron. The electron affinities of the noble gases have not been measured conclusively, so they may or may not have slightly negative values.[36]

Electron affinity generally increases across a period. This is caused by the filling of the valence shell of the atom; a group 17 atom releases more energy than a group 1 atom on gaining an electron because it obtains a filled valence shell and is therefore more stable.[36]

A trend of decreasing electron affinity going down groups would be expected. The additional electron will be entering an orbital farther away from the nucleus. As such this electron would be less attracted to the nucleus and would release less energy when added. However, in going down a group, around one-third of elements are anomalous, with heavier elements having higher electron affinities than their next lighter congeners. Largely, this is due to the poor shielding by d and f electrons. A uniform decrease in electron affinity only applies to group 1 atoms.[37]

5.3.6 Metallic character

The lower the values of ionization energy, electronegativity and electron affinity, the more metallic character the element has. Conversely, nonmetallic character increases with higher values of these properties.[38] Given the periodic trends of these three properties, metallic character tends to decrease going across a period (or row) and, with some irregularities (mostly) due to poor screening of the nucleus by d and f electrons, and relativistic effects,[39] tends to increase going down a group (or column or family). Thus,

the most metallic elements (such as caesium and francium) are found at the bottom left of traditional periodic tables and the most nonmetallic elements (oxygen, fluorine, chlorine) at the top right. The combination of horizontal and vertical trends in metallic character explains the stair-shaped dividing line between metals and nonmetals found on some periodic tables, and the practice of sometimes categorizing several elements adjacent to that line, or elements adjacent to those elements, as metalloids.[40][41]

5.4 History

Main article: History of the periodic table

5.4.1 First systemization attempts

The discovery of the elements mapped to significant periodic table development dates (pre-, per- and post-)

In 1789, Antoine Lavoisier published a list of 33 chemical elements, grouping them into gases, metals, nonmetals, and earths.[42] Chemists spent the following century searching for a more precise classification scheme. In 1829, Johann Wolfgang Döbereiner observed that many of the elements could be grouped into triads based on their chemical properties. Lithium, sodium, and potassium, for example, were grouped together in a triad as soft, reactive metals. Döbereiner also observed that, when arranged by atomic weight, the second member of each triad was roughly the average of the first and the third;[43] this became known as the Law of Triads.[44] German chemist Leopold Gmelin worked with this system, and by 1843 he had identified ten triads, three groups of four, and one group of five. Jean-Baptiste Dumas published work in 1857 describing relationships between various groups of metals. Although various chemists were able to identify relationships between small groups of elements, they had yet to build one scheme that encompassed them all.[43]

In 1858, German chemist August Kekulé observed that carbon often has four other atoms bonded to it. Methane, for example, has one carbon atom and four hydrogen atoms. This concept eventually became known as valency; different elements bond with different numbers of atoms.[45]

In 1862, Alexandre-Emile Béguyer de Chancourtois, a French geologist, published an early form of periodic table, which he called the telluric helix or screw. He was the first person to notice the periodicity of the elements. With the elements arranged in a spiral on a cylinder by order of increasing atomic weight, de Chancourtois showed that elements with similar properties seemed to occur at regular intervals. His chart included some ions and compounds in addition to elements. His paper also used geological rather than chemical terms and did not include a diagram; as a result, it received little attention until the work of Dmitri Mendeleev.[46]

In 1864, Julius Lothar Meyer, a German chemist, published a table with 44 elements arranged by valency. The table showed that elements with similar properties often shared the same valency.[47] Concurrently, William Odling (an English chemist) published an arrangement of 57 elements, ordered on the basis of their atomic weights. With some irregularities and gaps, he noticed what appeared to be a periodicity of atomic weights among the elements and that this accorded with 'their usually received groupings.'[48] Odling alluded to the idea of a periodic law but did not pursue it.[49] He subsequently proposed (in 1870) a valence-based classification of the elements.[50]

No.	No.	No.	No.	No.	No.	No.	No.
H 1	F 8	Cl 15	Co & Ni 22	Br 29	Pd 36	I 42	Pt & Ir 50
Li 2	Na 9	K 16	Cu 23	Rb 30	Ag 37	Cs 44	Os 51
G 3	Mg 10	Ca 17	Zn 24	Sr 31	Cd 38	Ba & V 45	Hg 52
Bo 4	Al 11	Cr 19	Y 25	Ce & La 33	U 40	Ta 46	Tl 53
C 5	Si 12	Ti 18	In 26	Zr 32	Sn 39	W 47	Pb 54
N 6	P 13	Mn 20	As 27	Di & Mo 34	Sb 41	Nb 48	Bi 55
O 7	S 14	Fe 21	Se 28	Ro & Ru 35	Te 43	Au 49	Th 56

Newlands's periodic table, as presented to the Chemical Society in 1866, and based on the law of octaves

English chemist John Newlands produced a series of papers from 1863 to 1866 noting that when the elements were listed in order of increasing atomic weight, similar physical and chemical properties recurred at intervals of eight; he likened such periodicity to the octaves of music.[51][52] This so termed Law of Octaves, however, was ridiculed by Newlands' contemporaries, and the Chemical Society refused to publish his work.[53] Newlands was nonetheless able to draft a table of the elements and used it to predict the existence of missing elements, such as germanium.[54] The Chemical Society only acknowledged the significance of his discoveries five years after they credited Mendeleev.[55]

In 1867, Gustavus Hinrichs, a Danish born academic chemist based in America, published a spiral periodic system based on atomic spectra and weights, and chemical similarities. His work was regarded as idiosyncratic, ostentatious and labyrinthine and this may have militated against its recognition and acceptance.[56][57]

5.4.2 Mendeleev's table

Dmitri Mendeleev

Russian chemistry professor Dmitri Mendeleev and German chemist Julius Lothar Meyer independently published their periodic tables in 1869 and 1870, respectively.[58] Mendeleev's table was his first published version; that of Meyer was an expanded version of his (Meyer's) table of 1864.[59] They both constructed their tables by listing the elements in rows or columns in order of atomic weight and starting a new row or column when the characteristics of the elements began to repeat.[60]

The recognition and acceptance afforded to Mendeleev's table came from two decisions he made. The first was to leave gaps in the table when it seemed that the corresponding element had not yet been discovered.[61] Mendeleev was not the first chemist to do so, but he was the first to be recognized as using the trends in his periodic table to predict the properties of those missing elements, such as gallium and germanium.[62] The second decision was to occasionally ignore the order suggested by the atomic weights and

ОПЫТЪ СИСТЕМЫ ЭЛЕМЕНТОВЪ,

ОСНОВАННОЙ НА ИХЪ АТОМНОМЪ ВѢСЬ И ХИМИЧЕСКОМЪ СХОДСТВѢ.

```
                    Ti=50      Zr=90      ?=180.
                    V=51       Nb=94      Ta=182.
                    Cr=52      Mo=96      W=186.
                    Mn=55      Rh=104,4   Pt=197,1.
                    Fe=56      Ru=104,4   Ir=198.
                  Ni=Co=59     Pd=106,6   Os=199.
  H=1                          Cu=63,4    Ag=108     Hg=200.
          Be= 9,4 Mg=24        Zn=65,2    Cd=112
          B=11    Al=27,3      ?=68       Ur=116     Au=197?
          C=12    Si=28        ?=70       Sn=118
          N=14    P=31         As=75      Sb=122     Bi=210?
          O=16    S=32         Se=79,4    Te=128?
          F=19    Cl=35,5      Br=80      I=127
  Li=7    Na=23   K=39         Rb=85,4    Cs=133     Tl=204.
                  Ca=40        Sr=87,6    Ba=137     Pb=207.
                  ?=45         Ce=92
                  ?Er=56       La=94
                  ?Yt=60       Di=95
                  ?In=75,6     Th=118?
```

Д. Менделѣевъ

A version of Mendeleev's 1869 periodic table: An experiment on a system of elements. Based on their atomic weights and chemical similarities. *This early arrangement presents the periods vertically, and the groups horizontally.*

Reihen	Gruppe I. — R²O	Gruppe II. — RO	Gruppe III. — R²O³	Gruppe IV. RH⁴ RO²	Gruppe V. RH³ R²O⁵	Gruppe VI. RH² RO³	Gruppe VII. RH R²O⁷	Gruppe VIII. — RO⁴
1	H=1							
2	Li=7	Be=9,4	B=11	C=12	N=14	O=16	F=19	
3	Na=23	Mg=24	Al=27,3	Si=28	P=31	S=32	Cl=35,5	
4	K=39	Ca=40	—=44	Ti=48	V=51	Cr=52	Mn=55	Fe=56, Co=59, Ni=59, Cu=63.
5	(Cu=63)	Zn=65	—=68	—=72	As=75	Se=78	Br=80	
6	Rb=85	Sr=87	?Yt=88	Zr=90	Nb=94	Mo=96	—=100	Ru=104, Rh=104, Pd=106, Ag=108.
7	(Ag=108)	Cd=112	In=113	Sn=118	Sb=122	Te=125	J=127	
8	Cs=133	Ba=137	?Di=138	?Ce=140	—	—	—	— — — —
9	(—)	—	—	—	—	—	—	
10	—	—	?Er=178	?La=180	Ta=182	W=184	—	Os=195, Ir=197, Pt=198, Au=199.
11	(Au=199)	Hg=200	Tl=204	Pb=207	Bi=208	—	—	
12	—	—	—	Th=231	—	U=240	—	— — — —

Mendeleev's 1871 periodic table with eight groups of elements. Dashes represented elements unknown in 1871.

Eight-column form of periodic table, updated with all elements discovered to 2015

switch adjacent elements, such as tellurium and iodine, to better classify them into chemical families. Later in 1913, Henry Moseley determined experimental values of the nuclear charge or atomic number of each element, and showed that Mendeleev's ordering actually corresponds to the order of increasing atomic number.[63]

The significance of atomic numbers to the organization of the periodic table was not appreciated until the existence and properties of protons and neutrons became understood. Mendeleev's periodic tables used atomic weight instead of atomic number to organize the elements, information determinable to fair precision in his time. Atomic weight worked well enough in most cases to (as noted) give a presentation that was able to predict the properties of missing elements more accurately than any other method then known. Substitution of atomic numbers, once understood, gave a definitive, integer-based sequence for the elements, and Moseley predicted that the only missing elements (in 1913) between aluminum (Z=13) and gold (Z=79) (in 1913) were Z = 43, 61, 72 and 75, which were all later discovered. The sequence of atomic numbers is still used today even as new synthetic elements are being produced and studied.[64]

5.4.3 Second version and further development

In 1871, Mendeleev published his periodic table in a new form, with groups of similar elements arranged in columns rather than in rows, and those columns numbered I to VIII corresponding with the element's oxidation state. He also gave detailed predictions for the properties of elements he had earlier noted were missing, but should exist.[65] These gaps were subsequently filled as chemists discovered additional naturally occurring elements.[66] It is often stated that the last naturally occurring element to be discovered was francium (referred to by Mendeleev as *eka-caesium*) in 1939.[67] However, plutonium, produced synthetically in 1940, was identified in trace quantities as a naturally occurring primordial element in 1971,[68] and by 2011 it was known that all the elements up to californium can occur naturally as trace amounts in uranium ores by neutron capture and beta decay.[1]

The popular[69] periodic table layout, also known as the common or standard form (as shown at various other points in this article), is attributable to Horace Groves Deming. In 1923, Deming, an American chemist, published short (Mendeleev style) and medium (18-column) form periodic tables.[70][n 5] Merck and Company prepared a hand-out form of Deming's 18-column medium table, in 1928, which was widely circulated in American schools. By the 1930s Deming's table was appearing in handbooks and encyclopaedias of chemistry. It was also distributed for many years by the Sargent-Welch Scientific Company.[71][72][73]

With the development of modern quantum mechanical theories of electron configurations within atoms, it became apparent that each period (row) in the table corresponded to the filling of a quantum shell of electrons. Larger atoms have more electron sub-shells, so later tables have required progressively longer periods.[74]

Although minute quantities of some transuranic elements occur naturally,[1] they were all first discovered in laboratories. Their production has expanded the periodic table significantly, the first of these being neptunium, synthesized in 1939.[77] Because many of the transuranic elements are highly unstable and decay quickly, they are challenging to detect and characterize when produced. There have been controversies concerning the acceptance of competing discovery claims for some elements, requiring independent review to determine which party has priority, and hence naming rights. The most recently accepted and named elements are flerovium (element 114) and livermorium (element 116), both named on 31 May 2012.[78] In 2010, a joint Russia–US collaboration at Dubna, Moscow Oblast, Russia, claimed to have synthesized six atoms of ununseptium (element 117), making it the most recently claimed discovery.[79]

5.5   Alternative structures

Main article: Alternative periodic tables
There are many periodic tables with structures other than

Glenn T. Seaborg who, in 1945, suggested a new periodic table showing the actinides as belonging to a second f-block series

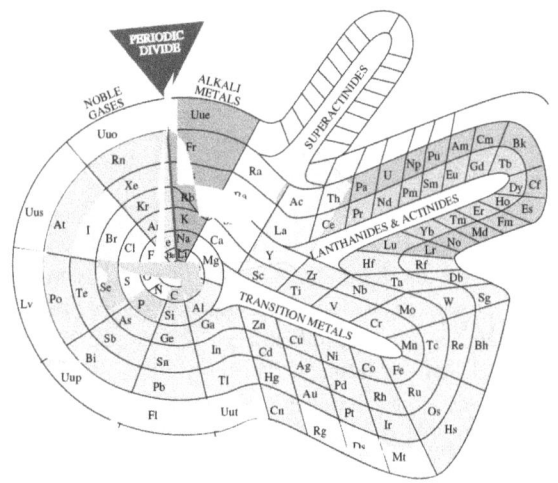

Theodor Benfey's spiral periodic table

In 1945, Glenn Seaborg, an American scientist, made the suggestion that the actinide elements, like the lanthanides, were filling an f sub-level. Before this time the actinides were thought to be forming a fourth d-block row. Seaborg's colleagues advised him not to publish such a radical suggestion as it would most likely ruin his career. As Seaborg considered he did not then have a career to bring into disrepute, he published anyway. Seaborg's suggestion was found to be correct and he subsequently went on to win the 1951 Nobel Prize in chemistry for his work in synthesizing actinide elements.[75][76][n 6]

that of the standard form. Within 100 years of the appearance of Mendeleev's table in 1869 it has been estimated that around 700 different periodic table versions were published.[80] As well as numerous rectangular variations, other periodic table formats have included, for example,[n 7] circular, cubic, cylindrical, edificial (building-like), helical, lemniscate, octagonal prismatic, pyramidal, separated, spherical, spiral, and triangular forms. Such alternatives are often developed to highlight or emphasize chemical or physical properties of the elements that are not as apparent in traditional periodic tables.[80]

A popular[81] alternative structure is that of Theodor Benfey (1960). The elements are arranged in a continuous spiral, with hydrogen at the center and the transition metals, lanthanides, and actinides occupying peninsulas.[82]

Most periodic tables are two-dimensional;[11] however, three-dimensional tables are known to as far back as at least 1862 (pre-dating Mendeleev's two-dimensional table of 1869). More recent examples include Courtines' Periodic Classification (1925),[83] Wringley's Lamina System (1949),[84] Giguère's Periodic helix (1965)[85][n 8] and Dufour's Periodic Tree (1996).[86] Going one better, Stowe's Physicist's Periodic Table (1989)[87] has been described as being four-dimensional (having three spatial dimensions and one colour dimension).[88]

The various forms of periodic tables can be thought of as lying on a chemistry–physics continuum.[89] Towards the chemistry end of the continuum can be found, as an example, Rayner-Canham's 'unruly'[90] Inorganic Chemist's Periodic Table (2002),[91] which emphasizes trends and patterns, and unusual chemical relationships and properties. Near the physics end of the continuum is Janet's Left-Step Periodic Table (1928). This has a structure that shows a closer connection to the order of electron-shell filling and, by association, quantum mechanics.[92] Somewhere in the middle of the continuum is the ubiquitous common or standard form of periodic table. This is regarded as better expressing empirical trends in physical state, electrical and thermal conductivity, and oxidation numbers, and other properties easily inferred from traditional techniques of the chemical laboratory.[93]

5.6 Open questions and controversies

5.6.1 Elements with unknown chemical properties

Although all elements up to ununoctium have been discovered, of the elements above hassium (element 108), only copernicium (element 112) and flerovium (element 114) have known chemical properties. The other elements may behave differently from what would be predicted by extrapolation, due to relativistic effects; for example, flerovium has been predicted to possibly exhibit some noble-gas-like properties, even though it is currently placed in the carbon group.[94] More recent experiments have suggested, however, that flerovium behaves chemically like lead, as expected from its periodic table position.[95]

5.6.2 Further periodic table extensions

Main article: Extended periodic table

It is unclear whether new elements will continue the pattern of the current periodic table as period 8, or require further adaptations or adjustments. Seaborg expected the eighth period to follow the previously established pattern exactly, so that it would include a two-element s-block for elements 119 and 120, a new g-block for the next 18 elements, and 30 additional elements continuing the current f-, d-, and p-blocks.[97] More recently, physicists such as Pekka Pyykkö have theorized that these additional elements do not follow the Madelung rule, which predicts how electron shells are filled and thus affects the appearance of the present periodic table.[98]

5.6.3 Element with the highest possible atomic number

The number of possible elements is not known. A very early suggestion made by Elliot Adams in 1911, and based on the arrangement of elements in each horizontal periodic table row, was that elements of atomic weight greater than 256± (which would equate to between elements 99 and 100 in modern-day terms) did not exist.[99] A higher—more recent—estimate is that the periodic table may end soon after the island of stability,[100] which is expected to center around element 126, as the extension of the periodic and nuclides tables is restricted by proton and neutron drip lines.[101] Other predictions of an end to the periodic table include at element 128 by John Emsley,[1] at element 137 by Richard Feynman,[102] and at element 155 by Albert Khazan.[1][n 9]

Bohr model

The Bohr model exhibits difficulty for atoms with atomic number greater than 137, as any element with an atomic number greater than 137 would require 1s electrons to be traveling faster than c, the speed of light.[103] Hence the non-relativistic Bohr model is inaccurate when applied to such an element.

Relativistic Dirac equation

The relativistic Dirac equation has problems for elements with more than 137 protons. For such elements, the wave function of the Dirac ground state is oscillatory rather than bound, and there is no gap between the positive and negative energy spectra, as in the Klein paradox.[104] More accurate calculations taking into account the effects of the finite size of the nucleus indicate that the binding energy first exceeds the limit for elements with more than 173 protons. For heavier elements, if the innermost orbital (1s) is not filled, the electric field of the nucleus will pull an electron out of the vacuum, resulting in the spontaneous emission of a positron;[105] however, this does not happen if the innermost orbital is filled, so that element 173 is not necessarily the end of the periodic table.[106]

5.6.4 Placement of hydrogen and helium

Hydrogen and helium are often placed in different places than their electron configurations would indicate; hydrogen is usually placed above lithium, in accordance with its electron configuration, but is sometimes placed above fluorine,[107] or even carbon,[107] as it also behaves somewhat similarly to them. Hydrogen is also sometimes placed in its own group, as it does not behave similarly enough to any element to be placed in a group with another.[108] Helium is almost always placed above neon, as they are very similar chemically, although it is occasionally placed above beryllium on account of having a comparable electron shell configuration (helium: $1s^2$; beryllium: [He] $2s^2$).[19]

5.6.5 Groups included in the transition metals

The definition of a transition metal, as given by IUPAC, is an element whose atom has an incomplete d sub-shell, or which can give rise to cations with an incomplete d sub-shell.[109] By this definition all of the elements in groups 3–11 are transition metals. The IUPAC definition therefore excludes group 12, comprising zinc, cadmium and mercury, from the transition metals category.

Some chemists treat the categories "d-block elements" and "transition metals" interchangeably, thereby including groups 3–12 among the transition metals. In this instance the group 12 elements are treated as a special case of transition metal in which the d electrons are not ordinarily involved in chemical bonding. The recent discovery that mercury can use its d electrons in the formation of mercury(IV) fluoride (HgF_4) has prompted some commentators to suggest that mercury can be regarded as a transition metal.[110] Other commentators, such as Jensen,[111] have argued that the formation of a compound like HgF_4 can occur only under highly abnormal conditions. As such, mercury could not be regarded as a transition metal by any reasonable interpretation of the ordinary meaning of the term.[111]

Still other chemists further exclude the group 3 elements from the definition of a transition metal. They do so on the basis that the group 3 elements do not form any ions having a partially occupied d shell and do not therefore exhibit any properties characteristic of transition metal chemistry.[112] In this case, only groups 4–11 are regarded as transition metals.

5.6.6 Period 6 and 7 elements in group 3

Although scandium and yttrium are always the first two elements in group 3 the identity of the next two elements is not settled. They are either lanthanum and actinium; or lutetium and lawrencium. There are strong chemical and physical arguments supporting the latter arrangement[113][114] but not all authors have been convinced.[115] Most working chemists are not aware there is any controversy.[116]

Lanthanum and actinium are traditionally depicted as the remaining group 3 members.[117][118] It has been suggested that this layout originated in the 1940s, with the appearance of periodic tables relying on the electron configurations of the elements and the notion of the differentiating electron. The configurations of caesium, barium and lanthanum are [Xe]$6s^1$, [Xe]$6s^2$ and [Xe]$5d^16s^2$. Lanthanum thus has a 5d differentiating electron and this establishes "it in group 3 as the first member of the d-block for period 6."[111] A consistent set of electron configurations is then seen in group 3: scandium [Ar]$3d^14s^2$, yttrium [Kr]$4d^15s^2$ and lanthanum [Xe]$5d^16s^2$. Still in period 6, ytterbium was assigned an electron configuration of [Xe]$4f^{13}5d^16s^2$ and lutetium [Xe]$4f^{14}5d^16s^2$, "resulting in a 4f differentiating electron for lutetium and firmly establishing it as the last member of the f-block for period 6."[111]

In other tables, lutetium and lawrencium are the remaining group 3 members.[119] It has been known since the early 20th century that, "yttrium and (to a lesser degree) scandium are closer in their chemical properties to lutetium and the other heavy rare earths [i.e. lanthanides] than they are to lanthanum."[111] Accordingly, lutetium rather than lanthanum was assigned to group 3 by some chemists in the 1920s and 30s. Later spectroscopic work found that the electron configuration of ytterbium was in fact [Xe]$4f^{14}6s^2$. This meant that ytterbium and lutetium—the latter with [Xe]$4f^{14}5d^16s^2$—both had 14 f electrons, "resulting in a d rather than an f differentiating electron" for lutetium and making it an "equally valid candidate" with [Xe]$5d^16s^2$ lanthanum, for the group 3 periodic table position below yttrium.[111] Several physicists in the 1950s and 60s opted for lutetium, in light of a comparison of several of its physical properties with those of lanthanum.[111] This arrangement, in which lanthanum is the first member of the f-block, is disputed by some authors since lanthanum lacks any f electrons. However, it has been argued that this is not valid concern given other periodic table anomalies—

thorium, for example, has no f electrons yet is part of the f-block.[120] As to lawrencium, its electron configuration was confirmed in 2015 as $[Rn]5f^{14}7s^27p^1$. Such a configuration represents another periodic table anomaly, regardless of whether lawrencium is located in the f-block or the d-block, as the only potentially applicable p-block position has been reserved for ununtrium with its predicted electron configuration of $[Rn]5f^{14}6d^{10}7s^27p^1$.[121]

Some tables, including the table on the IUPAC site,[122][n 10] place footnote markers in the two positions below scandium and yttrium, and show both lanthanum and lutetium, and actinium and lawrencium as being part of, respectively, the lanthanide series and the actinide series of elements. This arrangement emphasizes similarities in the chemistry of the 15 lanthanide elements (La–Lu) over electron configuration arguments. The actinides are more diverse in their behavior. Most early members show some similarities to transition metals; actinium and the later members are more like lanthanides.[123]

5.6.7 Optimal form

The many different forms of periodic table have prompted the question of whether there is an optimal or definitive form of periodic table. The answer to this question is thought to depend on whether the chemical periodicity seen to occur among the elements has an underlying truth, effectively hard-wired into the universe, or if any such periodicity is instead the product of subjective human interpretation, contingent upon the circumstances, beliefs and predilections of human observers. An objective basis for chemical periodicity would settle the questions about the location of hydrogen and helium, and the composition of group 3. Such an underlying truth, if it exists, is thought to have not yet been discovered. In its absence, the many different forms of periodic table can be regarded as variations on the theme of chemical periodicity, each of which

explores and emphasizes different aspects, properties, perspectives and relationships of and among the elements.[n 11] The ubiquity of the standard or medium-long periodic table is thought to be a result of this layout having a good balance of features in terms of ease of construction and size, and its depiction of atomic order and periodic trends.[49][124]

5.7 See also

- Abundance of the chemical elements
- Atomic electron configuration table
- Element collecting
- List of elements
- List of periodic table-related articles
- Table of nuclides
- The Mystery of Matter: Search for the Elements (PBS film)
- Timeline of chemical element discoveries

5.8 Notes

[1] The elements discovered initially by synthesis and later in nature are technetium (Z=43), promethium (61), astatine (85), francium (87), neptunium (93), plutonium (94), americium (95), curium (96), berkelium (97) and californium (98).

[2] There is an inconsistency and some irregularities in this convention. Thus, helium is shown in the p-block but is actually an s-block element, and (for example) the d-subshell in the d-block is actually filled by the time group 11 is reached, rather than group 12.

[3] The noble gases, astatine, francium, and all elements heavier than americium were left out as there is no data for them.

[4] While fluorine is the most electronegative of the elements under the Pauling scale, neon is the most electronegative element under other scales, such as the Allen scale.

[5] An antecedent of Deming's 18-column table may be seen in Adams' 16-column Periodic Table of 1911. Adams omits the rare earths and the 'radioactive elements' (i.e. the actinides) from the main body of his table and instead shows them as being 'careted in only to save space' (rare earths between Ba and eka-Yt; radioactive elements between eka-Te and eka-I). See: Elliot Q. A. (1911). "A modification of the periodic table". *Journal of the American Chemical Society.* **33**(5): 684–688 (687).

[6] A second extra-long periodic table row, to accommodate known and undiscovered elements with an atomic weight greater than bismuth (thorium, protactinium and uranium, for example), had been postulated as far back as 1892. Most investigators, however, considered that these elements were analogues of the third series transition elements, hafnium, tantalum and tungsten. The existence of a second inner transition series, in the form of the actinides, was not accepted until similarities with the electron structures of the lanthanides had been established. See: van Spronsen, J. W. (1969). *The periodic system of chemical elements*. Amsterdam: Elsevier. p. 315–316, ISBN 0-444-40776-6.

[7] See *The Internet database of periodic tables* for depictions of these kinds of variants.

[8] The animated depiction of Giguère's periodic table that is widely available on the internet (including from here) is erroneous, as it does not include hydrogen and helium. Giguère included hydrogen, above lithium, and helium, above beryllium. See: Giguère P.A. (1966). "The "new look" for the periodic system". *Chemistry in Canada* **18** (12): 36–39 (see p. 37).

[9] Karol (2002, p. 63) contends that gravitational effects would become significant when atomic numbers become astronomically large, thereby overcoming other super-massive nuclei instability phenomena, and that neutron stars (with atomic numbers on the order of 10^{21}) can arguably be regarded as representing the heaviest known elements in the universe. See: Karol P. J. (2002). "The Mendeleev–Seaborg periodic table: Through Z = 1138 and beyond". *Journal of Chemical Education* **79** (1): 60–63.

[10] Although this form of the table is sometimes referred to as the "approved" or "official" IUPAC periodic table, "IUPAC has not approved any specific form of the periodic table..." See: Leigh, G. J. (January–February 2009). "Periodic Tables and IUPAC". *Chemistry International* **31** (1).

[11] Scerri, one of the foremost authorities on the history of the periodic table (Sella 2013), favoured the concept of an optimal form of periodic table but has recently changed his mind and now supports the value of a plurality of periodic tables. See: Sella A. (2013). 'An elementary history lesson'. *New Scientist*. 2929, 13 August: 51, accessed 4 September 2013; and Scerri, E. (2013). 'Is there an optimal periodic table and other bigger questions in the philosophy of science.'. 9 August, accessed 4 September 2013.

5.9   References

[1] Emsley, John (2011). *Nature's Building Blocks: An A-Z Guide to the Elements* (New ed.). New York, NY: Oxford University Press. ISBN 978-0-19-960563-7.

[2] Greenwood, pp. 24–27

[3] Gray, p. 6

[4] Koppenol, W. H. (2002). "Naming of New Elements (IUPAC Recommendations 2002)" (PDF). *Pure and Applied Chemistry* **74** (5): 787–791. doi:10.1351/pac200274050787.

[5] Silva, Robert J. (2006). "Fermium, Mendelevium, Nobelium and Lawrencium". In Morss; Edelstein, Norman M.; Fuger, Jean. *The Chemistry of the Actinide and Transactinide Elements* (3rd ed.). Dordrecht, The Netherlands: Springer Science+Business Media. ISBN 1-4020-3555-1.

[6] Gray, p. 11

[7] Scerri 2007, p. 24

[8] Messler, R. W. (2010). *The essence of materials for engineers*. Sudbury, MA: Jones & Bartlett Publishers. p. 32. ISBN 0-7637-7833-8.

[9] Bagnall, K. W. (1967). "Recent advances in actinide and lanthanide chemistry". In Fields, P.R.; Moeller, T. *Advances in chemistry, Lanthanide/Actinide chemistry*. Advances in Chemistry **71**. American Chemical Society. pp. 1–12. doi:10.1021/ba-1967-0071. ISBN 0-8412-0072-6.

[10] Day, M. C., Jr.; Selbin, J. (1969). *Theoretical inorganic chemistry* (2nd ed.). New York: Nostrand-Rienhold Book Corporation. p. 103. ISBN 0-7637-7833-8.

[11] Holman, J.; Hill, G. C. (2000). *Chemistry in context* (5th ed.). Walton-on-Thames: Nelson Thornes. p. 40. ISBN 0-17-448276-0.

[12] Leigh, G. J. (1990). *Nomenclature of Inorganic Chemistry: Recommendations 1990*. Blackwell Science. ISBN 0-632-02494-1.

[13] Fluck, E. (1988). "New Notations in the Periodic Table" (PDF). *Pure Appl. Chem.* (IUPAC) **60** (3): 431–436. doi:10.1351/pac198860030431. Retrieved 24 March 2012.

[14] Moore, p. 111

[15] Greenwood, p. 30

[16] Stoker, Stephen H. (2007). *General, organic, and biological chemistry*. New York: Houghton Mifflin. p. 68. ISBN 978-0-618-73063-6. OCLC 52445586.

[17] Mascetta, Joseph (2003). *Chemistry The Easy Way* (4th ed.). New York: Hauppauge. p. 50. ISBN 978-0-7641-1978-1. OCLC 52047235.

[18] Kotz, John; Treichel, Paul; Townsend, John (2009). *Chemistry and Chemical Reactivity, Volume 2* (7th ed.). Belmont: Thomson Brooks/Cole. p. 324. ISBN 978-0-495-38712-1. OCLC 220756597.

[19] Gray, p. 12

[20] Jones, Chris (2002). *d- and f-block chemistry*. New York: J. Wiley & Sons. p. 2. ISBN 978-0-471-22476-1. OCLC 300468713.

[21] Silberberg, M. S. (2006). *Chemistry: The molecular nature of matter and change* (4th ed.). New York: McGraw-Hill. p. 536. ISBN 0-07-111658-3.

[22] Manson, S. S.; Halford, G. R. (2006). *Fatigue and durability of structural materials*. Materials Park, Ohio: ASM International. p. 376. ISBN 0-87170-825-6.

[23] Bullinger, Hans-Jörg (2009). *Technology guide: Principles, applications, trends*. Berlin: Springer-Verlag. p. 8. ISBN 978-3-540-88545-0.

[24] Jones, B. W. (2010). *Pluto: Sentinel of the outer solar system*. Cambridge: Cambridge University Press. pp. 169–71. ISBN 978-0-521-19436-5.

[25] Hinrichs, G. D. (1869). "On the classification and the atomic weights of the so-called chemical elements, with particular reference to Stas's determinations". *Proceedings of the American Association for the Advancement of Science* **18** (5): 112–124.

[26] Myers, R. (2003). *The basics of chemistry*. Westport, CT: Greenwood Publishing Group. pp. 61–67. ISBN 0-313-31664-3.

[27] Chang, Raymond (2002). *Chemistry* (7 ed.). New York: McGraw-Hill. pp. 289–310; 340–42. ISBN 0-07-112072-6.

[28] Greenwood, p. 27

[29] Jolly, W. L. (1991). *Modern Inorganic Chemistry* (2nd ed.). McGraw-Hill. p. 22. ISBN 978-0-07-112651-9.

[30] Greenwood, p. 28

[31] IUPAC, *Compendium of Chemical Terminology*, 2nd ed. (the "Gold Book") (1997). Online corrected version: (2006–) "Electronegativity".

[32] Pauling, L. (1932). "The Nature of the Chemical Bond. IV. The Energy of Single Bonds and the Relative Electronegativity of Atoms". *Journal of the American Chemical Society* **54** (9): 3570–3582. doi:10.1021/ja01348a011.

[33] Allred, A. L. (1960). "Electronegativity values from thermochemical data". *Journal of Inorganic and Nuclear Chemistry* (Northwestern University) **17** (3–4): 215–221. doi:10.1016/0022-1902(61)80142-5. Retrieved 11 June 2012.

[34] Huheey, Keiter & Keiter, p. 42

[35] Siekierski, Slawomir; Burgess, John (2002). *Concise chemistry of the elements*. Chichester: Horwood Publishing. pp. 35–36. ISBN 1-898563-71-3.

[36] Chang, pp. 307–309

[37] Huheey, Keiter & Keiter, pp. 42, 880–81

[38] Yoder, C. H.; Suydam, F. H.; Snavely, F. A. (1975). *Chemistry* (2nd ed.). Harcourt Brace Jovanovich. p. 58. ISBN 0-15-506465-7.

[39] Huheey, Keiter & Keiter, pp. 880–85

[40] Sacks, O (2009). *Uncle Tungsten: Memories of a chemical boyhood*. New York: Alfred A. Knopf. pp. 191, 194. ISBN 0-375-70404-3.

[41] Gray, p. 9

[42] Siegfried, Robert (2002). *From elements to atoms a history of chemical composition*. Philadelphia, Pennsylvania: Library of Congress Cataloging-in-Publication Data. p. 92. ISBN 0-87169-924-9.

[43] Ball, p. 100

[44] Horvitz, Leslie (2002). *Eureka!: Scientific Breakthroughs That Changed The World*. New York: John Wiley. p. 43. ISBN 978-0-471-23341-1. OCLC 50766822.

[45] van Spronsen, J. W. (1969). *The periodic system of chemical elements*. Amsterdam: Elsevier. p. 19. ISBN 0-444-40776-6.

[46] "Alexandre-Emile Bélguier de Chancourtois (1820-1886)" (in French). Annales des Mines history page. Retrieved 18 September 2014.

[47] Venable, pp. 85–86; 97

[48] Odling, W. (2002). "On the proportional numbers of the elements". *Quarterly Journal of Science* **1**: 642–648 (643).

[49] Scerri, Eric R. (2011). *The periodic table: A very short introduction*. Oxford: Oxford University Press. ISBN 978-0-19-958249-5.

[50] Kaji, M. (2004). "Discovery of the periodic law: Mendeleev and other researchers on element classification in the 1860s". In Rouvray, D. H.; King, R. Bruce. *The periodic table: Into the 21st Century*. Research Studies Press. pp. 91–122 (95). ISBN 0-86380-292-3.

[51] Newlands, John A. R. (20 August 1864). "On Relations Among the Equivalents". *Chemical News* **10**: 94–95.

[52] Newlands, John A. R. (18 August 1865). "On the Law of Octaves". *Chemical News* **12**: 83.

[53] Bryson, Bill (2004). *A Short History of Nearly Everything*. Black Swan. pp. 141–142. ISBN 978-0-552-15174-0.

[54] Scerri 2007, p. 306

[55] Brock, W. H.; Knight, D. M. (1965). "The Atomic Debates: 'Memorable and Interesting Evenings in the Life of the Chemical Society'". *Isis* (The University of Chicago Press) **56** (1): 5–25. doi:10.1086/349922.

[56] Scerri 2007, pp. 87, 92

[57] Kauffman, George B. (March 1969). "American forerunners of the periodic law". *Journal of Chemical Education* **46** (3): 128–135 (132). Bibcode:1969JChEd..46..128K. doi:10.1021/ed046p128.

[58] Mendelejew, Dimitri (1869). "Über die Beziehungen der Eigenschaften zu den Atomgewichten der Elemente". *Zeitschrift für Chemie* (in German): 405–406.

[59] Venable, pp. 96–97; 100–102

[60] Ball, pp. 100–102

[61] Pullman, Bernard (1998). *The Atom in the History of Human Thought.* Translated by Axel Reisinger. Oxford University Press. p. 227. ISBN 0-19-515040-6.

[62] Ball, p. 105

[63] Atkins, P. W. (1995). *The Periodic Kingdom.* HarperCollins Publishers, Inc. p. 87. ISBN 0-465-07265-8.

[64] Samanta, C.; Chowdhury, P. Roy; Basu, D.N. (2007). "Predictions of alpha decay half lives of heavy and superheavy elements". *Nucl. Phys. A* **789**: 142–154. arXiv:nucl-th/0703086. Bibcode:2007NuPhA.789..142S. doi:10.1016/j.nuclphysa.2007.04.001.

[65] Scerri 2007, p. 112

[66] Kaji, Masanori (2002). "D.I. Mendeleev's Concept of Chemical Elements and the Principle of Chemistry" (PDF). *Bull. Hist. Chem.* (Tokyo Institute of Technology) **27** (1): 4–16. Retrieved 11 June 2012.

[67] Adloff, Jean-Pierre; Kaufman, George B. (25 September 2005). "Francium (Atomic Number 87), the Last Discovered Natural Element". The Chemical Educator. Retrieved 26 March 2007.

[68] Hoffman, D. C.; Lawrence, F. O.; Mewherter, J. L.; Rourke, F. M. (1971). "Detection of Plutonium-244 in Nature". *Nature* **234** (5325): 132–134. Bibcode:1971Natur.234..132H. doi:10.1038/234132a0.

[69] Gray, p. 12

[70] Deming, Horace G (1923). *General chemistry: An elementary survey.* New York: J. Wiley & Sons. pp. 160, 165.

[71] Abraham, M; Coshow, D; Fix, W. *Periodicity:A source book module, version 1.0* (PDF). New York: Chemsource, Inc. p. 3.

[72] Emsley, J (7 March 1985). "Mendeleyev's dream table". *New Scientist*: 32–36(36).

[73] Fluck, E (1988). "New notations in the period table". *Pure & Applied Chemistry* **60** (3): 431–436 (432). doi:10.1351/pac198860030431.

[74] Ball, p. 111

[75] Scerri 2007, pp. 270–71

[76] Masterton, William L.; Hurley, Cecile N.; Neth, Edward J. *Chemistry: Principles and reactions* (7th ed.). Belmont, CA: Brooks/Cole Cengage Learning. p. 173. ISBN 1-111-42710-0.

[77] Ball, p. 123

[78] Barber, Robert C.; Karol, Paul J; Nakahara, Hiromichi; Vardaci, Emanuele; Vogt, Erich W. (2011). "Discovery of the elements with atomic numbers greater than or equal to 113 (IUPAC Technical Report)". *Pure Appl. Chem.* **83** (7): 1485. doi:10.1351/PAC-REP-10-05-01.

[79] Эксперимент по синтезу 117-го элемента получает продолжение[Experiment on sythesis of the 117th element is to be continued] (in Russian). JINR. 2012.

[80] Scerri 2007, p. 20

[81] Emsely, J; Sharp, R (21 June 2010). "The periodic table: Top of the charts". *The Independent.*

[82] Seaborg, Glenn (1964). "Plutonium: The Ornery Element". *Chemistry* **37** (6): 14.

[83] Mark R. Leach. "1925 Courtines' Periodic Classification". Retrieved 16 October 2012.

[84] Mark R. Leach. "1949 Wringley's Lamina System". Retrieved 16 October 2012.

[85] Mazurs, E.G. (1974). *Graphical Representations of the Periodic System During One Hundred Years.* Alabama: University of Alabama Press. p. 111. ISBN 978-0-8173-3200-6.

[86] Mark R. Leach. "1996 Dufour's Periodic Tree". Retrieved 16 October 2012.

[87] Mark R. Leach. "1989 Physicist's Periodic Table by Timothy Stowe". Retrieved 16 October 2012.

[88] Bradley, David (20 July 2011). "At last, a definitive periodic table?". *ChemViews Magazine.* doi:10.1002/chemv.201000107.

[89] Scerri 2007, pp. 285–86

[90] Scerri 2007, p. 285

[91] Mark R. Leach. "2002 Inorganic Chemist's Periodic Table". Retrieved 16 October 2012.

[92] Scerri, Eric (2008). "The role of triads in the evolution of the periodic table: Past and present". *Journal of Chemical Education* **85** (4): 585–89 (see p.589). Bibcode:2008JChEd..85..585S. doi:10.1021/ed085p585.

[93] Bent, H. A.; Weinhold, F (2007). "Supporting information: News from the periodic table: An introduction to "Periodicity symbols, tables, and models for higher-order valency and donor–acceptor kinships"". *Journal of Chemical Education* **84** (7): 3–4. doi:10.1021/ed084p1145.

[94] Schändel, Matthias (2003). *The Chemistry of Superheavy Elements.* Dordrecht: Kluwer Academic Publishers. p. 277. ISBN 1-4020-1250-0.

[95] Scerri 2011, pp. 142–143

[96] Fricke, B.; Greiner, W.; Waber, J. T. (1971). "The continuation of the periodic table up to Z = 172. The chemistry of superheavy elements". *Theoretica chimica acta* (Springer-Verlag) **21** (3): 235–260. doi:10.1007/BF01172015. Retrieved 28 November 2012.

[97] Frazier, K. (1978). "Superheavy Elements". *Science News* **113** (15): 236–238. doi:10.2307/3963006. JSTOR 3963006.

[98] Pyykkö, Pekka (2011). "A suggested periodic table up to Z ≤ 172, based on Dirac–Fock calculations on atoms and ions". *Physical Chemistry Chemical Physics* **13** (1): 161–168. Bibcode:2011PCCP...13..161P. doi:10.1039/c0cp01575j. PMID 20967377.

[99] Elliot, Q. A. (1911). "A modification of the periodic table". *Journal of the American Chemical Society* **33** (5): 684–688 (688). doi:10.1021/ja02218a004.

[100] Glenn Seaborg (c. 2006). "transuranium element (chemical element)". Encyclopædia Britannica. Retrieved 16 March 2010.

[101] Cwiok, S.; Heenen, P.-H.; Nazarewicz, W. (2005). "Shape coexistence and triaxiality in the superheavy nuclei". *Nature* **433** (7027): 705–9. Bibcode:2005Natur.433..705C. doi:10.1038/nature03336. PMID 15716943.

[102] Column: The crucible Ball, Philip in Chemistry World, Royal Society of Chemistry, Nov. 2010

[103] Eisberg, R.; Resnick, R. (1985). *Quantum Physics of Atoms, Molecules, Solids, Nuclei and Particles*. Wiley.

[104] Bjorken, J. D.; Drell, S. D. (1964). *Relativistic Quantum Mechanics*. McGraw-Hill.

[105] Greiner, W.; Schramm, S. (2008). "American Journal of Physics" **76**. p. 509., and references therein.

[106] Ball, Philip (November 2010). "Would Element 137 Really Spell the End of the Periodic Table? Philip Ball Examines the Evidence". Royal Society of Chemistry. Retrieved 30 September 2012.

[107] Cronyn, Marshall W. (August 2003). "The Proper Place for Hydrogen in the Periodic Table". *Journal of Chemical Education* **80** (8): 947–951. Bibcode:2003JChEd..80..947C. doi:10.1021/ed080p947.

[108] Gray, p. 14

[109] IUPAC, *Compendium of Chemical Terminology*, 2nd ed. (the "Gold Book") (1997). Online corrected version: (2006–) "transition element".

[110] Xuefang Wang; Lester Andrews; Sebastian Riedel; Martin Kaupp (2007). "Mercury Is a Transition Metal: The First Experimental Evidence for HgF$_4$". *Angew. Chem. Int. Ed.* **46** (44): 8371–8375. doi:10.1002/anie.200703710. PMID 17899620.

[111] William B. Jensen (2008). "Is Mercury Now a Transition Element?". *J. Chem. Educ.* **85** (9): 1182–1183. Bibcode:2008JChEd..85.1182J. doi:10.1021/ed085p1182.

[112] Rayner-Canham, G; Overton, T. *Descriptive inorganic chemistry* (4th ed.). New York: W H Freeman. pp. 484–485. ISBN 0-7167-8963-9.

[113] Thyssen, P.; Binnemanns, K. (2011). "1: Accommodation of the rare earths in the periodic table: A historical analysis". In Gschneidner Jr., K. A.; Büzli, J-C. J.; Pecharsky, V. K. *Handbook on the Physics and Chemistry of Rare Earths* **41**. Amsterdam: Elsevier. pp. 80–81. ISBN 978-0-444-53590-0.

[114] Keeler, J.; Wothers, P. (2014). *Chemical Structure and Reactivity: An Integrated Approach*. Oxford: Oxford University. p. 259. ISBN 978-0-19-9604135.

[115] Scerri, E. (2012). "Mendeleev's Periodic Table Is Finally Completed and What To Do about Group 3?". *Chemistry International* **34** (4).

[116] Castelvecchi, Davide (8 April 2015). "Exotic atom struggles to find its place in the periodic table". *Nature News*. Retrieved 20 Sep 2015.

[117] Emsley, J. (2011). *Nature's Building Blocks* (new ed.). Oxford: Oxford University. p. 651. ISBN 978-0-19-960563-7.

[118] See, for example: "Periodic Table". Royal Society of Chemistry. Retrieved 20 Sep 2015.

[119] See, for example: Brown, T. L.; LeMay Jr., H. E.; Bursten, B. E.; Murphy, C. J. (2009). *Chemistry: The Central Science* (11th ed.). Upper Saddle River, New Jersey: Pearson Education. p. endpapers. ISBN 0-13-235848-4.

[120] Scerri, E (2015). "Five ideas in chemical education that must die - part five". *educationinchemistryblog*. Royal Society of Chemistry. Retrieved Sep 19, 2015. It is high time that the idea of group 3 consisting of Sc, Y, La and Ac is abandoned

[121] Jensen, W. B. (2015). "Some Comments on the Position of Lawrencium in the Periodic Table" (PDF). Retrieved 20 Sep 2015.

[122] "Periodic Table of the Elements". International Union of Pure and Applied Chemistry. Retrieved 3 April 2010.

[123] Owen, S. M. (1991). *A Guide to Modern Inorganic Chemistry*. Harlow, Essex: Longman Scientific & Technical. p. 190. ISBN 0-58-206439-2.

[124] Francl, Michelle (May 2009). "Table manners" (PDF). *Nature Chemistry* **1** (2): 97–98. Bibcode:2009NatCh...1...97F. doi:10.1038/nchem.183. PMID 21378810.

5.10   Bibliography

- Ball, Philip (2002). *The Ingredients: A Guided Tour of the Elements.* Oxford: Oxford University Press. ISBN 0-19-284100-9.

- Chang, Raymond (2002). *Chemistry* (7th ed.). New York: McGraw-Hill Higher Education. ISBN 978-0-19-284100-1.

- Gray, Theodore (2009). *The Elements: A Visual Exploration of Every Known Atom in the Universe.* New York: Black Dog & Leventhal Publishers. ISBN 978-1-57912-814-2.

- Greenwood, Norman N.; Earnshaw, Alan (1984). *Chemistry of the Elements.* Oxford: Pergamon Press. ISBN 0-08-022057-6.

- Huheey, JE; Keiter, EA; Keiter, RL. *Principles of structure and reactivity* (4th ed.). New York: Harper Collins College Publishers. ISBN 0-06-042995-X.

- Moore, John (2003). *Chemistry For Dummies.* New York: Wiley Publications. p. 111. ISBN 978-0-7645-5430-8. OCLC 51168057.

- Scerri, Eric (2007). *The periodic table: Its story and its significance.* Oxford: Oxford University Press. ISBN 0-19-530573-6.

- Scerri, Eric R. (2011). *The periodic table: A very short introduction.* Oxford: Oxford University Press. ISBN 978-0-19-958249-5.

- Venable, F P (1896). *The Development of the Periodic Law.* Easton PA: Chemical Publishing Company.

5.11   External links

- M. Dayah. "Dynamic Periodic Table". Retrieved 14 May 2012.

- Brady Haran. "The Periodic Table of Videos". University of Nottingham. Retrieved 14 May 2012.

- Mark Winter. "WebElements: the periodic table on the web". University of Sheffield. Retrieved 14 May 2012.

- Mark R. Leach. "The INTERNET Database of Periodic Tables". Retrieved 14 May 2012.

Chapter 6

Period (periodic table)

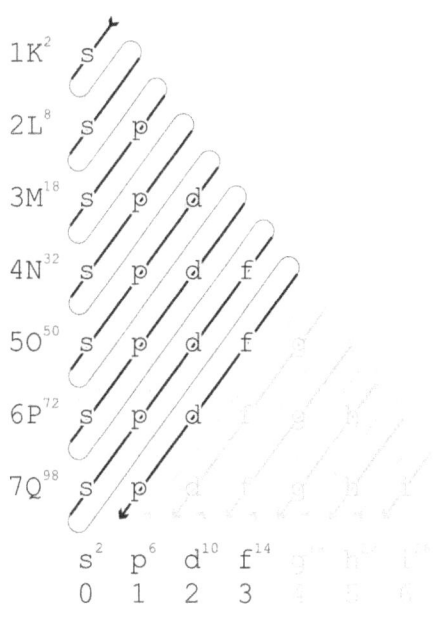

In the periodic table of the elements, each numbered row is a period.

In the periodic table of the elements, elements are arranged in a series of rows (or **periods**) so that those with similar properties appear in a column. Elements of the same period have the same number of electron shells; with each group across a period, the elements have one more proton and electron and become less metallic. This arrangement reflects the *periodic* recurrence of similar properties as the atomic number increases. For example, the alkaline metals lie in one group (group 1) and share similar properties, such as high reactivity and the tendency to lose one electron to arrive at a noble-gas electronic configuration. The periodic table of elements has a total of 118 elements.

Modern quantum mechanics explains these periodic trends in properties in terms of electron shells. As atomic number increases, shells fill with electrons in approximately the order shown at right. The filling of each shell corresponds to a row in the table.

In the s-block and p-block of the periodic table, elements within the same period generally do not exhibit trends and similarities in properties (vertical trends down groups are more significant). However in the d-block, trends across periods become significant, and in the f-block elements show a high degree of similarity across periods.

The Madelung energy ordering rule describes the order in which orbitals are arranged by increasing energy according to the Madelung rule. Each diagonal corresponds to a different value of $n + l$.

6.1 Periods

Seven periods of elements occur naturally on Earth. For period 8, which includes elements which may be synthesized after 2013, see the extended periodic table.

A group in chemistry means a family of objects with similarities like different families. There are 7 periods, going horizontally across the periodic table.

6.1.1 Period 1

The first period contains fewer elements than any other, with only two, hydrogen and helium. They therefore do not follow the octet rule. Chemically, helium behaves as a noble gas, and thus is taken to be part of the group 18 elements. However, in terms of its nuclear structure it belongs to the s block, and is therefore sometimes classified as a group 2 element, or simultaneously both 2 and 18. Hydrogen readily loses and gains an electron, and so behaves chemically as both a group 1 and a group 17 element.

- Hydrogen (H) is the most abundant of the chemical elements, constituting roughly 75% of the universe's elemental mass.[1] Ionized hydrogen is just a proton. Stars in the main sequence are mainly composed of hydrogen in its plasma state. Elemental hydrogen is relatively rare on Earth, and is industrially produced from hydrocarbons such as methane. Hydrogen can form compounds with most elements and is present in water and most organic compounds.[2]

- Helium (He) exists only as a gas except in extreme conditions.[3] It is the second lightest element and is the second most abundant in the universe.[4] Most helium was formed during the Big Bang, but new helium is created through nuclear fusion of hydrogen in stars.[5] On Earth, helium is relatively rare, only occurring as a byproduct of the natural decay of some radioactive elements.[6] Such 'radiogenic' helium is trapped within natural gas in concentrations of up to seven percent by volume.[7]

6.1.2   Period 2

Period 2 elements involve the 2s and 2p orbitals. They include the biologically most essential elements besides hydrogen: carbon, nitrogen, and oxygen.

- Lithium (Li) is the lightest metal and the least dense solid element.[8] In its non-ionized state it is one of the most reactive elements, and so is only ever found naturally in compounds. It is the heaviest primordial element forged in large quantities during the Big Bang.

- Beryllium (Be) has one of the highest melting points of all the light metals. Small amounts of beryllium were synthesised during the Big Bang, although most of it decayed or reacted further within stars to create larger nucleii, like carbon, nitrogen or oxygen. Beryllium is classified by the International Agency for Research on Cancer as a group 1 carcinogen.[9] Between 1% and 15% of people are sensitive to beryllium and may develop an inflammatory reaction in their respiratory system and skin, called chronic beryllium disease.[10]

- Boron (B) does not occur naturally as a free element, but in compounds such as borates. It is

an essential plant micronutrient, required for cell wall strength and development, cell division, seed and fruit development, sugar transport and hormone development,[11][12] though high levels are toxic.

- Carbon (C) is the fourth most abundant element in the universe by mass after hydrogen, helium and oxygen[13] and is the second most abundant element in the human body by mass after oxygen,[14] the third most abundant by number of atoms.[15] There are an almost infinite number of compounds that contain carbon due to carbon's ability to form long stable chains of C—C bonds.[16][17] All organic compounds, those essential for life, contain at least one atom of carbon;[16][17] combined with hydrogen, oxygen, nitrogen, sulfur, and phosphorus, carbon is the basis of every important biological compound.[17]

- Nitrogen (N) is found mainly as mostly inert diatomic gas, N_2, which makes up 78% of the Earth's atmosphere. It is an essential component of proteins and therefore of life.

- Oxygen (O) comprising 21% of the atmosphere and is required for respiration by all (or nearly all) animals, as well as being the principal component of water. Oxygen is the third most abundant element in the universe, and oxygen compounds dominate the Earth's crust.

- Fluorine (F) is the most reactive element in its non-ionized state, and so is never found that way in nature.

- Neon (Ne) is a noble gas used in neon lighting.

6.1.3   Period 3

All period three elements occur in nature and have at least one stable isotope. All but the noble gas argon are essential to basic geology and biology.

- Sodium (Na) is an alkali metal. It is present in Earth's oceans in large quantities in the form of sodium chloride (table salt).

- Magnesium (Mg) is an alkaline earth metal. Magnesium ions are found in chlorophyll.

- Aluminium (Al) is a post-transition metal. It is the most abundant metal in the Earth's crust.

- Silicon (Si) is a metalloid. It is a semiconductor, making it the principal component in many integrated circuits. Silicon dioxide is the principal constituent of sand. As Carbon is to Biology, Silicon is to Geology.

- Phosphorus (P) is a nonmetal essential to DNA. It is highly reactive, and as such is never found in nature as a free element.

- Sulfur (S) is a nonmetal. It is found in two amino acids: cysteine and methionine.

- Chlorine (Cl) is a halogen. It is used as a disinfectant, especially in swimming pools.

- Argon (Ar) is a noble gas, making it almost entirely nonreactive. Incandescent lamps are often filled with noble gases such as argon in order to preserve the filaments at high temperatures.

6.1.4 Period 4

From left to right, aqueous solutions of: $Co(NO_3)_2$ (red); $K_2Cr_2O_7$ (orange); K_2CrO_4 (yellow); $NiCl_2$ (green); $CuSO_4$ (blue); $KMnO_4$ (purple).

Period 4 includes the biologically essential elements potassium and calcium, and is the first period in the d-block with the lighter transition metals. These include iron, the heaviest element forged in main-sequence stars and a principal component of the earth, as well as other important metals such as cobalt, nickel, copper, and zinc. Almost all have biological roles.

6.1.5 Period 5

Period 5 contains the heaviest few elements that have biological roles, molybdenum and iodine. (Tungsten, a period 6 element, is the only heavier element that has a biological role.) It includes technetium, the lightest exclusively radioactive element.

6.1.6 Period 6

Period 6 is the first period to include the f-block, with the lanthanides (also known as the rare earth elements), and includes the heaviest stable elements. Many of these heavy metals are toxic and some are radioactive, but platinum and gold are largely inert.

6.1.7 Period 7

All elements of period 7 are radioactive. This period contains the heaviest element which occurs naturally on earth, californium. All of the subsequent elements in the period have been synthesized artificially. Whilst one of these (einsteinium) is now available in macroscopic quantities, most are extremely rare, having only been prepared in microgram amounts or less. Some of the later elements have only ever been identified in laboratories in quantities of a few atoms at a time.

Although the rarity of many of these elements means that experimental results are not very extensive, periodic and group trends in behaviour appear to be less well defined for period 7 than for other periods. Whilst francium and radium do show typical properties of Groups 1 and 2 respectively, the actinides display a much greater variety of behaviour and oxidation states than the lanthanides. These peculiarities of period 7 may be due to a variety of factors, including a large degree of spin-orbit coupling and relativistic effects, ultimately caused by the very high positive electrical charge from their massive atomic nuclei.

6.1.8 Period 8

Main article: Extended periodic table

No element of the eighth period has yet been synthesized. A g-block is predicted. It is not clear if all elements predicted for the eighth period are in fact physically possible. There may therefore be no ninth period.

6.2 References

[1] Palmer, David (November 13, 1997). "Hydrogen in the Universe". NASA. Retrieved 2008-02-05.

[2] "hydrogen". *Encyclopædia Britannica*. 2008.

[3] "Helium: physical properties". WebElements. Retrieved 2008-07-15.

[4] "Helium: geological information". WebElements. Retrieved 2008-07-15.

[5] Cox, Tony (1990-02-03). "Origin of the chemical elements". *New Scientist*. Retrieved 2008-07-15.

[6] "Helium supply deflated: production shortages mean some industries and partygoers must squeak by.". Houston Chronicle. 2006-11-05.

[7] Brown, David (2008-02-02). "Helium a New Target in New Mexico". American Association of Petroleum Geologists. Retrieved 2008-07-15.

[8] Lithium at WebElements.

[9] "IARC Monograph, Volume 58". International Agency for Research on Cancer. 1993. Retrieved 2008-09-18.

[10] Information about chronic beryllium disease.

[11] "Functions of Boron in Plant Nutrition" (PDF). U.S. Borax Inc.

[12] Blevins, Dale G.; Lukaszewski, Krystyna M. (1998). "Functions of Boron in Plant Nutrition". *Annual Review of Plant Physiology and Plant Molecular Biology* **49**: 481–500. doi:10.1146/annurev.arplant.49.1.481. PMID 15012243.

[13] Ten most abundant elements in the universe, taken from *The Top 10 of Everything*, 2006, Russell Ash, page 10. Retrieved October 15, 2008.

[14] Chang, Raymond (2007). *Chemistry, Ninth Edition*. McGraw-Hill. p. 52. ISBN 0-07-110595-6.

[15] Freitas Jr., Robert A. (1999). *Nanomedicine*. Landes Bioscience. Tables 3-1 & 3-2. ISBN 1-57059-680-8.

[16] "Structure and Nomenclature of Hydrocarbons". Purdue University. Retrieved 2008-03-23.

[17] Alberts, Bruce; Alexander Johnson; Julian Lewis; Martin Raff; Keith Roberts; Peter Walter. *Molecular Biology of the Cell*. Garland Science.

Chapter 7

Main sequence

For the racehorse, see Main Sequence (horse).

In astronomy, the **main sequence** is a continuous and dis-

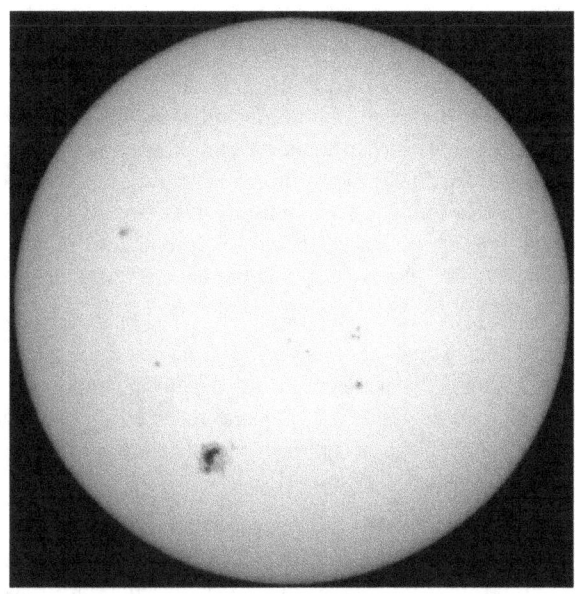

The Sun is the most familiar example of a main sequence star

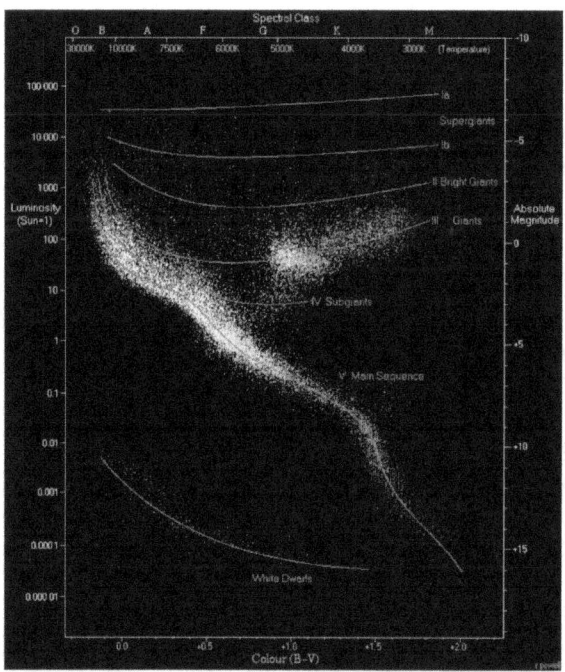

A Hertzsprung–Russell diagram plots the actual brightness (or absolute magnitude) of a star against its color index (represented as B-V). The main sequence is visible as a prominent diagonal band that runs from the upper left to the lower right. This plot shows 22,000 stars from the Hipparcos Catalogue together with 1,000 low-luminosity stars (red and white dwarfs) from the Gliese Catalogue of Nearby Stars.

tinctive band of stars that appears on plots of stellar color versus brightness. These color-magnitude plots are known as Hertzsprung–Russell diagrams after their co-developers, Ejnar Hertzsprung and Henry Norris Russell. Stars on this band are known as **main-sequence stars** or "dwarf" stars.[1][2]

After a star has formed, it generates thermal energy in the dense core region through the nuclear fusion of hydrogen atoms into helium. During this stage of the star's lifetime, it is located along the main sequence at a position determined primarily by its mass, but also based upon its chemical composition and other factors. All main-sequence stars are in hydrostatic equilibrium, where outward thermal pressure from the hot core is balanced by the inward pressure of gravitational collapse from the overlying layers. The strong dependence of the rate of energy generation in the core on the temperature and pressure helps to sustain this balance.

Energy generated at the core makes its way to the surface and is radiated away at the photosphere. The energy is carried by either radiation or convection, with the latter occurring in regions with steeper temperature gradients, higher opacity or both.

The main sequence is sometimes divided into upper and lower parts, based on the dominant process that a star uses to generate energy. Stars below about 1.5 times the mass of the Sun (or 1.5 solar masses ($M\odot$)) primarily fuse hydrogen atoms together in a series of stages to form helium, a sequence called the proton–proton chain. Above this mass, in the upper main sequence, the nuclear fusion process mainly uses atoms of carbon, nitrogen and oxygen as intermediaries in the CNO cycle that produces helium from hydrogen atoms. Main-sequence stars with more than two solar masses undergo convection in their core regions, which acts to stir up the newly created helium and maintain the proportion of fuel needed for fusion to occur. Below this mass, stars have cores that are entirely radiative with convective zones near the surface. With decreasing stellar mass, the proportion of the star forming a convective envelope steadily increases, while main-sequence stars below 0.4 $M_\odot$ undergo convection throughout their mass. When core convection does not occur, a helium-rich core develops surrounded by an outer layer of hydrogen.

In general, the more massive a star is, the shorter its lifespan on the main sequence. After the hydrogen fuel at the core has been consumed, the star evolves away from the main sequence on the HR diagram. The behavior of a star now depends on its mass, with stars below 0.23 $M_\odot$ becoming white dwarfs directly, while stars with up to ten solar masses pass through a red giant stage.[3] More massive stars can explode as a supernova,[4] or collapse directly into a black hole.

7.1 History

Hot and brilliant O-type main-sequence stars in star-forming regions. These are all regions of star formation that contain many hot young stars including several bright stars of spectral type O.[5]

In the early part of the 20th century, information about the types and distances of stars became more readily available. The spectra of stars were shown to have distinctive features, which allowed them to be categorized. Annie Jump Cannon and Edward C. Pickering at Harvard College Observatory developed a method of categorization that became known as the Harvard Classification Scheme, published in the *Harvard Annals* in 1901.[6]

In Potsdam in 1906, the Danish astronomer Ejnar Hertzsprung noticed that the reddest stars—classified as K and M in the Harvard scheme—could be divided into two distinct groups. These stars are either much brighter than the Sun, or much fainter. To distinguish these groups, he called them "giant" and "dwarf" stars. The following year he began studying star clusters; large groupings of stars that are co-located at approximately the same distance. He published the first plots of color versus luminosity for these stars. These plots showed a prominent and continuous sequence of stars, which he named the Main Sequence.[7]

At Princeton University, Henry Norris Russell was following a similar course of research. He was studying the relationship between the spectral classification of stars and their actual brightness as corrected for distance—their absolute magnitude. For this purpose he used a set of stars that had reliable parallaxes and many of which had been categorized at Harvard. When he plotted the spectral types of these stars against their absolute magnitude, he found that dwarf stars followed a distinct relationship. This allowed the real brightness of a dwarf star to be predicted with reasonable accuracy.[8]

Of the red stars observed by Hertzsprung, the dwarf stars also followed the spectra-luminosity relationship discovered by Russell. However, the giant stars are much brighter than dwarfs and so, do not follow the same relationship. Russell proposed that the "giant stars must have low density or great surface-brightness, and the reverse is true of dwarf stars". The same curve also showed that there were very few faint white stars.[8]

In 1933, Bengt Strömgren introduced the term Hertzsprung–Russell diagram to denote a luminosity-spectral class diagram.[9] This name reflected the parallel development of this technique by both Hertzsprung and Russell earlier in the century.[7]

As evolutionary models of stars were developed during the 1930s, it was shown that, for stars of a uniform chemical composition, a relationship exists between a star's mass and its luminosity and radius. That is, for a given mass and composition, there is a unique solution for determining the star's radius and luminosity. This became known as the Vogt-Russell theorem; named after Heinrich Vogt and Henry Norris Russell. By this theorem, once a star's chemical composition and its position on the main sequence is known, so too is the star's mass and radius. (However, it was subsequently discovered that the theorem breaks down somewhat for stars of non-uniform composition.)[10]

A refined scheme for stellar classification was published in 1943 by W. W. Morgan and P. C. Keenan.[11] The MK classification assigned each star a spectral type—based on the Harvard classification—and a luminosity class. The Harvard classification had been developed by assigning a different letter to each star based on the strength of the hydrogen spectral line, before the relationship between spectra and temperature was known. When ordered by temperature and when duplicate classes were removed, the spectral types of stars followed, in order of decreasing temperature with colors ranging from blue to red, the sequence O, B, A, F, G, K and M. (A popular mnemonic for memorizing this sequence of stellar classes is "Oh Be A Fine Girl/Guy, Kiss Me".) The luminosity class ranged from I to V, in order of decreasing luminosity. Stars of luminosity class V belonged to the main sequence.[12]

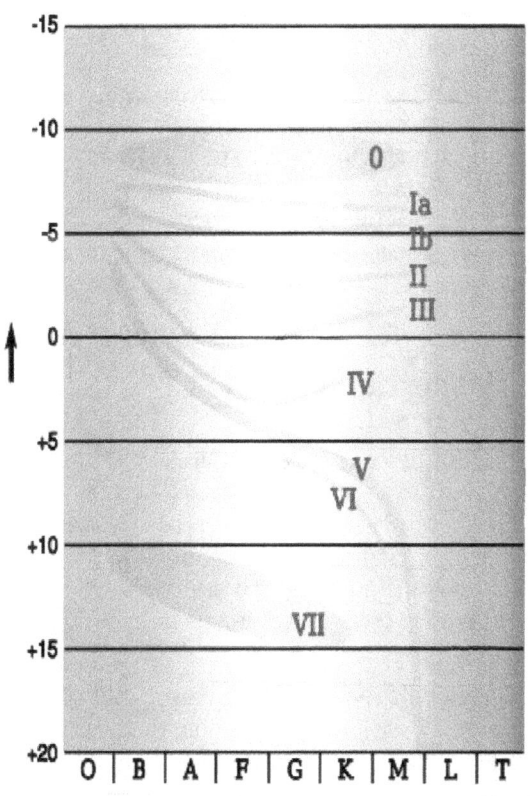

Hertzsprung–Russell diagram
Spectral type
Brown dwarfs
White dwarfs
Red dwarfs
Subdwarfs
Main sequence
("dwarfs")
Subgiants
Giants
Bright giants
Supergiants
Hypergiants
absolute
magni-
tude
(MV)

7.2 Formation

Main article: Star formation

When a protostar is formed from the collapse of a giant molecular cloud of gas and dust in the local interstellar medium, the initial composition is homogeneous throughout, consisting of about 70% hydrogen, 28% helium and trace amounts of other elements, by mass.[13] The initial mass of the star depends on the local conditions within the cloud. (The mass distribution of newly formed stars is described empirically by the initial mass function.)[14] During the initial collapse, this pre-main-sequence star generates energy through gravitational contraction. Upon reaching a suitable density, energy generation is begun at the core using an exothermic nuclear fusion process that converts hydrogen into helium.[12]

Once nuclear fusion of hydrogen becomes the dominant energy production process and the excess energy gained from gravitational contraction has been lost,[15] the star lies along a curve on the Hertzsprung–Russell diagram (or HR diagram) called the standard main sequence. Astronomers will sometimes refer to this stage as "zero age main sequence", or ZAMS.[16] The ZAMS curve can be calculated using computer models of stellar properties at the point when stars begin hydrogen fusion. From this point, the brightness and surface temperature of stars typically increase with age.[17]

A star remains near its initial position on the main sequence until a significant amount of hydrogen in the core has been consumed, then begins to evolve into a more luminous star. (On the HR diagram, the evolving star moves up and to the right of the main sequence.) Thus the main sequence represents the primary hydrogen-burning stage of a star's lifetime.[12]

7.3   Properties

The majority of stars on a typical HR diagram lie along the main-sequence curve. This line is pronounced because both the spectral type and the luminosity depend only on a star's mass, at least to zeroth-order approximation, as long as it is fusing hydrogen at its core—and that is what almost all stars spend most of their "active" lives doing.[18]

The temperature of a star determines its spectral type via its effect on the physical properties of plasma in its photosphere. A star's energy emission as a function of wavelength is influenced by both its temperature and composition. A key indicator of this energy distribution is given by the color index, $B - V$, which measures the star's magnitude in blue (B) and green-yellow (V) light by means of filters.[note 1] This difference in magnitude provides a measure of a star's temperature.

7.4   Dwarf terminology

Main-sequence stars are called dwarf stars, but this terminology is partly historical and can be somewhat confusing. For the cooler stars, dwarfs such as red dwarfs, orange dwarfs, and yellow dwarfs are indeed much smaller and dimmer than other stars of those colors. However, for hotter blue and white stars, the size and brightness difference between so-called *dwarf* stars that are on the main sequence and the so-called *giant* stars that are not becomes smaller; for the hottest stars it is not directly observable. For those stars the terms *dwarf* and *giant* refer to differences in spectral lines which indicate if a star is on the main sequence or off it. Nevertheless, very hot main-sequence stars are still sometimes called dwarfs, even though they have roughly the same size and brightness as the "giant" stars of that temperature.[19]

The common use of *dwarf* to mean main sequence is confusing in another way, because there are dwarf stars which are not main-sequence stars. For example, a white dwarf is the dead core of a star that is left after the star has shed

its outer layers, that is much smaller than a main-sequence star—roughly the size of the Earth. These represent the final evolutionary stage of many main-sequence stars.[20]

7.5   Parameters

By treating the star as an idealized energy radiator known as a black body, the luminosity L and radius R can be related to the effective temperature T_{eff} by the Stefan–Boltzmann law:

$$L = 4\pi\sigma R^2 Teff^4$$

where σ is the Stefan–Boltzmann constant. As the position of a star on the HR diagram shows its approximate luminosity, this relation can be used to estimate its radius.[21]

The mass, radius and luminosity of a star are closely interlinked, and their respective values can be approximated by three relations. First is the Stefan–Boltzmann law, which relates the luminosity L, the radius R and the surface temperature *Teff*. Second is the mass–luminosity relation, which relates the luminosity L and the mass M. Finally, the relationship between M and R is close to linear. The ratio of M to R increases by a factor of only three over 2.5 orders of magnitude of M. This relation is roughly proportional to the star's inner temperature TI, and its extremely slow increase reflects the fact that the rate of energy generation in the core strongly depends on this temperature, while it has to fit the mass–luminosity relation. Thus, a too high or too low temperature will result in stellar instability.

A better approximation is to take $\epsilon = L/M$, the energy generation rate per unit mass, as ε is proportional to TI^{15}, where TI is the core temperature. This is suitable for stars at least as massive as the Sun, exhibiting the CNO cycle, and gives the better fit $R \propto M^{0.78}$.[22]

7.5.1   Sample parameters

The table below shows typical values for stars along the main sequence. The values of luminosity (L), radius (R) and mass (M) are relative to the Sun—a dwarf star with a spectral classification of G2 V. The actual values for a star may vary by as much as 20–30% from the values listed below.[23]

7.6 Energy generation

See also: Stellar nucleosynthesis

All main-sequence stars have a core region where energy is generated by nuclear fusion. The temperature and density of this core are at the levels necessary to sustain the energy production that will support the remainder of the star. A reduction of energy production would cause the overlaying mass to compress the core, resulting in an increase in the fusion rate because of higher temperature and pressure. Likewise an increase in energy production would cause the star to expand, lowering the pressure at the core. Thus the star forms a self-regulating system in hydrostatic equilibrium that is stable over the course of its main sequence lifetime.[29]

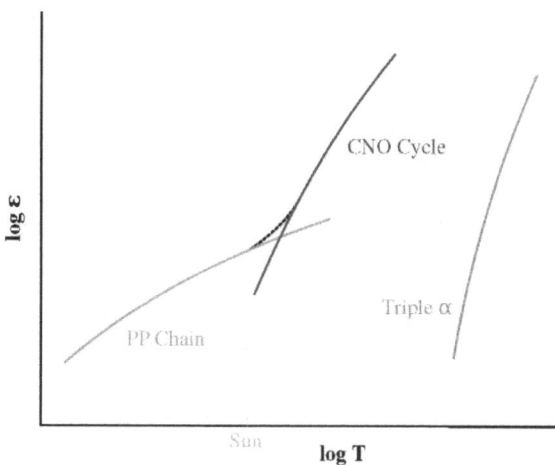

This graph shows the logarithm of the relative energy output (ε) for the proton-proton (PP), CNO and triple-α fusion processes at different temperatures. The dashed line shows the combined energy generation of the PP and CNO processes within a star. At the Sun's core temperature, the PP process is more efficient.

Main-sequence stars employ two types of hydrogen fusion processes, and the rate of energy generation from each type depends on the temperature in the core region. Astronomers divide the main sequence into upper and lower parts, based on which of the two is the dominant fusion process. In the lower main sequence, energy is primarily generated as the result of the proton-proton chain, which directly fuses hydrogen together in a series of stages to produce helium.[30] Stars in the upper main sequence have sufficiently high core temperatures to efficiently use the CNO cycle. (See the chart.) This process uses atoms of carbon, nitrogen and oxygen as intermediaries in the process of fusing hydrogen into helium.

At a stellar core temperature of 18 Million Kelvin, the PP process and CNO cycle are equally efficient, and each type

generates half of the star's net luminosity. As this is the core temperature of a star with about 1.5 $M\odot$, the upper main sequence consists of stars above this mass. Thus, roughly speaking, stars of spectral class F or cooler belong to the lower main sequence, while class A stars or hotter are upper main-sequence stars.[17] The transition in primary energy production from one form to the other spans a range difference of less than a single solar mass. In the Sun, a one solar mass star, only 1.5% of the energy is generated by the CNO cycle.[31] By contrast, stars with 1.8 $M\odot$ or above generate almost their entire energy output through the CNO cycle.[32]

The observed upper limit for a main-sequence star is 120–200 $M\odot$.[33] The theoretical explanation for this limit is that stars above this mass can not radiate energy fast enough to remain stable, so any additional mass will be ejected in a series of pulsations until the star reaches a stable limit.[34] The lower limit for sustained proton–proton nuclear fusion is about 0.08 $M\odot$ or 80 times the mass of Jupiter.[30] Below this threshold are sub-stellar objects that can not sustain hydrogen fusion, known as brown dwarfs.[35]

7.7 Structure

Main article: Stellar structure

Because there is a temperature difference between the

is transported by radiation, is stable against convection and there is very little mixing of the plasma. By contrast, in a convection zone the energy is transported by bulk movement of plasma, with hotter material rising and cooler material descending. Convection is a more efficient mode for carrying energy than radiation, but it will only occur under conditions that create a steep temperature gradient.[29][36]

In massive stars (above 10 $M_\odot$)[37] the rate of energy generation by the CNO cycle is very sensitive to temperature, so the fusion is highly concentrated at the core. Consequently, there is a high temperature gradient in the core region, which results in a convection zone for more efficient energy transport.[30] This mixing of material around the core removes the helium ash from the hydrogen-burning region, allowing more of the hydrogen in the star to be consumed during the main-sequence lifetime. The outer regions of a massive star transport energy by radiation, with little or no convection.[29]

Intermediate mass stars such as Sirius may transport energy primarily by radiation, with a small core convection region.[38] Medium-sized, low mass stars like the Sun have a core region that is stable against convection, with a convection zone near the surface that mixes the outer layers. This results in a steady buildup of a helium-rich core, surrounded by a hydrogen-rich outer region. By contrast, cool, very low-mass stars (below 0.4 $M_\odot$) are convective throughout.[14] Thus the helium produced at the core is distributed across the star, producing a relatively uniform atmosphere and a proportionately longer main sequence lifespan.[29]

7.8   Luminosity-color variation

As non-fusing helium ash accumulates in the core of a main-sequence star, the reduction in the abundance of hydrogen per unit mass results in a gradual lowering of the fusion rate within that mass. Since it is the outflow of fusion-supplied energy that supports the higher layers of the star, the core is compressed, producing higher temperatures and pressures. Both factors increase the rate of fusion thus moving the equilibrium towards a smaller, denser, hotter core producing more energy whose increased outflow pushes the higher layers further out. Thus there is a steady increase in the luminosity and radius of the star over time.[17] For example, the luminosity of the early Sun was only about 70% of its current value.[39] As a star ages this luminosity increase changes its position on the HR diagram. This effect results in a broadening of the main sequence band because stars are observed at random stages in their lifetime. That is, the main sequence band develops a thickness on the HR diagram; it is not simply a narrow line.[40]

Other factors that broaden the main sequence band on the HR diagram include uncertainty in the distance to stars and the presence of unresolved binary stars that can alter the observed stellar parameters. However, even perfect observation would show a fuzzy main sequence because mass is not the only parameter that affects a star's color and luminosity. Variations in chemical composition caused by the initial abundances, the star's evolutionary status,[41] interaction with a close companion,[42] rapid rotation,[43] or a magnetic field can all slightly change a main-sequence star's HR diagram position, to name just a few factors. As an example, there are metal-poor stars (with a very low abundance of elements with higher atomic numbers than helium) that lie just below the main sequence and are known as subdwarfs. These stars are fusing hydrogen in their cores and so they mark the lower edge of main sequence fuzziness caused by variance in chemical composition.[44]

A nearly vertical region of the HR diagram, known as the instability strip, is occupied by pulsating variable stars known as Cepheid variables. These stars vary in magnitude at regular intervals, giving them a pulsating appearance. The strip intersects the upper part of the main sequence in the region of class *A* and *F* stars, which are between one and two solar masses. Pulsating stars in this part of the instability strip that intersects the upper part of the main sequence are called Delta Scuti variables. Main-sequence stars in this region experience only small changes in magnitude and so this variation is difficult to detect.[45] Other classes of unstable main-sequence stars, like Beta Cephei variables, are unrelated to this instability strip.

7.9   Lifetime

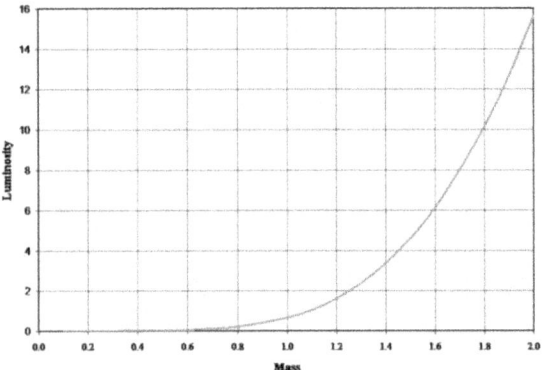

This plot gives an example of the mass-luminosity relationship for zero-age main-sequence stars. The mass and luminosity are relative to the present-day Sun.

The total amount of energy that a star can generate through nuclear fusion of hydrogen is limited by the amount of hydrogen fuel that can be consumed at the core. For a star in equilibrium, the energy generated at the core must be at least equal to the energy radiated at the surface. Since the luminosity gives the amount of energy radiated per unit time, the total life span can be estimated, to first approximation, as the total energy produced divided by the star's luminosity.[46]

For a star with at least 0.5 $M_\odot$, once the hydrogen supply in its core is exhausted and it expands to become a red giant, it can start to fuse helium atoms to form carbon. The energy output of the helium fusion process per unit mass is only about a tenth the energy output of the hydrogen process, and the luminosity of the star increases.[47] This results in a much shorter length of time in this stage compared to the main sequence lifetime. (For example, the Sun is predicted to spend 130 million years burning helium, compared to about 12 billion years burning hydrogen.)[48] Thus, about 90% of the observed stars above 0.5 $M_\odot$ will be on the main sequence.[49] On average, main-sequence stars are known to follow an empirical mass-luminosity relationship.[50] The luminosity (*L*) of the star is roughly proportional to the total mass (*M*) as the following power law:

$$L \propto M^{3.5}$$

This relationship applies to main-sequence stars in the range 0.1–50 $M_\odot$.[51]

The amount of fuel available for nuclear fusion is proportional to the mass of the star. Thus, the lifetime of a star on the main sequence can be estimated by comparing it to solar evolutionary models. The Sun has been a main-sequence star for about 4.5 billion years and it will become a red giant in 6.5 billion years,[52] for a total main sequence lifetime of roughly 10^{10} years. Hence:[53]

$$\tau_{MS} \approx 10^{10} \text{years} \cdot \left[\frac{M}{M_\odot} \right] \cdot \left[\frac{L_\odot}{L} \right] = 10^{10} \text{years} \cdot \left[\frac{M}{M_\odot} \right]^{-2.5}$$

where *M* and *L* are the mass and luminosity of the star, respectively, $M_\odot$ is a solar mass, $L_\odot$ is the solar luminosity and τ_{MS} is the star's estimated main sequence lifetime.

Although more massive stars have more fuel to burn and might be expected to last longer, they also must radiate a proportionately greater amount with increased mass. Thus, the most massive stars may remain on the main sequence for only a few million years, while stars with less than a tenth of a solar mass may last for over a trillion years.[54]

The exact mass-luminosity relationship depends on how efficiently energy can be transported from the core to the surface. A higher opacity has an insulating effect that retains more energy at the core, so the star does not need to

produce as much energy to remain in hydrostatic equilibrium. By contrast, a lower opacity means energy escapes more rapidly and the star must burn more fuel to remain in equilibrium.[55] Note, however, that a sufficiently high opacity can result in energy transport via convection, which changes the conditions needed to remain in equilibrium.[17]

In high-mass main-sequence stars, the opacity is dominated by electron scattering, which is nearly constant with increasing temperature. Thus the luminosity only increases as the cube of the star's mass.[47] For stars below 10 $M_\odot$, the opacity becomes dependent on temperature, resulting in the luminosity varying approximately as the fourth power of the star's mass.[51] For very low mass stars, molecules in the atmosphere also contribute to the opacity. Below about 0.5 $M_\odot$, the luminosity of the star varies as the mass to the power of 2.3, producing a flattening of the slope on a graph of mass versus luminosity. Even these refinements are only an approximation, however, and the mass-luminosity relation can vary depending on a star's composition.[14]

7.10 Evolutionary tracks

See also: Stellar evolution

Once a main-sequence star consumes the hydrogen at its

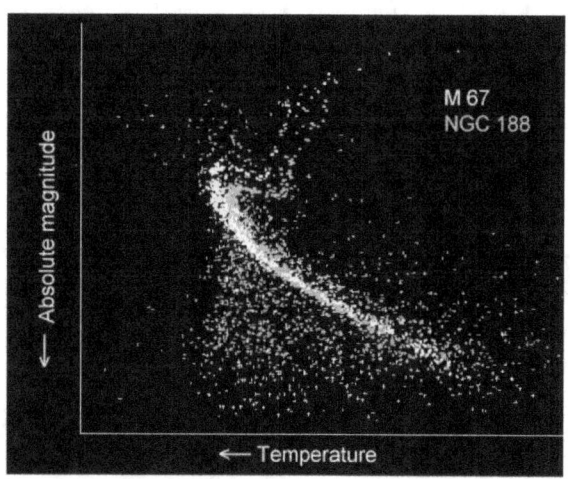

This shows the Hertzsprung–Russell diagrams for two open clusters. NGC 188 (blue) is older, and shows a lower turn off from the main sequence than that seen in M67 (yellow). The dots outside the two sequences are mostly foreground and background stars with no relation to the clusters.

core, the loss of energy generation causes its gravitational collapse to resume. Stars with less than 0.23 $M_\odot$,[3] are predicted to directly become white dwarfs once energy generation by nuclear fusion of hydrogen at their core comes to a halt. In stars between this threshold and 10 $M_\odot$, the hydrogen surrounding the helium core reaches sufficient temperature and pressure to undergo fusion, forming a hydrogen-burning shell. In consequence of this change, the outer envelope of the star expands and decreases in temperature, turning it into a red giant. At this point the star is evolving off the main sequence and entering the giant branch. The path which the star now follows across the HR diagram, to the upper right of the main sequence, is called an evolutionary track.

The helium core of a red giant continues to collapse until it is entirely supported by electron degeneracy pressure—a quantum mechanical effect that restricts how closely matter can be compacted. For stars of more than about 0.5 $M\odot$,[56] the core eventually reaches a temperature where it becomes hot enough to burn helium into carbon via the triple alpha process.[57][58] Stars with more than 5–7.5 $M\odot$ can additionally fuse elements with higher atomic numbers.[59][60] For stars with ten or more solar masses, this process can lead to an increasingly dense core that finally collapses, ejecting the star's overlying layers in a Type II supernova explosion,[4] Type Ib supernova or Type Ic supernova.

When a cluster of stars is formed at about the same time, the life span of these stars will depend on their individual masses. The most massive stars will leave the main sequence first, followed steadily in sequence by stars of ever lower masses. Thus the stars will evolve in order of their position on the main sequence, proceeding from the most massive at the left toward the right of the HR diagram. The current position where stars in this cluster are leaving the main sequence is known as the turn-off point. By knowing the main sequence lifespan of stars at this point, it becomes possible to estimate the age of the cluster.[61]

7.11   See also

- Hydrogen-burning process

- Red dwarf

- Supergiant

7.12   Notes

[1] By measuring the difference between these values, this eliminates the need to correct the magnitudes for distance. However, see extinction.

[2] The Sun is a typical type G2V star.

7.13   References

[1] Harding E. Smith (1999-04-21). "The Hertzsprung-Russell Diagram". *Gene Smith's Astronomy Tutorial*. Center for Astrophysics & Space Sciences, University of California, San Diego. Retrieved 2009-10-29.

[2] Richard Powell (2006). "The Hertzsprung Russell Diagram". *An Atlas of the Universe*. Retrieved 2009-10-29.

[3] Adams, Fred C.; Laughlin, Gregory (April 1997). "A Dying Universe: The Long Term Fate and Evolution of Astrophysical Objects". *Reviews of Modern Physics* **69** (2): 337–372. arXiv:astro-ph/9701131. Bibcode:1997RvMP...69..337A. doi:10.1103/RevModPhys.69.337.

[4] Gilmore, Gerry (2004). "The Short Spectacular Life of a Superstar". *Science* **304** (5697): 1915–1916. doi:10.1126/science.1100370. PMID 15218132. Retrieved 2007-05-01.

[5] "The Brightest Stars Don't Live Alone". *ESO Press Release*. Retrieved 27 July 2012.

[6] Longair, Malcolm S. (2006). *The Cosmic Century: A History of Astrophysics and Cosmology*. Cambridge University Press. pp. 25–26. ISBN 0-521-47436-1.

[7] Brown, Laurie M.; Pais, Abraham; Pippard, A. B., eds. (1995). *Twentieth Century Physics*. Bristol; New York: Institute of Physics, American Institute of Physics. p. 1696. ISBN 0-7503-0310-7. OCLC 33102501.

[8] Russell, H. N. (1913). ""Giant" and "dwarf" stars". *The Observatory* **36**: 324–329. Bibcode:1913Obs....36..324R.

[9] Strömgren, Bengt (1933). "On the Interpretation of the Hertzsprung-Russell-Diagram". *Zeitschrift für Astrophysik* **7**: 222–248. Bibcode:1933ZA......7..222S.

[10] Schatzman, Evry L.; Praderie, Francoise (1993). *The Stars*. Springer. pp. 96–97. ISBN 3-540-54196-9.

[11] Morgan, W. W.; Keenan, P. C.; Kellman, E. (1943). *An atlas of stellar spectra, with an outline of spectral classification*. Chicago, Illinois: The University of Chicago press. Retrieved 2008-08-12.

[12] Unsöld, Albrecht (1969). *The New Cosmos*. Springer-Verlag New York Inc. p. 268. ISBN 0-387-90886-2.

[13] Gloeckler, George; Geiss, Johannes (2004). "Composition of the local interstellar medium as diagnosed with pickup ions". *Advances in Space Research* **34** (1): 53–60. Bibcode:2004AdSpR..34...53G. doi:10.1016/j.asr.2003.02.054.

[14] Kroupa, Pavel (2002). "The Initial Mass Function of Stars: Evidence for Uniformity in Variable Systems". *Science* **295** (5552): 82–91. arXiv:astro-ph/0201098. Bibcode:2002Sci...295...82K. doi:10.1126/science.1067524. PMID 11778039. Retrieved 2007-12-03.

[15] Schilling, Govert (2001). "New Model Shows Sun Was a Hot Young Star". *Science* **293** (5538): 2188–2189. doi:10.1126/science.293.5538.2188. PMID 11567116. Retrieved 2007-02-04.

[16] "Zero Age Main Sequence". *The SAO Encyclopedia of Astronomy*. Swinburne University. Retrieved 2007-12-09.

[17] Clayton, Donald D. (1983). *Principles of Stellar Evolution and Nucleosynthesis*. University of Chicago Press. ISBN 0-226-10953-4.

[18] "Main Sequence Stars". Australia Telescope Outreach and Education. Retrieved 2007-12-04.

[19] Moore, Patrick (2006). *The Amateur Astronomer*. Springer. ISBN 1-85233-878-4.

[20] "White Dwarf". *COSMOS—The SAO Encyclopedia of Astronomy*. Swinburne University. Retrieved 2007-12-04.

[21] "Origin of the Hertzsprung-Russell Diagram". University of Nebraska. Retrieved 2007-12-06.

[22] "A course on stars' physical properties, formation and evolution" (PDF). University of St. Andrews. Retrieved 2010-05-18.

[23] Siess, Lionel (2000). "Computation of Isochrones". Institut d'Astronomie et d'Astrophysique, Université libre de Bruxelles. Retrieved 2007-12-06.—Compare, for example, the model isochrones generated for a ZAMS of 1.1 solar masses. This is listed in the table as 1.26 times the solar luminosity. At metallicity Z=0.01 the luminosity is 1.34 times solar luminosity. At metallicity Z=0.04 the luminosity is 0.89 times the solar luminosity.

[24] Zombeck, Martin V. (1990). *Handbook of Space Astronomy and Astrophysics* (2nd ed.). Cambridge University Press. ISBN 0-521-34787-4. Retrieved 2007-12-06.

[25] "SIMBAD Astronomical Database". Centre de Données astronomiques de Strasbourg. Retrieved 2008-11-21.

[26] Luck, R. Earle; Heiter, Ulrike (2005). "Stars within 15 Parsecs: Abundances for a Northern Sample". *The Astronomical Journal* **129** (2): 1063–1083. Bibcode:2005AJ....129.1063L. doi:10.1086/427250.

[27] "LTT 2151 – High proper-motion Star". Centre de Données astronomiques de Strasbourg. Retrieved 2008-08-12.

[28] Staff (2008-01-01). "List of the Nearest Hundred Nearest Star Systems". Research Consortium on Nearby Stars. Retrieved 2008-08-12.

[29] Brainerd, Jerome James (2005-02-16). "Main-Sequence Stars". The Astrophysics Spectator. Retrieved 2007-12-04.

[30] Karttunen, Hannu (2003). *Fundamental Astronomy*. Springer. ISBN 3-540-00179-4.

[31] Bahcall, John N.; Pinsonneault, M. H.; Basu, Sarbani (2001-07-10). "Solar Models: Current Epoch and Time Dependences, Neutrinos, and Helioseismological Properties". *The Astrophysical Journal* **555** (2): 990–1012. arXiv:astro-ph/0212331. Bibcode:2003PhRvL..90m1301B. doi:10.1086/321493.

[32] Salaris, Maurizio; Cassisi, Santi (2005). *Evolution of Stars and Stellar Populations*. John Wiley and Sons. p. 128. ISBN 0-470-09220-3.

[33] Oey, M. S.; Clarke, C. J. (2005). "Statistical Confirmation of a Stellar Upper Mass Limit". *The Astrophysical Journal* **620** (1): L43–L46. arXiv:astro-ph/0501135. Bibcode:2005ApJ...620L..43O. doi:10.1086/428396.

[34] Ziebarth, Kenneth (1970). "On the Upper Mass Limit for Main-Sequence Stars". *Astrophysical Journal* **162**: 947–962. Bibcode:1970ApJ...162..947Z. doi:10.1086/150726.

[35] Burrows, A.; Hubbard, W. B.; Saumon, D.; Lunine, J. I. (March 1993). "An expanded set of brown dwarf and very low mass star models". *Astrophysical Journal, Part 1* **406** (1): 158–171. Bibcode:1993ApJ...406..158B. doi:10.1086/172427.

[36] Aller, Lawrence H. (1991). *Atoms, Stars, and Nebulae*. Cambridge University Press. ISBN 0-521-31040-7.

[37] Bressan, A. G.; Chiosi, C.; Bertelli, G. (1981). "Mass loss and overshooting in massive stars". *Astronomy and Astrophysics* **102** (1): 25–30. Bibcode:1981A&A...102...25B.

[38] Lochner, Jim; Gibb, Meredith; Newman, Phil (2006-09-06). "Stars". NASA. Retrieved 2007-12-05.

[39] Gough, D. O. (1981). "Solar interior structure and luminosity variations". *Solar Physics* **74** (1): 21–34. Bibcode:1981SoPh...74...21G. doi:10.1007/BF00151270.

[40] Padmanabhan, Thanu (2001). *Theoretical Astrophysics*. Cambridge University Press. ISBN 0-521-56241-4.

[41] Wright, J. T. (2004). "Do We Know of Any Maunder Minimum Stars?". *The Astronomical Journal* **128** (3): 1273–1278. arXiv:astro-ph/0406338. Bibcode:2004AJ....128.1273W. doi:10.1086/423221. Retrieved 2007-12-06.

[42] Tayler, Roger John (1994). *The Stars: Their Structure and Evolution*. Cambridge University Press. ISBN 0-521-45885-4.

[43] Sweet, I. P. A.; Roy, A. E. (1953). "The structure of rotating stars". *Monthly Notices of the Royal Astronomical Society* **113**: 701–715. Bibcode:1953MNRAS.113..701S. doi:10.1093/mnras/113.6.701.

[44] Burgasser, Adam J.; Kirkpatrick, J. Davy; Lepine, Sebastien (July 5–9, 2004). *Spitzer Studies of Ultracool Subdwarfs: Metal-poor Late-type M, L and T Dwarfs*. Proceedings of the 13th Cambridge Workshop on Cool Stars, Stellar Systems and the Sun (Hamburg, Germany: Dordrecht, D. Reidel Publishing Co): 237. Retrieved 2007-12-06.

[45] Green, S. F.; Jones, Mark Henry; Burnell, S. Jocelyn (2004). *An Introduction to the Sun and Stars*. Cambridge University Press. ISBN 0-521-54622-2.

[46] Richmond, Michael W. (2004-11-10). "Stellar evolution on the main sequence". Rochester Institute of Technology. Retrieved 2007-12-03.

[47] Prialnik, Dina (2000). *An Introduction to the Theory of Stellar Structure and Evolution*. Cambridge University Press. ISBN 0-521-65937-X.

[48] Schröder, K.-P.; Connon Smith, Robert (May 2008). "Distant future of the Sun and Earth revisited". *Monthly Notices of the Royal Astronomical Society* **386** (1): 155–163. arXiv:0801.4031. Bibcode:2008MNRAS.386..155S. doi:10.1111/j.1365-2966.2008.13022.x.

[49] Arnett, David (1996). *Supernovae and Nucleosynthesis: An Investigation of the History of Matter, from the Big Bang to the Present*. Princeton University Press. ISBN 0-691-01147-8.—Hydrogen fusion produces 8×10^{18} erg/g while helium fusion produces 8×10^{17} erg/g.

[50] For a detailed historical reconstruction of the theoretical derivation of this relationship by Eddington in 1924, see: Lecchini, Stefano (2007). *How Dwarfs Became Giants. The Discovery of the Mass-Luminosity Relation*. Bern Studies in the History and Philosophy of Science. ISBN 3-9522882-6-8.

[51] Rolfs, Claus E.; Rodney, William S. (1988). *Cauldrons in the Cosmos: Nuclear Astrophysics*. University of Chicago Press. ISBN 0-226-72457-3.

[52] Sackmann, I.-Juliana; Boothroyd, Arnold I.; Kraemer, Kathleen E. (November 1993). "Our Sun. III. Present and Future". *Astrophysical Journal* **418**: 457–468. Bibcode:1993ApJ...418..457S. doi:10.1086/173407.

[53] Hansen, Carl J.; Kawaler, Steven D. (1994). *Stellar Interiors: Physical Principles, Structure, and Evolution*. Birkhäuser. p. 28. ISBN 0-387-94138-X.

[54] Laughlin, Gregory; Bodenheimer, Peter; Adams, Fred C. (1997). "The End of the Main Sequence". *The Astrophysical Journal* **482** (1): 420–432. Bibcode:1997ApJ...482..420L. doi:10.1086/304125.

[55] Imamura, James N. (1995-02-07). "Mass-Luminosity Relationship". University of Oregon. Archived from the original on December 14, 2006. Retrieved 2007-01-08.

[56] Fynbo, Hans O. U. et al. (2004). "Revised rates for the stellar triple-α process from measurement of 12C nuclear resonances". *Nature* **433** (7022): 136–139. doi:10.1038/nature03219. PMID 15650733.

[57] Sitko, Michael L. (2000-03-24). "Stellar Structure and Evolution". University of Cincinnati. Archived from the original on March 26, 2005. Retrieved 2007-12-05.

[58] Staff (2006-10-12). "Post-Main Sequence Stars". Australia Telescope Outreach and Education. Retrieved 2008-01-08.

[59] Girardi, L.; Bressan, A.; Bertelli, G.; Chiosi, C. (2000). "Evolutionary tracks and isochrones for low- and intermediate-mass stars: From 0.15 to 7 M_{sun}, and from Z=0.0004 to 0.03". *Astronomy and Astrophysics Supplement* **141** (3): 371–383. arXiv:astro-ph/9910164. Bibcode:2000A&AS..141..371G. doi:10.1051/aas:2000126.

[60] Poelarends, A. J. T.; Herwig, F.; Langer, N.; Heger, A. (March 2008). "The Supernova Channel of Super-AGB Stars". *The Astrophysical Journal* **675** (1): 614–625. arXiv:0705.4643. Bibcode:2008ApJ...675..614P. doi:10.1086/520872.

[61] Krauss, Lawrence M.; Chaboyer, Brian (2003). "Age Estimates of Globular Clusters in the Milky Way: Constraints on Cosmology". *Science* **299** (5603): 65–69. Bibcode:2003Sci...299...65K. doi:10.1126/science.1075631. PMID 12511641.

7.14   Further reading

7.14.1   General

- Kippenhahn, Rudolf, *100 Billion Suns*, Basic Books, New York, 1983.

7.14.2   Technical

- Arnett, David, *Supernovae and Nucleosynthesis*, Princeton University Press, Princeton, 1996.

- Bahcall, John N., *Neutrino Astrophysics*, Cambridge University Press, Cambridge, 1989.

- Bahcall, John N., Pinsonneault, M.H., and Basu, Sarbani, *"Solar Models: Current Epoch and Time Dependences, Neutrinos, and Helioseismological Properties,"* The Astrophysical Journal, 555, 990, 2001.

- Barnes, C. A., Clayton, D. D., and Schramm, D. N.(eds.), *Essays in Nuclear Astrophysics*, Cambridge University Press, Cambridge, 1982.

- Bowers, Richard L., and Deeming, Terry, *Astrophysics I: Stars*, Jones and Bartlett, Publishers, Boston, 1984.

- Bradley W. Carroll and Dale A. Ostlie (2007). *An Introduction to Modern Astrophysics*. Person Education Addison-Wesley San Francisco. ISBN 0-80530402-9.

- Chabrier, Gilles, and Baraffe, Isabelle, *"Theory of Low-Mass Stars and Substellar Objects,"* Annual Review of Astronomy and Astrophysics, 38, 337, 2000.

- Chandrasekhar, S., *An Introduction to the study of stellar Structure,* Dover Publications, Inc., New York, 1967.

- Clayton, Donald D., *Principles of Stellar Evolution and Nucleosynthesis,* University of Chicago Press, Chicago, 1983.

- Cox, J. P., and Giuli, R. T., *Principles of Stellar Structure,* Gordon and Breach, New York, 1968.

- Fowler, William ., Caughlan, Georgeanne R., and Zimmerman, Barbara A., *"Thermonuclear Reaction Rates, I,"* Annual Review of Astronomy and Astrophysics, 5, 525, 1967.

- Fowler, William A., Caughlan, Georgeanne R., and Zimmerman, Barbara A., *"Thermonuclear Reaction Rates, II, "* Annual Review of Astronomy and Astrophysics, 13, 69, 1975.

- Hansen, Carl J., Kawaler, Steven D., and Trimble, *Virginia Stellar Interiors: Physical Principles, Structure, and Evolution, Second Edition,* Springer-Verlag, New York, 2004.

- Harris, Michael J., Fowler, William A., Caughlan, Georgeanne R., and Zimmerman, Barbara A., *"Thermonuclear Reaction Rates, III,"* Annual Review of Astronomy and Astrophysics, 21, 165, 1983.

- Iben, Icko, Jr, *"Stellar Evolution Within and Off the Main Sequence,"* Annual Review of Astronomy and Astrophysics, 5, 571, 1967.

- Iglesias, Carlos A, and Rogers, Forrest J., *"Updated Opal Opacities,"* The Astrophysical Journal, 464, 943, 1996.

- Kippenhahn, Rudolf, and Weigert, Alfred, *Stellar Structure and Evolution,* Springer-Verlag, Berlin, 1990.

- Liebert, James, and Probst, Ronald G., *"Very Low Mass Stars,"* Annual Review of Astronomy and Astrophysics, 25, 437, 1987.

- Padmanabhan, T., *Theoretical Astrophysics,* Cambridge University Press, Cambridge, 2002.

- Prialnik, Dina, *An Introduction to the Theory of Stellar Structure and Evolution,* Cambridge University Press, Cambridge, 2000.

- Novotny, Eva, *Introduction to Stellar Atmospheres and Interior,* Oxford University Press, New York, 1973.

- Shore, Steven N., *The Tapestry of Modern Astrophysics,* John Wiley and Sons, Hoboken, 2003.

Chapter 8

Stellar nucleosynthesis

Stellar nucleosynthesis is the process by which the natural abundances of the chemical elements within stars vary due to nuclear fusion reactions in the cores and overlying mantles of stars. Stars are said to evolve (age) with changes in the abundances of the elements within. Core fusion increases the atomic weight of its gaseous elements, causing pressure loss and contraction accompanied by increase of temperature.[1] Structural changes of the star (evolution) become necessary to stabilize it. Stars lose most of their mass when it is ejected late in their stellar lifetimes, thereby increasing the abundance of elements heavier than helium in the interstellar medium. The term supernova nucleosynthesis is used to describe the creation of elements during the evolution and explosion of a presupernova star, as Fred Hoyle advocated presciently in 1954.[2] One stimulus to the development of the theory of nucleosynthesis was the variations in the abundances of elements found in the universe. Those abundances, when plotted on a graph as a function of atomic number of the element, have a jagged sawtooth shape that varies by factors of tens of millions. This suggested a natural process other than a random distribution. Such a graph of the abundances can be seen at History of nucleosynthesis theory. Stellar nucleosynthesis is the dominating contributor to several processes that also occur under the collective term nucleosynthesis.

A second stimulus to understanding the processes of stellar nucleosynthesis occurred during the 20th century, when it was realized that the energy released from nuclear fusion reactions accounted for the longevity of the Sun as a source[3] of heat and light. The fusion of nuclei in a star, starting from its initial hydrogen and helium abundance, provides that energy and synthesizes new nuclei as a byproduct of that fusion process. This became clear during the decade prior to World War II. The fusion product nuclei are restricted to those only slightly heavier than the fusing nuclei; thus they do not contribute heavily to the natural abundances of the elements. Nonetheless, this insight raised the plausibility of explaining all of the natural abundances of elements in this way. The prime energy producer in the sun is the fusion of hydrogen to form helium, which occurs at a solar-core temperature of 14 million kelvin.

8.1 History

In 1920 Arthur Eddington proposed that stars obtained their energy from nuclear fusion of hydrogen to form helium and raised the possibility that the heavier elements are produced in stars.

In 1920, Arthur Eddington, on the basis of the precise measurements of atomic masses by F.W. Aston and a preliminary suggestion by Jean Perrin, proposed that stars obtained their energy from nuclear fusion of hydrogen to form helium and raised the possibility that the heavier elements are produced in stars.[4][5][6] This was a preliminary step toward the idea of nucleosynthesis. In 1928, George

Gamow derived what is now called the Gamow factor, a quantum-mechanical formula that gave the probability of bringing two nuclei sufficiently close for the strong nuclear force to overcome the Coulomb barrier. The Gamow factor was used in the decade that followed by Atkinson and Houtermans and later by Gamow himself and Edward Teller to derive the rate at which nuclear reactions would proceed at the high temperatures believed to exist in stellar interiors.

In 1939, in a paper entitled "Energy Production in Stars", Hans Bethe analyzed the different possibilities for reactions by which hydrogen is fused into helium.[7] He defined two processes that he believed to be the sources of energy in stars. The first one, the proton–proton chain reaction, is the dominant energy source in stars with masses up to about the mass of the Sun. The second process, the carbon-nitrogen-oxygen cycle, which was also considered by Carl Friedrich von Weizsäcker in 1938, is most important in more massive stars. These works concerned the energy generation capable of keeping stars hot. A clear physical description of the p-p chain and of the CNO cycle appears in a 1968 textbook.[8] Bethe's two papers did not address the creation of heavier nuclei, however. That theory was begun by Fred Hoyle in 1946 with his argument that a collection of very hot nuclei would assemble into iron.[9] Hoyle followed that in 1954 with a large paper describing how advanced fusion stages within stars would synthesize elements between carbon and iron in mass.[10] This is the dominant work in stellar nucleosynthesis.[11] It provided the roadmap to how the most abundant elements on earth had been synthesized from initial hydrogen and helium, making clear how those abundant elements increased their galactic abundances as the galaxy aged.

Quickly, Hoyle's theory was expanded to other processes, beginning with the publication of a celebrated review paper in 1957 by Burbidge, Burbidge, Fowler and Hoyle (commonly referred to as the B^{2}FH paper).[12] This review paper collected and refined earlier research into a heavily cited picture that gave promise of accounting for the observed relative abundances of the elements; but it did not itself enlarge Hoyle's 1954 picture for the origin of primary nuclei as much as many assumed, except in the understanding of nucleosynthesis of those elements heavier than iron. Significant improvements were made by Alastair GW Cameron and by Donald D. Clayton. Cameron presented his own independent approach[13] (following Hoyle's approach for the most part) of nucleosynthesis. He introduced computers into time-dependent calculations of evolution of nuclear systems. Clayton calculated the first time-dependent models of the S-process[14] and of the R-process,[15] as well as of the burning of silicon into the abundant alpha-particle nuclei and iron-group elements,[16] and discovered radiogenic chronologies[17] for determining the age of the elements.

The entire research field expanded rapidly in the 1970s.

8.2 Key reactions

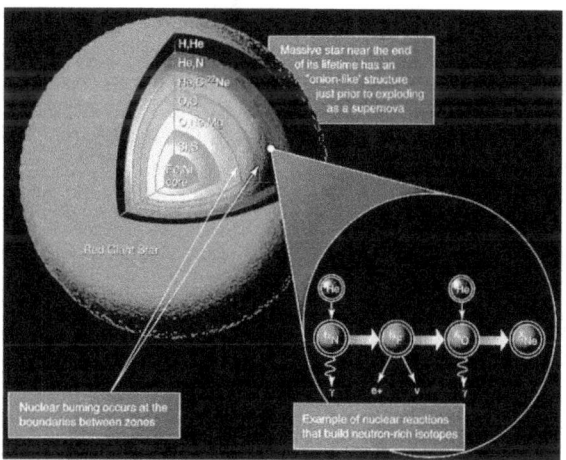

Cross section of a red giant showing nucleosynthesis and elements formed.

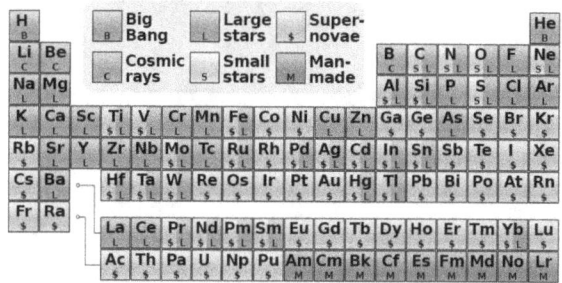

A version of the periodic table indicating the origins – including stellar nucleosynthesis – of the elements. All elements above 103 (lawrencium) are also manmade and are not included.

The most important reactions in stellar nucleosynthesis:

- Hydrogen fusing:
  - Deuterium burning
  - The proton-proton chain
  - The carbon-nitrogen-oxygen cycle
- Helium burning:
  - The triple-alpha process
  - The alpha process
- Burning of heavier elements:
  - Lithium burning: a process found most commonly in brown dwarfs

- Carbon burning process

- Neon burning process

- Oxygen burning process

- Silicon burning process

- Production of elements heavier than iron:

  - Neutron capture:

    - The R-process

    - The S-process

  - Proton capture:

    - The Rp-process

  - Photo-disintegration:

    - The P-process

8.2.1 Hydrogen burning

Main articles: Proton-proton chain reaction, CNO cycle
and Deuterium burning

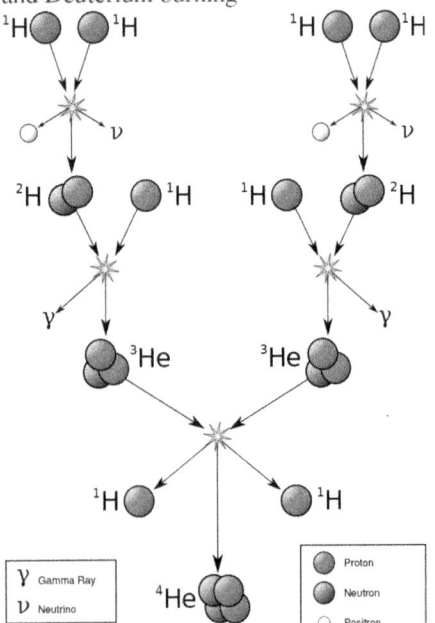

Illustration of the proton–proton chain reaction sequence

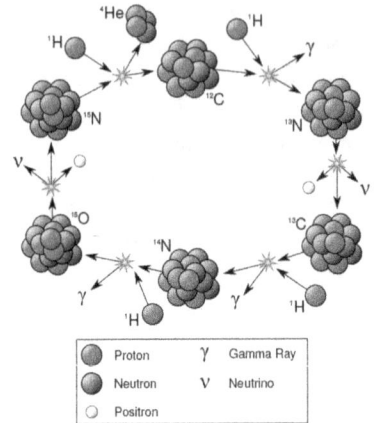

Overview of the CNO-I cycle. The helium nucleus is
released at the top-left step.

"Hydrogen burning" is an expression that astronomers
sometimes use for the stellar process that results in the nuclear fusion of four protons to form a nucleus of helium-
4.[18] (This should not be confused with the chemical
combustion of hydrogen in an oxidizing atmosphere.)
There are two predominant processes by which stellar hydrogen burning occurs.

In the cores of lower mass main sequence stars such as the
Sun, the dominant process is the proton-proton chain reaction (pp-chain reaction). This creates a helium-4 nucleus
through a sequence of chain reactions that begin with the fusion of two protons to form a nucleus of deuterium.[19] The
subsequent process of deuterium burning will consume any
pre-existing deuterium found at the core. The pp-chain reaction cycle is relatively insensitive to temperature, so this
hydrogen burning process can occur in up to a third of the
star's radius and occupy half the star's mass. As a result,
for stars above 35% of the Sun's mass,[20] the energy flux
toward the surface is sufficiently low that the core region remains a radiative zone, rather than becoming convective.[21]
In each complete fusion cycle, the p-p chain reaction releases about 26.2 MeV.[19]

In higher mass stars, the dominant process is the CNO cycle, which is a catalytic cycle that uses nuclei of carbon,
nitrogen and oxygen as intermediaries to produce a helium
nucleus.[19] During a complete CNO cycle, 25.0 MeV of
energy is released. The difference in energy compared to
the p-p chain reaction is accounted for by the energy lost
through neutrino emission.[19] The CNO cycle is very temperature sensitive, so it is strongly concentrated at the core.
About 90% of the CNO cycle energy generation occurs
within the inner 15% of the star's mass.[22] This results in
an intense outward energy flux that can not be sustained by
radiative transfer. As a result, the core region becomes a
convection zone, which stirs the hydrogen burning region
and keeps it well mixed with the surrounding proton-rich

region.[23] This core convection occurs in stars where the CNO cycle contributes more than 20% of the total energy. As the star ages and the core temperature increases, the region occupied by the convection zone slowly shrinks from 20% of the mass down to the inner 8% of the mass.[22]

The type of hydrogen burning process that dominates inside a star is determined by the temperature dependency differences between the two reactions. The pp-chain reaction starts at temperatures around 4×10^6 K,[24] making it the dominant mechanism in smaller stars. A self-maintaining CNO chain requires a higher temperature of approximately 15×10^6 K, but thereafter it increases more rapidly in efficiency than the pp-chain reaction as the temperature grows.[25] Above approximately 17×10^6 K, the CNO cycle becomes the dominant source of energy. This temperature is achieved in the cores of main sequence stars with at least 1.3 times the mass of the Sun.[26] The Sun itself has a core temperature of around 15.7×10^6 K and only 0.8% of the energy being produced in the Sun comes from the CNO cycle.[27] As a main sequence star ages, the core temperature will rise, resulting in a steadily increasing contribution from its CNO cycle.[22]

Once a star with about 0.5–10 times the mass of the Sun has consumed nearly all the hydrogen at its core, it begins to evolve up the red giant branch. Hydrogen burning occurs in a shell surrounding an inert helium core until the steadily increasing core temperature exceeds 1×10^8 K. At that point helium burning begins with a thermal runaway process called the helium flash with hydrogen burning continuing in a thin shell surrounding the now active helium core.[21]

8.3   References

[1] Donald D. Clayton, Principles of Stellar Evolution and Nucleosynthesis, Mc-Graw Hill, New York (1968) Chapter 6

[2] F. Hoyle, Synthesis of the elements between carbon and nickel, Astrophys. J. Suppl., 1, 121 (1954)

[3] Donald D. Clayton, *Principles of stellar Evolution and Nucleosynthesis*. McGraw-Hill, New York (1968); reissued by University of Chicago Press (1983)

[4] A.S. Eddington, The Internal Constitution of the Stars, *The Observatory*, **43**, 341 (1920) http://adsabs.harvard.edu/abs/1920Obs....43..341E

[5] A.S. Eddington, The Internal Constitution of the Stars, *Nature*, **106**, 106 (1920) http://adsabs.harvard.edu/abs/1920Natur.106...14E

[6] Why the Stars Shine D.Selle, Guidestar (Houston Astronomical Society), October 2012, p.6-8

[7] Energy Production in Stars by Hans Bethe

[8] Donald D. Clayton, Principles of Stellar Evolution and Nucleosynthesis, McGraw-Hill, New York (1968)

[9] F. Hoyle (1946). "The synthesis of the elements from hydrogen". *Monthly Notices of the Royal Astronomical Society* **106**: 343–383. Bibcode:1946MNRAS.106..343H. doi:10.1093/mnras/106.5.343.

[10] F. Hoyle, Synthesis of the elements between carbon and nickel, *Astrophys. J. Suppl.*, **1**, 121 (1954)

[11] D. D. Clayton, Hoyle's equation, *Science*, **318**, 1876–77 (2007)

[12] E. M. Burbidge; G. R. Burbidge; W. A. Fowler; F. Hoyle (1957). "Synthesis of the Elements in Stars". *Reviews of Modern Physics* **29** (4): 547–650. Bibcode:1957RvMP...29..547B. doi:10.1103/RevModPhys.29.547.

[13] A. G. W. Cameron, Stellar Evolution, Nuclear astrophysics and nucleogenesis, Chalk River (Canada) Report CRL-41 (1957)

[14] Donald D. Clayton, W. A. Fowler, T. E. Hull, and B. A. Zimmerman, "Neutron capture chains in heavy element synthesis", *Annals of Physics*, **12**, 331–408, (1961)

[15] Seeger, P. A., W. A. Fowler, and Donald D. Clayton, "Nucleosynthesis of heavy elements by neutron capture", *Astrophys. J. Suppl*, **XI**, 121–66, (1965)

[16] Bodansky, D., Donald D. Clayton, and W. A. Fowler, "Nucleosynthesis during silicon burning", *Phys. Rev. Letters*, **20**, 161–64, (1968); Bodansky, D., Donald D. Clayton, and W. A. Fowler, Nuclear quasi-equilibrium during silicon burning, *Astrophys. J. Suppl.* No. 148, **16**, 299–371, (1968)

[17] Donald D. Clayton, "Cosmoradiogenic chronologies of nucleosynthesis", *Astrophys. J.*, **139**, 637–63, (1964)

[18] Jones, Lauren V. (2009), *Stars and galaxies*, Greenwood guides to the universe, ABC-CLIO, pp. 65–67, ISBN 0-313-34075-7

[19] Böhm-Vitense, Erika (1992), *Introduction to Stellar Astrophysics* **3**, Cambridge University Press, pp. 93–100, ISBN 0-521-34871-4

[20] Reiners, A.; Basri, G. (March 2009). "On the magnetic topology of partially and fully convective stars". *Astronomy and Astrophysics* **496** (3): 787–790. arXiv:0901.1659. Bibcode:2009A&A...496..787R. doi:10.1051/0004-6361:200811450.

[21] de Loore, Camiel W. H.; Doom, C. (1992), *Structure and evolution of single and binary stars*, Astrophysics and space science library **179**, Springer, pp. 200–214, ISBN 0-7923-1768-8

[22] Jeffrey, C. Simon (2010), "Stellar structure and evolution: an introduction", in Goswami, A.; Reddy, B. E., *Principles and Perspectives in Cosmochemistry*, Springer, pp. 64–66, ISBN 3-642-10368-5

[23] Karttunen, Hannu; Oja, Heikki (2007), *Fundamental astronomy* (5th ed.), Springer, p. 247, ISBN 3-540-34143-9

[24] Reid, I. Neill; Hawley, Suzanne L. (2005), *New light on dark stars: red dwarfs, low-mass stars, brown dwarfs*, Springer-Praxis books in astrophysics and astronomy (2nd ed.), Springer, p. 108, ISBN 3-540-25124-3

[25] Salaris, Maurizio; Cassisi, Santi (2005), *Evolution of stars and stellar populations*, John Wiley and Sons, pp. 119–123, ISBN 0-470-09220-3

[26] Schuler, S. C.; King, J. R.; The, L.-S. (2009), "Stellar Nucleosynthesis in the Hyades Open Cluster", *The Astrophysical Journal* **701** (1): 837–849, arXiv:0906.4812, Bibcode:2009ApJ...701..837S, doi:10.1088/0004-637X/701/1/837

[27] Goupil, M. J.; Lebreton, Y.; Marques, J. P.; Samadi, R.; Baudin, F. (January 2011), "Open issues in probing interiors of solar-like oscillating main sequence stars 1. From the Sun to nearly suns", *Journal of Physics: Conference Series* **271** (1): 012031, arXiv:1102.0247, Bibcode:2011JPhCS.271a2031G, doi:10.1088/1742-6596/271/1/012031

8.4   Further reading

- Bethe, H. A. (1939). "Energy Production in Stars". *Physical Review* **55** (1): 103. Bibcode:1939PhRv...55..103B. doi:10.1103/PhysRev.55.103.

- Bethe, H. A. (1939). "Energy Production in Stars". *Physical Review* **55** (5): 434–456. Bibcode:1939PhRv...55..434B. doi:10.1103/PhysRev.55.434.

- Hoyle, F. (1954). "On Nuclear Reactions occurring in very hot stars: Synthesis of elements from carbon to nickel". *Astrophys. J.* **1** (Supplement 1): 121–146. Bibcode:1954ApJS....1..121H. doi:10.1086/190005.

- Clayton, Donald D. (1968). *Principles of Stellar Evolution and Nucleosynthesis*. New York: McGraw-Hill.

- Ray (2004). "Stars as thermonuclear reactors: Their fuels and ashes". arXiv:astro-ph/0405568 [astro-ph].

- G. Wallerstein; I. Iben Jr.; P. Parker; A.M. Boesgaard; G.M. Hale; A. E. Champagne et al. (1997). "Synthesis of the elements in stars: forty years of progress" (pdf). *Reviews of Modern Physics* **69**

(4): 995–1084. Bibcode:1997RvMP...69..995W. doi:10.1103/RevModPhys.69.995. Retrieved 2006-08-04.

- Woosley, S. E.; A. Heger; T. A. Weaver (2002). "The evolution and explosion of massive stars". *Reviews of Modern Physics* **74** (4): 1015–1071. Bibcode:2002RvMP...74.1015W. doi:10.1103/RevModPhys.74.1015.

- Clayton, Donald D. (2003). *Handbook of Isotopes in the Cosmos*. Cambridge: Cambridge University Press. ISBN 0-521-82381-1.

8.5   External links

- How the Sun Shines by John N. Bahcall

- Nucleosynthesis in NASA's Cosmicopia

Chapter 9

Big Bang nucleosynthesis

In physical cosmology, **Big Bang nucleosynthesis** (abbreviated BBN, also known as **primordial nucleosynthesis**) refers to the production of nuclei other than those of the lightest isotope of hydrogen (hydrogen-1, ^{1}H, having a single proton as a nucleus) during the early phases of the universe. Primordial nucleosynthesis is believed by most cosmologists to have taken place from 10 seconds to 20 minutes after the Big Bang, and is calculated to be responsible for the formation of most of the universe's helium as the isotope helium-4 (^{4}He), along with small amounts of the hydrogen isotope deuterium (^{2}H or D), the helium isotope helium-3 (^{3}He), and a very small amount of the lithium isotope lithium-7 (^{7}Li). In addition to these stable nuclei, two unstable or radioactive isotopes were also produced: the heavy hydrogen isotope tritium (^{3}H or T); and the beryllium isotope beryllium-7 (^{7}Be); but these unstable isotopes later decayed into ^{3}He and ^{7}Li, as above.

Essentially all of the elements that are heavier than lithium and beryllium were created much later, by stellar nucleosynthesis in evolving and exploding stars.

9.1 Characteristics

There are two important characteristics of Big Bang nucleosynthesis (BBN):

- The era began at temperatures of around 10 MeV (116 gigakelvin) and ended at temperatures below 100 keV (1.16 gigakelvin).[1] The corresponding time interval was from a few tenths of a second to up to 10^3 seconds.[2] The temperature/time relation in this era can be given by the equation:

$$tT^2 = (0.74 \text{ s MeV}^2) \times (10.75/g_*)^{1/2}$$

  where t is time in seconds, T is temperature in MeV and g* is the effective number of particle species.[3] (g* includes contributions of 2 from photons, 7/2 from electron-positron pairs and 7/4

from each neutrino flavor. In the standard model g* is 10.75). This expression also shows how a different number of neutrino flavors will change the rate of cooling of the early universe.

- It was widespread, encompassing the entire observable universe.

The key parameter which allows one to calculate the effects of BBN is the baryon/photon number ratio, which is a small number of order 6 x 10^{-10}. This parameter corresponds to the baryon density and controls the rate at which nucleons collide and react; from this we can derive elemental abundances. Although the baryon per photon ratio is important in determining elemental abundances, the precise value makes little difference to the overall picture. Without major changes to the Big Bang theory itself, BBN will result in mass abundances of about 75% of hydrogen-1, about 25% helium-4, about 0.01% of deuterium and helium-3, trace amounts (on the order of 10^{-10}) of lithium, and negligible heavier elements. (Traces of boron have been found in some old stars, giving rise to the question whether some boron, not really predicted by the theory, might have been produced in the Big Bang. The question is not presently resolved.[4]) That the observed abundances in the universe are generally consistent with these abundance numbers is considered strong evidence for the Big Bang theory.

In this field, for historical reasons it is customary to quote the Helium-4 fraction *by mass*, symbol Y, so that 25% helium-4 means that helium-4 atoms account for 25% of the mass, but only about 8% of the nuclei would be helium-4 nuclei. Other (trace) nuclei are usually expressed as number ratios to hydrogen.

9.2 Important parameters

The creation of light elements during BBN was dependent on a number of parameters; among those was the neutron-

proton ratio (calculable from Standard Model physics) and
the baryon-photon ratio.

9.2.1   Neutron-proton ratio

Neutrons can react with positrons or electron neutrinos to
create protons and other products in one of the following
reactions:

$$n + e^+ \leftrightarrow \text{anti-}\nu_e + p$$

$$n + \nu_e \leftrightarrow p + e^-$$

At times much earlier than 1 sec, the n/p ratio was close to
1:1. As the temperature dropped, the equilibrium shifted
in favour of protons due to their slightly lower mass, and
the n/p ratio decreased. These reactions continue until expansion of the universe outpaces the reactions, which occurs at about T = 0.7 MeV and is called the freeze out
temperature.[5] At freeze out, the neutron-proton ratio was
about 1/5. However, free neutrons are unstable with a mean
life of 880 sec; some neutrons decayed in the next few minutes before fusing into any nucleus, so the ratio of total
neutrons to protons after nucleosynthesis ends is about 1/7.
Almost all neutrons that fused instead of decaying ended
up combined into helium-4, due to the fact that helium-4
has the highest binding energy per nucleon among light elements. This predicts that about 8% of all atoms should be
helium-4, leading to a mass fraction of helium-4 of about
25%, which is in line with observations. Small traces of
deuterium and helium-3 remained as there was insufficient
time and density for them to react and form helium-4.

9.2.2   Baryon-photon ratio

The baryon-photon ratio, η, is the key parameter determining the abundance of light elements present in the early universe. Baryons can react with light elements in the following
reactions:

$$(p,n) + {}^2H \rightarrow ({}^3He, {}^3H)$$

$$({}^3He, {}^3H) + (n,p) \rightarrow {}^4He$$

It is evident that reactions with baryons during BBN would
ultimately result in helium-4, and also that the abundance of
primordial deuterium is indirectly related to the baryon density or baryon-photon ratio. That is, the larger the baryon-
photon ratio the more reactions there will be and the more
efficiently deuterium will be eventually transformed into
helium-4. This result makes deuterium a very useful tool
in measuring the baryon-to-photon ratio.

9.3   Sequence

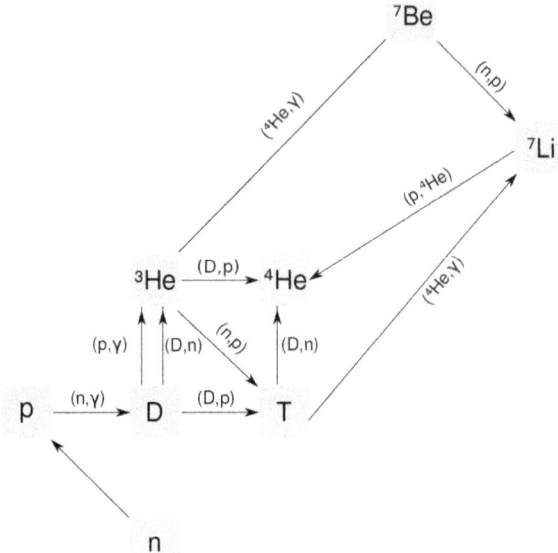

The main nuclear reaction chains for Big Bang nucleosynthesis

Big Bang nucleosynthesis began a few seconds after the big
bang, when the universe had cooled sufficiently to allow
deuterium nuclei to survive disruption by high-energy photons. This time is essentially independent of dark matter
content, since the universe was highly radiation dominated
until much later, and this dominant component controls the
temperature/time relation. The relative abundances of protons and neutrons follow from simple thermodynamical arguments, combined with the way that the mean temperature
of the universe changes over time. If the reactions needed to
reach the thermodynamically favoured equilibrium values
are too slow compared to the temperature change brought
about by the expansion, abundances would have remained
at some specific non-equilibrium value. Combining thermodynamics and the changes brought about by cosmic expansion, one can calculate the fraction of protons and neutrons based on the temperature at this point. The answer is
that there are about seven protons for every neutron at the
beginning of nucleosynthesis. This fraction is in favour of
protons, primarily because their lower mass with respect to
the neutron favors their production. Free neutrons decay
to protons with a half-life of about 10.2 minutes, but this
time-scale is longer than the first three minutes of nucleogenesis, during which time a substantial fraction of them were
combined with protons into deuterium and then He-4. The
sequence of these reaction chains is shown on the image.[6]

One feature of BBN is that the physical laws and constants
that govern the behavior of matter at these energies are very
well understood, and hence BBN lacks some of the speculative uncertainties that characterize earlier periods in the

life of the universe. Another feature is that the process of nucleosynthesis is determined by conditions at the start of this phase of the life of the universe, and proceeds independently of what happened before.

As the universe expands, it cools. Free neutrons and protons are less stable than helium nuclei, and the protons and neutrons have a strong tendency to form helium-4. However, forming helium-4 requires the intermediate step of forming deuterium. Before nucleosynthesis began, the temperature was high enough for many photons to have energy greater than the binding energy of deuterium; therefore any deuterium that was formed was immediately destroyed (a situation known as the **deuterium bottleneck**). Hence, the formation of helium-4 is delayed until the universe became cool enough for deuterium to survive (at about T = 0.1 MeV); after which there was a sudden burst of element formation. However, very shortly thereafter, at twenty minutes after the Big Bang, the universe became too cool for any further nuclear fusion and nucleosynthesis to occur. At this point, the elemental abundances were nearly fixed, and only change was the result of the radioactive decay of some products of BBN (such as tritium).[7]

9.3.1 History of theory

The history of Big Bang nucleosynthesis began with the calculations of Ralph Alpher in the 1940s. Alpher published the Alpher–Bethe–Gamow paper that outlined the theory of light-element production in the early universe.

During the 1970s, there was a major puzzle in that the density of baryons as calculated by Big Bang nucleosynthesis was much less than the observed mass of the universe based on calculations of the expansion rate. This puzzle was resolved in large part by postulating the existence of dark matter.

9.3.2 Heavy elements

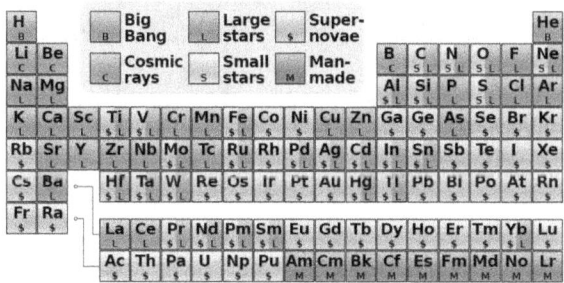

A version of the periodic table indicating the origins – including big bang nucleosynthesis – of the elements. All elements above 103 (lawrencium) are also manmade and are not included.

Big Bang nucleosynthesis produced no elements heavier than beryllium, due to a bottleneck: the absence of a stable nucleus with 8 or 5 nucleons. This deficit of larger atoms also limited the amounts of lithium-7 and beryllium-9 produced during BBN. In stars, the bottleneck is passed by triple collisions of helium-4 nuclei, producing carbon (the triple-alpha process). However, this process is very slow, taking tens of thousands of years to convert a significant amount of helium to carbon in stars, and therefore it made a negligible contribution in the minutes following the Big Bang.

9.3.3 Helium-4

Main article: Helium-4

Big Bang nucleo-synthesis predicts a primordial abundance of about 25% helium-4 by mass, irrespective of the initial conditions of the universe. As long as the universe was hot enough for protons and neutrons to transform into each other easily, their ratio, determined solely by their relative masses, was about 1 neutron to 7 protons (allowing for some decay of neutrons into protons). Once it was cool enough, the neutrons quickly bound with an equal number of protons to form first deuterium, then helium-4. Helium-4 is very stable and is nearly the end of this chain if it runs for only a short time, since helium neither decays nor combines easily to form heavier nuclei (since there are no stable nuclei with mass numbers of 5 or 8, helium does not combine easily with either protons, or with itself). Once temperatures are lowered, out of every 16 nucleons (2 neutrons and 14 protons), 4 of these (25% of the total particles and total mass) combine quickly into one helium-4 nucleus. This produces one helium for every 12 hydrogens, resulting in a universe that is a little over 8% helium by number of atoms, and 25% helium by mass.

One analogy is to think of helium-4 as ash, and the amount of ash that one forms when one completely burns a piece of wood is insensitive to how one burns it. The resort to the BBN theory of the helium-4 abundance is necessary as there is far more helium-4 in the universe than can be explained by stellar nucleosynthesis. In addition, it provides an important test for the Big Bang theory. If the observed helium abundance is much different from 25%, then this would pose a serious challenge to the theory. This would particularly be the case if the early helium-4 abundance was much smaller than 25% because it is hard to destroy helium-4. For a few years during the mid-1990s, observations suggested that this might be the case, causing astrophysicists to talk about a Big Bang nucleosynthetic crisis, but further observations were consistent with the Big Bang theory.[8]

9.3.4 Deuterium

Main article: Deuterium

Deuterium is in some ways the opposite of helium-4 in that while helium-4 is very stable and very difficult to destroy, deuterium is only marginally stable and easy to destroy. The temperatures, time, and densities were sufficient to combine a substantial fraction of the deuterium nuclei to form helium-4 but insufficient to carry the process further using helium-4 in the next fusion step. BBN did not convert all of the deuterium in the universe to helium-4 due to the expansion that cooled the universe and reduced the density and so, cut that conversion short before it could proceed any further. One consequence of this is that unlike helium-4, the amount of deuterium is very sensitive to initial conditions. The denser the initial universe was, the more deuterium would be converted to helium-4 before time ran out, and the less deuterium would remain.

There are no known post-Big Bang processes which can produce significant amounts of deuterium. Hence observations about deuterium abundance suggest that the universe is not infinitely old, which is in accordance with the Big Bang theory.

During the 1970s, there were major efforts to find processes that could produce deuterium, but those revealed ways of producing isotopes other than deuterium. The problem was that while the concentration of deuterium in the universe is consistent with the Big Bang model as a whole, it is too high to be consistent with a model that presumes that most of the universe is composed of protons and neutrons. If one assumes that all of the universe consists of protons and neutrons, the density of the universe is such that much of the currently observed deuterium would have been burned into helium-4. The standard explanation now used for the abundance of deuterium is that the universe does not consist mostly of baryons, but that non-baryonic matter (also known as dark matter) makes up most of the mass of the universe. This explanation is also consistent with calculations that show that a universe made mostly of protons and neutrons would be far more *clumpy* than is observed.

It is very hard to come up with another process that would produce deuterium other than by nuclear fusion. Such a process would require that the temperature be hot enough to produce deuterium, but not hot enough to produce helium-4, and that this process should immediately cool to non-nuclear temperatures after no more than a few minutes. It would also be necessary for the deuterium to be swept away before it reoccurs.

Producing deuterium by fission is also difficult. The problem here again is that deuterium is very unlikely due to nuclear processes, and that collisions between atomic nuclei are likely to result either in the fusion of the nuclei, or in the release of free neutrons or alpha particles. During the 1970s, cosmic ray spallation was proposed as a source of deuterium. That theory failed to account for the abundance of deuterium, but led to explanations of the source of other light elements.

9.4 Measurements and status of theory

The theory of BBN gives a detailed mathematical description of the production of the light "elements" deuterium, helium-3, helium-4, and lithium-7. Specifically, the theory yields precise quantitative predictions for the mixture of these elements, that is, the primordial abundances at the end of the big-bang.

In order to test these predictions, it is necessary to reconstruct the primordial abundances as faithfully as possible, for instance by observing astronomical objects in which very little stellar nucleosynthesis has taken place (such as certain dwarf galaxies) or by observing objects that are very far away, and thus can be seen in a very early stage of their evolution (such as distant quasars).

As noted above, in the standard picture of BBN, all of the light element abundances depend on the amount of ordinary matter (baryons) relative to radiation (photons). Since the universe is presumed to be homogeneous, it has one unique value of the baryon-to-photon ratio. For a long time, this meant that to test BBN theory against observations one had to ask: can *all* of the light element observations be explained with a *single value* of the baryon-to-photon ratio? Or more precisely, allowing for the finite precision of both the predictions and the observations, one asks: is there some *range* of baryon-to-photon values which can account for all of the observations?

More recently, the question has changed: Precision observations of the cosmic microwave background radiation[9][10] with the Wilkinson Microwave Anisotropy Probe (WMAP) and Planck give an independent value for the baryon-to-photon ratio. Using this value, are the BBN predictions for the abundances of light elements in agreement with the observations?

The present measurement of helium-4 indicates good agreement, and yet better agreement for helium-3. But for lithium-7, there is a significant discrepancy between BBN and WMAP/Planck, and the abundance derived from Population II stars. The discrepancy is a factor of 2.4—4.3 below the theoretically predicted value and is considered a problem for the original models,[11] that have resulted in revised calculations of the standard BBN based on new nuclear data, and to various reevaluation proposals for primordial proton-proton nuclear reactions, especially the abundances of $^7Be(n,p)^7Li$ versus $^7Be(d,p)^8Be$.[12]

9.5 Non-standard scenarios

In addition to the standard BBN scenario there are numerous non-standard BBN scenarios. These should not be confused with non-standard cosmology: a non-standard BBN scenario assumes that the Big Bang occurred, but inserts additional physics in order to see how this affects elemental abundances. These pieces of additional physics include relaxing or removing the assumption of homogeneity, or inserting new particles such as massive neutrinos.

There have been, and continue to be, various reasons for researching non-standard BBN. The first, which is largely of historical interest, is to resolve inconsistencies between BBN predictions and observations. This has proved to be of limited usefulness in that the inconsistencies were resolved by better observations, and in most cases trying to change BBN resulted in abundances that were more inconsistent with observations rather than less. The second reason for researching non-standard BBN, and largely the focus of non-standard BBN in the early 21st century, is to use BBN to place limits on unknown or speculative physics. For example, standard BBN assumes that no exotic hypothetical particles were involved in BBN. One can insert a hypothetical particle (such as a massive neutrino) and see what has to happen before BBN predicts abundances which are very different from observations. This has been usefully done to put limits on the mass of a stable tau neutrino.

9.6 See also

- Nucleosynthesis
- Stellar nucleosynthesis
- Ultimate fate of the universe
- Chronology of the universe
- Big Bang

9.7 References

[1] Doglov, A. D. "Big Bang Nucleosynthesis." Nucl.Phys.Proc.Suppl. (2002): 137-43. ArXiv. 17 Jan. 2002. Web. 14 Jan. 2013.

[2] Grupen, Claus. "Big Bang Nucleosynthesis." Astroparticle Physics. Berlin: Springer, 2005. 213-28. Print.

[3] J. Beringer et al. (Particle Data Group), "Big-Bang cosmology" Phys. Rev. D86, 010001 (2012): (21.43)

[4] "Hubble Observations Bring Some Surprises". *The New York Times*. 1992-01-14. Retrieved 2010-04-26.

[5] Gary Steigman (December 2007). "Primordial Nucleosynthesis in the Precision Cosmology Era". *Annual Review of Nuclear and Particle Science*: 463–491. arXiv:0712.1100. Bibcode:2007ARNPS..57..463S. doi:10.1146/annurev.nucl.56.080805.140437.

[6] Bertulani, Carlos A. (2013). *Nuclei in the Cosmos*. World Scientific. ISBN 978-981-4417-66-2.

[7] Weiss, Achim. "Equilibrium and change: The physics behind Big Bang Nucleosynthesis". *Einstein Online*. Archived from the original on 8 February 2007. Retrieved 2007-02-24.

[8] Bludman, S. A. (December 1998). "Baryonic Mass Fraction in Rich Clusters and the Total Mass Density in the Cosmos". *Astrophysical Journal* **508** (2): 535–538. arXiv:astro-ph/9706047. Bibcode:1998ApJ...508..535B. doi:10.1086/306412.

[9] David Toback (2009). "Chapter 12: Cosmic Background Radiation"

[10] David Toback (2009). "Unit 4: The Evolution Of The Universe"

[11] R. H. Cyburt, B. D. Fields & K. A. Olive (2008). "A Bitter Pill: The Primordial Lithium Problem Worsens". arXiv:0808.2818.

[12] Weiss, Achim. "Elements of the past: Big Bang Nucleosynthesis and observation". *Einstein Online*. Archived from the original on 8 February 2007. Retrieved 2007-02-24. For a recent calculation of BBN predictions, see A. Coc et al. (2004). "Updated Big Bang Nucleosynthesis confronted to WMAP observations and to the Abundance of Light Elements". *Astrophysical Journal* **600** (2): 544. arXiv:astro-ph/0309480. Bibcode:2004ApJ...600..544C. doi:10.1086/380121.
For the observational values, see the following articles:

- Helium-4: K. A. Olive & E. A. Skillman (2004). "A Realistic Determination of the Error on the Primordial Helium Abundance". *Astrophysical Journal* **617** (1): 29. arXiv:astro-ph/0405588. Bibcode:2004ApJ...617...29O. doi:10.1086/425170.

- Helium-3: T. M. Bania, R. T. Rood & D. S. Balser (2002). "The cosmological density of baryons from observations of 3He+ in the Milky Way". *Nature* **415** (6867): 54–7. Bibcode:2002Natur.415...54B. doi:10.1038/415054a. PMID 11780112.

- Deuterium: J. M. O'Meara et al. (2001). "The Deuterium to Hydrogen Abundance Ratio Towards a Fourth QSO: HS0105+1619". *Astrophysical Journal* **552** (2): 718. arXiv:astro-ph/0011179. Bibcode:2001ApJ...552..718O. doi:10.1086/320579.

• Lithium-7: C. Charbonnel & F. Primas (2005). "The Lithium Content of the Galactic Halo Stars". *Astronomy & Astrophysics* **442** (3): 961. arXiv:astro-ph/0505247. Bibcode:2005A&A...442..961C. doi:10.1051/0004-6361:20042491. A. Korn et al. (2006). "A probable stellar solution to the cosmological lithium discrepancy". *Nature* **442** (7103): 657–9. arXiv:astro-ph/0608201. Bibcode:2006Natur.442..657K. doi:10.1038/nature05011. PMID 16900193.

9.8   External links

9.8.1   For a general audience

• Weiss, Achim. "Big Bang Nucleosynthesis: Cooking up the first light elements". *Einstein Online*. Archived from the original on 8 February 2007. Retrieved 2007-02-24.

• White, Martin: Overview of BBN

• Wright, Ned: BBN (cosmology tutorial)

• Big Bang nucleosynthesis on arxiv.org

• Burles, Scott; Nollett, Kenneth M.; Turner, Michael S. (1999-03-19). "Big-Bang Nucleosynthesis: Linking Inner Space and Outer Space". arXiv:astro-ph/9903300.

9.8.2   Technical articles

• Burles, Scott, and Kenneth M. Nollett, Michael S. Turner (2001). "What Is The BBN Prediction for the Baryon Density and How Reliable Is It?". *Phys. Rev. D* **63** (6): 063512. arXiv:astro-ph/0008495. Bibcode:2001PhRvD..63f3512B. doi:10.1103/PhysRevD.63.063512. **Report-no**: FERMILAB-Pub-00-239-A

• Jedamzik, Karsten, "*Non-Standard Big Bang Nucleosynthesis Scenarios*". Max-Planck-Institut für Astrophysik, Garching.

• Steigman, Gary, Primordial Nucleosynthesis: Successes And Challenges arXiv:astro-ph/0511534; Forensic Cosmology: Probing Baryons and Neutrinos With BBN and the CBR arXiv:hep-ph/0309347; and Big Bang Nucleosynthesis: Probing the First 20 Minutes arXiv:astro-ph/0307244

• R. A. Alpher, H. A. Bethe, G. Gamow, *The Origin of Chemical Elements*, *Physical Review* **73** (1948), 803. The so-called $\alpha\beta\gamma$ paper, in which Alpher and Gamow suggested that the light elements were created by hydrogen ions capturing neutrons in the hot, dense early universe. Bethe's name was added for symmetry

• Gamow, G. (1948). "The Origin of Elements and the Separation of Galaxies". *Physical Review* **74**: 505. Bibcode:1948PhRv...74..505G. doi:10.1103/physrev.74.505.2. These two 1948 papers of Gamow laid the foundation for our present understanding of big-bang nucleosynthesis

• Gamow, G. (1948). "The Evolution of the Universe". *Nature* **162**: 680. Bibcode:1948Natur.162..680G. doi:10.1038/162680a0.

• Alpher, R. A. (1948). "A Neutron-Capture Theory of the Formation and Relative Abundance of the Elements". *Physical Review* **74**: 1737. Bibcode:1948PhRv...74.1737A. doi:10.1103/PhysRev.74.1737.

• R. A. Alpher and R. Herman, "On the Relative Abundance of the Elements," *Physical Review* **74** (1948), 1577. This paper contains the first estimate of the present temperature of the universe

• Alpher, R. A.; Herman, R.; Gamow, G. (1948). "Evolution of the Universe". *Nature* **162**: 774. Bibcode:1948Natur.162..774A. doi:10.1038/162774b0.

• Java Big Bang element abundance calculator

Chapter 10

Supernova nucleosynthesis

Supernova nucleosynthesis is a theory of the production of many different chemical elements in supernova explosions, first advanced by Fred Hoyle in 1954.[1] The nucleosynthesis, or fusion of lighter elements into heavier ones, occurs during explosive oxygen burning and silicon burning.[2] Those fusion reactions create the elements silicon, sulfur, chlorine, argon, sodium, potassium, calcium, scandium, titanium and iron peak elements: vanadium, chromium, manganese, iron, cobalt, and nickel. These are called "primary elements", in that they can be fused from pure hydrogen and helium in massive stars. As a result of their ejection from supernovae, their abundances increase within the interstellar medium. Elements heavier than nickel are created primarily by a rapid capture of neutrons in a process called the r-process. However, these are much less abundant than the primary chemical elements. Other processes thought to be responsible for some of the nucleosynthesis of underabundant heavy elements, notably a proton capture process known as the rp-process and a photodisintegration process known as the gamma (or p) process. The latter synthesizes the lightest, most neutron-poor, isotopes of the heavy elements.

10.1 Cause

Main article: Supernova

A supernova is a massive explosion of a star that occurs under two principal scenarios. The first is that a white dwarf star undergoes a nuclear-based explosion after it reaches its Chandrasekhar limit after absorbing mass from a neighboring star (usually a red giant). The second, and more common, cause is when a massive star, usually a supergiant, reaches nickel-56 in its nuclear fusion (or burning) processes. This isotope undergoes radioactive decay into iron-56, which has one of the highest binding energies of all of the isotopes, and is the last element that produces a net release of energy by nuclear fusion, exothermically.

All nuclear fusion reactions that produce heavier elements cause the star to lose energy and are said to be endothermic reactions. The pressure that supports the star's outer layers drops sharply. As the outer envelope is no longer sufficiently supported by the radiation pressure, the star's gravity pulls its outer layers rapidly inward. As the star collapses, these outer layers collide with the incompressible stellar core, producing a shockwave that expands outward through the unfused material of the outer shell. The pressures and densities in the shockwave are sufficient to induce fusion in that material, and the energy released leads to the star's explosion, dispersing material from the star into interstellar space.

10.1.1 Nuclear fusion sequence and the alpha process

After a star completes the oxygen burning process, its core is composed primarily of silicon and sulfur.[3] If it has sufficiently high mass, it further contracts until its core reaches temperatures in the range of 2.7–3.5 GK (230–300 keV). At these temperatures, silicon and other elements can photodisintegrate, emitting a proton or alpha particle.[3] Silicon burning entails the *alpha process*, which creates new elements by adding one of these alpha particles[3] (the equivalent of a helium nucleus, two protons plus two neutrons) per step in the following sequence:

The entire silicon-burning sequence lasts about one day and stops when nickel-56 has been produced. The star can no longer release energy via nuclear fusion because a nucleus with 56 nucleons has the lowest mass per nucleon (any proton or neutron) of all the elements in the alpha process sequence. Although iron-58 and nickel-62 have slightly higher binding energies per nucleon than iron-56,[4] the next step up in the alpha process would be zinc−60, which has slightly *more* mass per nucleon and thus, is less thermodynamically favorable. Nickel-56 (which has 28 protons) has a half-life of 6.02 days and decays via β^+ decay to cobalt−56 (27 protons), which in turn has a half-life of 77.3 days as it decays to iron-56 (26 protons). However, only minutes are available for the nickel-56 to decay within the core of a massive star. The star has run out of nuclear fuel and within minutes begins to contract.

During this phase of the contraction, the potential energy of gravitational contraction heats the interior to 5 GK (430 keV) and this opposes and delays the contraction. However, since no additional heat energy can be generated via new fusion reactions, the final unopposed contraction rapidly accelerates into a collapse lasting only a few seconds. The central portion of the star is now crushed into either a neutron star or, if the star is massive enough, a black hole. The outer layers of the star are blown off in an explosion known as a Type II supernova that lasts days to months. The supernova explosion releases a large burst of neutrons, which synthesizes, in about one second while-inside the star, roughly half of the supply of elements in the universe that are heavier than iron, via a neutron-capture mechanism known as the *r-process* (where the "r" stands for rapid neutron capture).

10.2   Products

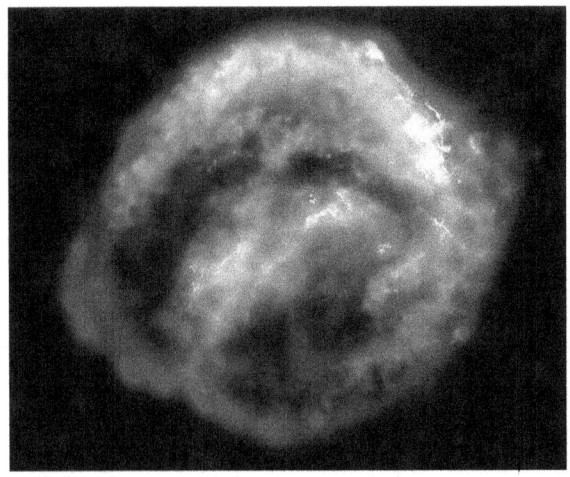

Composite image of Kepler's supernova from pictures by the Spitzer Space Telescope, Hubble Space Telescope, and Chandra X-ray Observatory.

The maximum weight for an element produced by fusion in a normal star is that of iron, reaching an isotope with an atomic mass of 56 (see Stellar nucleosynthesis). Prior to a supernova, fusion of elements between silicon and iron occurs only in the largest of stars, in the silicon burning process. (A slow neutron capture process, known as the s-process which also occurs during normal stellar nucleosynthesis can create elements up to bismuth with an atomic

mass of approximately 209. However, the s-process occurs primarily in low-mass stars that evolve more slowly.) Once the core fails to produce enough energy to support the outer envelope of gases the star explodes as a supernova producing the bulk of elements beyond iron. Production of elements from iron to uranium occurs within seconds in a supernova explosion. Due to the large amounts of energy released, much higher temperatures and densities are reached than at normal stellar temperatures. These conditions allow for an environment where transuranium elements might be formed.

10.3   The r-process

Main article: r-process

During supernova nucleosynthesis, the r-process (r for

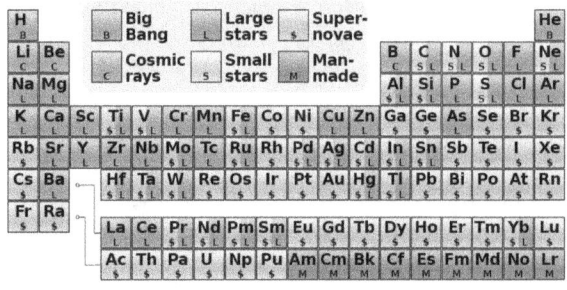

A version of the periodic table indicating the origins – including supernova nucleosynthesis – of the elements. All elements above 103 (lawrencium) are also manmade and are not included.

rapid) creates very neutron-rich heavy isotopes, which decay after the event to the first stable isotope, thereby creating the neutron-rich stable isotopes of all heavy elements. This neutron capture process occurs in high neutron density with high temperature conditions. In the r-process, any heavy nuclei are bombarded with a large neutron flux to form highly unstable neutron rich nuclei which very rapidly undergo beta decay to form more stable nuclei with higher atomic number and the same atomic weight. The neutron flux is astonishingly high, about 10^{22} neutrons per square centimeter per second. First calculation of a dynamic r-process, showing the evolution of calculated results with time,[5] also suggested that the r-process abundances are a superposition of differing neutron fluences. Small fluence produces the first r-process abundance peak near atomic weight A=130 but no actinides, whereas large fluence produces the actinides uranium and thorium but no longer contains the A=130 abundance peak. These processes occur in a fraction of a second to a few seconds, depending on details. Hundreds of subsequent papers published have utilized this time-dependent approach. Interestingly, the only modern nearby supernova, 1987A, has not revealed r-process enrichments. Modern thinking is that the r-process yield may be ejected from some supernovae but swallowed up in others as part of the residual neutron star or black hole.

10.4 See also

- Big bang nucleosynthesis
- Critical mass
- Nuclear decay
- Nuclear fission
- Nuclear fusion
- Nucleosynthesis
- Primordial nuclide
- Stellar nucleosynthesis
- Supernova

10.5 Notes

[1] Energy is produced in the isolated fusion reaction of nickel-56 with helium-4, but production of the latter (by photodisintegration of heavier nuclei) is costly, and consumes energy, causing alpha buildup of nickel to be shut off due to the essential fact that nickel-56 has nucleon binding energy less zinc-60.

10.6 References

[1] "Synthesis of the laments from carbon to nickel" Astrophys. J. Suppl. 1, 121 (1954)

[2] Woosley, S.E.; W. D. Arnett & D. D. Clayton (1973). "Explosive burning of oxygen and silicon". *The Astrophysical Journal Supplement* **26**: 231–312. Bibcode:1973ApJS...26..231W. doi:10.1086/190282.

[3] Clayton, Donald D. (1983). *Principles of Stellar Evolution and Nucleosynthesis*. University of Chicago Press. pp. 519–524. ISBN 9780226109534.

[4] Citation: *The atomic nuclide with the highest mean binding energy*, Fewell, M. P., American Journal of Physics, Volume 63, Issue 7, pp. 653–658 (1995). Click here for a high-resolution graph, *The Most Tightly Bound Nuclei*, which is part of the Hyperphysics project at Georgia State University.

[5] P. A. Seeger; W.A. Fowler; D. D. Clayton (1965). "Nucleosynthesis of heavy elements by neutron capture". *The Astrophysical Journal Supplement* **11**: 121–166. Bibcode:1965ApJS...11..121S. doi:10.1086/190111.

10.7 Other reading

- E. M. Burbidge, G. R. Burbidge, W. A. Fowler, F. Hoyle, *Synthesis of the Elements in Stars*, Rev. Mod. Phys. 29 (1957) 547 (article at the Physical Review Online Archive).

- D. D. Clayton, "Handbook of Isotopes in the Cosmos", Cambridge University Press, 2003, ISBN 0-521-82381-1.

10.8 External links

- Atom Smashers Shed Light on Supernovae, Big Bang *Sky & Telescope Online*, April 22, 2005

- G. Gonzalez; D. Brownlee; P. Ward (2001). "The Galactic Habitable Zone: Galactic Chemical Evolution" (PDF). *Icarus* **152**: 185–200. arXiv:astro-ph/0103165. Bibcode:2001Icar..152..185G. doi:10.1006/icar.2001.6617.

Chapter 11

Nucleosynthesis

For the song by Vangelis, see Albedo 0.39.

Nucleosynthesis is the process that creates new atomic nuclei from pre-existing nucleons, primarily protons and neutrons. The first nuclei were formed about three minutes after the Big Bang, through the process called Big Bang nucleosynthesis. It was then that hydrogen and helium formed to become the content of the first stars, and this primeval process is responsible for the present hydrogen/helium ratio of the cosmos.

With the formation of stars, heavier nuclei were created from hydrogen and helium by stellar nucleosynthesis, a process that continues today. Some of these elements, particularly those lighter than iron, continue to be delivered to the interstellar medium when low mass stars eject their outer envelope before they collapse to form white dwarfs. The remains of their ejected mass form the planetary nebulae observable throughout our galaxy.

Supernova nucleosynthesis within exploding stars by fusing carbon and oxygen is responsible for the abundances of elements between magnesium (atomic number 12) and nickel (atomic number 28).[1] Supernova nucleosynthesis is also thought to be responsible for the creation of rarer elements heavier than iron and nickel, in the last few seconds of a type II supernova event. The synthesis of these heavier elements absorbs energy (endothermic) as they are created, from the energy produced during the supernova explosion. Some of those elements are created from the absorption of multiple neutrons (the R process) in the period of a few seconds during the explosion. The elements formed in supernovas include the heaviest elements known, such as the long-lived elements uranium and thorium.

Cosmic ray spallation, caused when cosmic rays impact the interstellar medium and fragment larger atomic species, is a significant source of the lighter nuclei, particularly ^{3}He, ^{9}Be and ^{10,11}B, that are not created by stellar nucleosynthesis.

In addition to the fusion processes responsible for the growing abundances of elements in the universe, a few minor natural processes continue to produce very small numbers of new nuclides on Earth. These nuclides contribute little to their abundances, but may account for the presence of specific new nuclei. These nuclides are produced via radiogenesis (decay) of long-lived, heavy, primordial radionuclides such as uranium and thorium. Cosmic ray bombardment of elements on Earth also contribute to the presence of rare, short-lived atomic species called cosmogenic nuclides.

11.1 Timeline

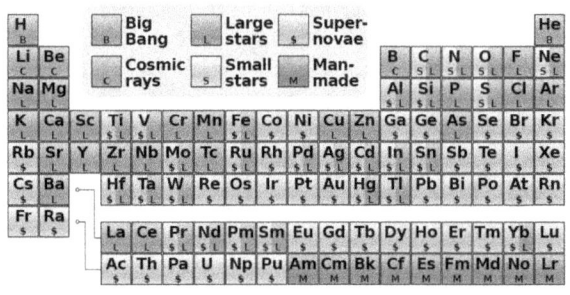

Periodic table showing the cosmogenic origin of each element. Elements from carbon up to sulfur may be made in small stars by the alpha process. Elements beyond iron are made in large stars with slow neutron capture (s-process), followed by expulsion to space in gas ejections (see planetary nebulae). Elements heavier than iron may be made in supernovae after the r-process, involving a dense burst of neutrons and rapid capture by the element.

It is thought that the primordial nucleons themselves were formed from the quark–gluon plasma during the Big Bang as it cooled below two trillion degrees. A few minutes afterward, starting with only protons and neutrons, nuclei up to lithium and beryllium (both with mass number 7) were formed, but the abundances of other elements dropped sharply with growing atomic mass. Some boron may have been formed at this time, but the process stopped before significant carbon could be formed, as this element requires a far higher product of helium density and time than were present in the short nucleosynthesis period of the Big Bang. That fusion process essentially shut down at about 20 minutes, due to drops in temperature and density as the universe continued to expand. This first process, Big Bang nucleosynthesis, was the first type of nucleogenesis to occur in the universe.

The subsequent nucleosynthesis of the heavier elements requires the extreme temperatures and pressures found within stars and supernovas. These processes began as hydrogen and helium from the Big Bang collapsed into the first stars at 500 million years. Star formation has occurred continuously in the galaxy since that time. The elements found on Earth, the so-called primordial elements, were created prior to Earth's formation by stellar nucleosynthesis and by supernova nucleosynthesis. They range in atomic numbers from $Z=6$ (carbon) to $Z=94$ (plutonium). Synthesis of these elements occurred either by nuclear fusion (including both rapid and slow multiple neutron capture) or to a lesser degree by nuclear fission followed by beta decay.

A star gains heavier elements by combining its lighter nuclei, hydrogen, deuterium, beryllium, lithium, and boron, which were found in the initial composition of the interstellar medium and hence the star. Interstellar gas therefore contains declining abundances of these light elements, which are present only by virtue of their nucleosynthesis during the Big Bang. Larger quantities of these lighter elements in the present universe are therefore thought to have been restored through billions of years of cosmic ray (mostly high-energy proton) mediated breakup of heavier elements in interstellar gas and dust. The fragments of these cosmic-ray collisions include the light elements Li, Be and B.

11.2 History of nucleosynthesis theory

The first ideas on nucleosynthesis were simply that the chemical elements were created at the beginning of the universe, but no rational physical scenario for this could be identified. Gradually it became clear that hydrogen and helium are much more abundant than any of the other elements. All the rest constitute less than 2% of the mass of the Solar System, and of other star systems as well. At the same time it was clear that oxygen and carbon were the next two most common elements, and also that there was a general trend toward high abundance of the light elements, especially those composed of whole numbers of helium-4 nuclei.

Arthur Stanley Eddington first suggested in 1920, that stars obtain their energy by fusing hydrogen into helium and

raised the possibility that the heavier elements may also form in stars.[2][3] This idea was not generally accepted, as the nuclear mechanism was not understood. In the years immediately before World War II, Hans Bethe first elucidated those nuclear mechanisms by which hydrogen is fused into helium.

Fred Hoyle's original work on nucleosynthesis of heavier elements in stars, occurred just after World War II.[4] His work explained the production of all heavier elements, starting from hydrogen. Hoyle proposed that hydrogen is continuously created in the universe from vacuum and energy, without need for universal beginning.

Hoyle's work explained how the abundances of the elements increased with time as the galaxy aged. Subsequently, Hoyle's picture was expanded during the 1960s by contributions from William A. Fowler, Alastair G. W. Cameron, and Donald D. Clayton, followed by many others. In the seminal 1957 review paper by E. M. Burbidge, G. R. Burbidge, Fowler and Hoyle (see Ref. list) is a well-known summary of the state of the field in 1957. That paper defined new processes for the transformation of one heavy nucleus into others within stars, processes that could be documented by astronomers.

The Big Bang itself had been proposed in 1931, long before this period, by Georges Lemaître, a Belgian physicist, who suggested that the evident expansion of the Universe in time required that the Universe, if contracted backwards in time, would continue to do so until it could contract no further. This would bring all the mass of the Universe to a single point, a "primeval atom", to a state before which time and space did not exist. Hoyle later gave Lemaître's model the derisive term of Big Bang, not realizing that Lemaître's model was needed to explain the existence of deuterium and nuclides between helium and carbon, as well as the fundamentally high amount of helium present, not only in stars but also in interstellar space. As it happened, both Lemaître and Hoyle's models of nucleosynthesis would be needed to explain the elemental abundances in the universe.

The goal of the theory of nucleosynthesis is to explain the vastly differing abundances of the chemical elements and their several isotopes from the perspective of natural processes. The primary stimulus to the development of this theory was the shape of a plot of the abundances versus the atomic number of the elements. Those abundances, when plotted on a graph as a function of atomic number, have a jagged sawtooth structure that varies by factors up to ten million. A very influential stimulus to nucleosynthesis research was an abundance table created by Hans Suess and Harold Urey that was based on the unfractionated abundances of the non-volatile elements found within unevolved meteorites.[5] Such a graph of the abundances is displayed on a logarithmic scale below, where the dramatically jagged structure is visually suppressed by the many powers of ten spanned in the vertical scale of this graph. See *Handbook of Isotopes in the Cosmos* for more data and discussion of abundances of the isotopes.[6]

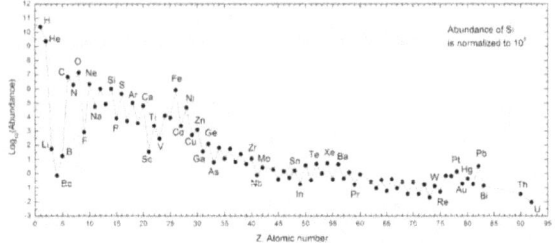

Abundances of the chemical elements in the Solar System. Hydrogen and helium are most common, residuals within the paradigm of the Big Bang.[7] The next three elements (Li, Be, B) are rare because they are poorly synthesized in the Big Bang and also in stars. The two general trends in the remaining stellar-produced elements are: (1) an alternation of abundance of elements according to whether they have even or odd atomic numbers, and (2) a general decrease in abundance, as elements become heavier. Within this trend is a peak at abundances of iron and nickel, which is especially visible on a logarithmic graph spanning fewer powers of ten, say between $\log A = 2$ (A=100) and $\log A = 6$ (A=1,000,000).

11.3   Processes

There are a number of astrophysical processes which are believed to be responsible for nucleosynthesis. The majority of these occur in shells within stars, and the chain of those nuclear fusion processes are known as hydrogen burning (via the proton-proton chain or the CNO cycle), helium burning, carbon burning, neon burning, oxygen burning and silicon burning. These processes are able to create elements up to and including iron and nickel. This is the region of nucleosynthesis within which the isotopes with the highest binding energy per nucleon are created. Heavier elements can be assembled within stars by a neutron capture process known as the s-process or in explosive environments, such as supernovae, by a number of other processes. Some of those others include the r-process, which involves rapid neutron captures, the rp-process, and the p-process (sometimes known as the gamma process), which results in the photodisintegration of existing nuclei.

11.4   The major types of nucleosynthesis

11.4.1   Big Bang nucleosynthesis

Main article: Big Bang nucleosynthesis

Big Bang nucleosynthesis occurred within the first three minutes of the beginning of the universe and is responsible for much of the abundance of ^{1}H (protium), ^{2}H (D, deuterium), ^{3}He (helium-3), and ^{4}He (helium-4). Although ^{4}He continues to be produced by stellar fusion and alpha decays and trace amounts of ^{1}H continue to be produced by spallation and certain types of radioactive decay, most of the mass of the isotopes in the universe are thought to have been produced in the Big Bang. The nuclei of these elements, along with some ^{7}Li and ^{7}Be are considered to have been formed between 100 and 300 seconds after the Big Bang when the primordial quark–gluon plasma froze out to form protons and neutrons. Because of the very short period in which nucleosynthesis occurred before it was stopped by expansion and cooling (about 20 minutes), no elements heavier than beryllium (or possibly boron) could be formed. Elements formed during this time were in the plasma state, and did not cool to the state of neutral atoms until much later.

$$n^0 \longrightarrow p^+ + e^- + \bar{\nu}_e \qquad p^+ + n^0 \longrightarrow {}^2_1D + \gamma$$
$${}^2_1D + p^+ \longrightarrow {}^3_2He + \gamma \qquad {}^2_1D + {}^2_1D \longrightarrow {}^3_2He + n^0$$
$${}^2_1D + {}^2_1D \longrightarrow {}^3_1T + p^+ \qquad {}^3_1T + {}^2_1D \longrightarrow {}^4_2He + n^0$$
$${}^3_1T + {}^4_2He \longrightarrow {}^7_3Li + \gamma \qquad {}^3_2He + n^0 \longrightarrow {}^3_1T + p^+$$
$${}^3_2He + {}^2_1D \longrightarrow {}^4_2He + p^+ \qquad {}^3_2He + {}^4_2He \longrightarrow {}^7_4Be + \gamma$$
$${}^7_3Li + p^+ \longrightarrow {}^4_2He + {}^4_2He \qquad {}^7_4Be + n^0 \longrightarrow {}^7_3Li + p^+$$

Chief nuclear reactions responsible for the relative abundances of light atomic nuclei observed throughout the universe.

11.4.2   Stellar nucleosynthesis

Main articles: Stellar nucleosynthesis, Proton-proton chain, Triple-alpha process, CNO cycle, s-process, p-process and photodisintegration

Stellar nucleosynthesis is the nuclear process by which new nuclei are produced. It occurs in stars during stellar evolution. It is responsible for the galactic abundances of elements from carbon to iron. Stars are thermonuclear furnaces in which H and He are fused into heavier nuclei by increasingly high temperatures as the composition of the core evolves.[8] Of particular importance is carbon, because its formation from He is a bottleneck in the entire process. Carbon is produced by the triple-alpha process in all stars. Carbon is also the main element that causes the release of free neutrons within stars, giving rise to the s-process, in which the slow absorption of neutrons converts iron into elements heavier than iron and nickel.[9]

The products of stellar nucleosynthesis are generally dispersed into the interstellar gas through mass loss episodes and the stellar winds of low mass stars. The mass loss events can be witnessed today in the planetary nebulae phase of low-mass star evolution, and the explosive ending of stars, called supernovae, of those with more than eight times the mass of the Sun.

The first direct proof that nucleosynthesis occurs in stars was the astronomical observation that interstellar gas has become enriched with heavy elements as time passed. As a result, stars that were born from it late in the galaxy, formed with much higher initial heavy element abundances than those that had formed earlier. The detection of technetium in the atmosphere of a red giant star in 1952,[10] by spectroscopy, provided the first evidence of nuclear activity within stars. Because technetium is radioactive, with a half-life much less than the age of the star, its abundance must reflect its recent creation within that star. Equally convincing evidence of the stellar origin of heavy elements, is the large overabundances of specific stable elements found in stellar atmospheres of asymptotic giant branch stars. Observation of barium abundances some 20-50 times greater than found in unevolved stars is evidence of the operation of the s-process within such stars. Many modern proofs of stellar nucleosynthesis are provided by the isotopic compositions of stardust, solid grains that have condensed from the gases of individual stars and which have been extracted from meteorites. Stardust is one component of cosmic dust, and is frequently called presolar grains. The measured isotopic compositions in stardust grains demonstrate many aspects of nucleosynthesis within the stars from which the grains condensed during the star's late-life mass-loss episodes.[11]

11.4.3 Explosive nucleosynthesis

Main articles: r-process, rp-process and Supernova nucleosynthesis

Supernova nucleosynthesis occurs in the energetic environment in supernovae, in which the elements between silicon and nickel are synthesized in quasiequilibrium[12] established during fast fusion that attaches by reciprocating balanced nuclear reactions to ^{28}Si. Quasiequilibrium can be thought of as *almost equilibrium* except for a high abundance of the ^{28}Si nuclei in the feverishly burning mix. This concept[13] was the most important discovery in nucleosynthesis theory of the intermediate-mass elements since Hoyle's 1954 paper because it provided an overarching understanding of the abundant and chemically important elements between silicon (A=28) and nickel (A=60). It re-

placed the incorrect although much cited alpha process of the B2FH paper, which inadvertently obscured Hoyle's better 1954 theory.[14] Further nucleosynthesis processes can occur, in particular the r-process (rapid process) described by the B2FH paper and first calculated by Seeger, Fowler and Clayton,[15] in which the most neutron-rich isotopes of elements heavier than nickel are produced by rapid absorption of free neutrons. The creation of free neutrons by electron capture during the rapid compression of the supernova core along with assembly of some neutron-rich seed nuclei makes the r-process a *primary process*, and one that can occur even in a star of pure H and He. This is in contrast to the B2FH designation of the process as a *secondary process*. This promising scenario, though generally supported by supernova experts, has yet to achieve a totally satisfactory calculation of r-process abundances. The primary r-process has been confirmed by astronomers who have observed old stars born when galactic metallicity was still small, that nonetheless contain their complement of r-process nuclei; thereby demonstrating that the metallicity is a product of an internal process. The r-process is responsible for our natural cohort of radioactive elements, such as uranium and thorium, as well as the most neutron-rich isotopes of each heavy element.

The rp-process (rapid proton) involves the rapid absorption of free protons as well as neutrons, but its role and its existence are less certain.

Explosive nucleosynthesis occurs too rapidly for radioactive decay to decrease the number of neutrons, so that many abundant isotopes with equal and even numbers of protons and neutrons are synthesized by the silicon quasiequilibrium process.[16] During this process, the burning of oxygen and silicon fuses nuclei that themselves have equal numbers of protons and neutrons to produce nuclides which consist of whole numbers of helium nuclei, up to 15 (representing ^{60}Ni). Such multiple-alpha-particle nuclides are totally stable up to ^{40}Ca (made of 10 helium nuclei), but heavier nuclei with equal and even numbers of protons and neutrons are tightly bound but unstable. The quasiequilibrium produces radioactive isobars ^{44}Ti, ^{48}Cr, ^{52}Fe, and ^{56}Ni, which (except ^{44}Ti) are created in abundance but decay after the explosion and leave the most stable isotope of the corresponding element at the same atomic weight. The most abundant and extant isotopes of elements produced in this way are ^{48}Ti, ^{52}Cr, and ^{56}Fe. These decays are accompanied by the emission of gamma-rays (radiation from the nucleus), whose spectroscopic lines can be used to identify the isotope created by the decay. The detection of these emission lines were an important early product of gamma-ray astronomy.[17]

The most convincing proof of explosive nucleosynthesis in supernovae occurred in 1987 when those gamma-ray lines were detected emerging from supernova 1987A. Gamma ray lines identifying ^{56}Co and ^{57}Co nuclei, whose radioactive half-lives limit their age to about a year, proved that they were created by their radioactive cobalt parents. This nuclear astronomy observation was predicted in 1969[18] as a way to confirm explosive nucleosynthesis of the elements, and that prediction played an important role in the planning for NASA's Compton Gamma-Ray Observatory.

Other proofs of explosive nucleosynthesis are found within the stardust grains that condensed within the interiors of supernovae as they expanded and cooled. Stardust grains are one component of cosmic dust. In particular, radioactive ^{44}Ti was measured to be very abundant within supernova stardust grains at the time they condensed during the supernova expansion.[19] This confirmed a 1975 prediction of the identification of supernova stardust (SUNOCONs), which became part of the pantheon of presolar grains. Other unusual isotopic ratios within these grains reveal many specific aspects of explosive nucleosynthesis.

11.4.4 Cosmic ray spallation

Main article: Cosmic ray spallation

Cosmic ray spallation process reduces the atomic weight of interstellar matter by the impact with cosmic rays, to produce some of the lightest elements present in the universe (though not a significant amount of deuterium). Most notably spallation is believed to be responsible for the generation of almost all of ^{3}He and the elements lithium, beryllium, and boron, although some 7Li and 7Be are thought to have been produced in the Big Bang. The spallation process results from the impact of cosmic rays (mostly fast protons) against the interstellar medium. These impacts fragment carbon, nitrogen, and oxygen nuclei present. The process results in the light elements beryllium, boron, and lithium in cosmos at much greater abundances than they are within solar atmospheres. The light elements ^{1}H and ^{4}He nuclei are not a product of spallation and are represented in the cosmos with approximately primordial abundance.

Beryllium and boron are not significantly produced by stellar fusion processes, due to the instability of any ^{8}Be formed from two ^{4}He nuclei.

11.5 Empirical evidence

Theories of nucleosynthesis are tested by calculating isotope abundances and comparing those results with observed results. Isotope abundances are typically calculated from the transition rates between isotopes in a network. Often these calculations can be simplified as a few key reac-

tions control the rate of other reactions.

11.6 Minor mechanisms and processes

Very small amounts of certain nuclides are produced on Earth by artificial means. Those are our primary source, for example, of technetium. However, some nuclides are also produced by a number of natural means that have continued after primordial elements were in place. These often act to produce new elements in ways that can be used to date rocks or to trace the source of geological processes. Although these processes do not produce the nuclides in abundance, they are assumed to be the entire source of the existing natural supply of those nuclides.

These mechanisms include:

- Radioactive decay may lead to radiogenic daughter nuclides. The nuclear decay of many long-lived primordial isotopes, especially uranium-235, uranium-238, and thorium-232 produce many intermediate daughter nuclides, before they too finally decay to isotopes of lead. The Earth's natural supply of elements like radon and polonium is via this mechanism. The atmosphere's supply of argon-40 is due mostly to the radioactive decay of potassium-40 in the time since the formation of the Earth. Little of the atmospheric argon is primordial. Helium-4 is produced by alpha-decay, and the helium trapped in Earth's crust is also mostly non-primordial. In other types of radioactive decay, such as cluster decay, larger species of nuclei are ejected (for example, neon-20), and these eventually become newly formed stable atoms.

- Radioactive decay may lead to spontaneous fission. This is not cluster decay, as the fission products may be split among nearly any type of atom. Thorium-232, uranium-235, and uranium-238 are primordial isotopes that undergo spontaneous fission. Natural technetium and promethium are produced in this manner.

- Nuclear reactions. Naturally-occurring nuclear reactions powered by radioactive decay give rise to so-called nucleogenic nuclides. This process happens when an energetic particle from a radioactive decay, often an alpha particle, reacts with a nucleus of another atom to change the nucleus into another nuclide. This process may also cause the production of further subatomic particles, such as neutrons. Neutrons can also be produced in spontaneous fission and by neutron emission. These neutrons can then go on to produce other nuclides via neutron-induced fission, or by neutron capture. For example, some stable isotopes such as neon-21 and neon-22 are produced by several routes of nucleogenic synthesis, and thus only part of their abundance is primordial.

- Nuclear reactions due to cosmic rays. By convention, these reaction-products are not termed "nucleogenic" nuclides, but rather cosmogenic nuclides. Cosmic rays continue to produce new elements on Earth by the same cosmogenic processes discussed above that produce primordial beryllium and boron. One important example is carbon-14, produced from nitrogen-14 in the atmosphere by cosmic rays. Iodine-129 is another example.

In addition to artificial processes, it is postulated that neutron star collision is the main source of elements heavier than iron.[20]

11.7   See also

- Stellar evolution

- Supernova nucleosynthesis

- Cosmic dust

- Metallicity

11.8   References

[1] Donald D. Clayton, *Handbook of isotopes in the cosmos*, Cambridge University Press (Cambridge 2003)

[2] A.S. Eddington, The Internal Constitution of the Stars, *The Observatory*, **43**, 341 (1920) http://adsabs.harvard.edu/abs/1920Obs....43..341E

[3] A.S. Eddington, The Internal Constitution of the Stars, *Nature*, **106**, 106 (1920) http://adsabs.harvard.edu/abs/1920Natur.106...14E

[4] Actually, before the war ended, he learned abut the problem of spherical implosion of plutonium in the Manhattan project. He saw an analogy between the plutonium fission reaction and the newly discovered supernovae, and he was able to show that exploding super novae produced all of the elements in the same proportion as existed on Earth. He felt that he had accidentally fallen into a subject that would make his career. Autobiography William A. Fowler

[5] H.E. Suess and H.C. Urey, Abundances of the elements, *Revs. Mod. Phys.*, **28**, 53 (1957)

[6] Donald D. Clayton, *Handbook of isotopes in the cosmos*, Cambridge University Press (Cambridge U.K. 2003)

[7] Massimo S. Stiavelli. From First Light to Reionization. John Wiley & Sons, Apr 22, 2009. Pg 8.

[8] Donald D. Clayton, *Principles of Stellar Evolution and Nucleosynthesis*, McGraw-Hill (New York 1968) Chapter 5; reissued by University of Chicago Press (Chicago 1883)

[9] D.D. Clayton, W.A. Fowler, T. Hull and B. Zimmerman, Neutron capture chains in heavy element synthesis, *Ann. Phys.*, **12**, 331-408 (1961); Donald D. Clayton, *Principles of Stellar Evolution and Nucleosynthesis*, McGraw-Hill (New York 1968) Chapter 7

[10] S. Paul W. Merrill (1952). "Spectroscopic Observations of Stars of Class S". *The Astrophysical Journal* **116**: 21. Bibcode:1952ApJ...116...21M. doi:10.1086/145589.

[11] Donald D. Clayton and L. R. Nittler (2004). "Astrophysics with Presolar Stardust". *Annual Review of Astronomy and Astrophysics* **42** (1): 39–78. Bibcode:2004ARA&A..42...39C. doi:10.1146/annurev.astro.42.053102.134022.

[12] D. Bodansky, Donald D. Clayton, and W. A. Fowler, (1968) Nuclear quasi-equilibrium during silicon burning, *Astrophys. J. Suppl.* No. 148, **16**, 299-371

[13] See also Chapter 7 of Donald D. Clayton, *Principles of Stellar Evolution and Nucleosynthesis*, McGraw-Hill, New York (1968)

[14] Donald D. Clayton, Hoyle's Equation, *Science*, **318**, 1876-77 (2007)

[15] P.A.Seeger, W. A. Fowler, and Donald D. Clayton, Nucleosynthesis of heavy elements by neutron capture, *Astrophys. J. Suppl*, **11**, 121-66, (1965)

[16] D. Bodansky, Donald D. Clayton, and W. A. Fowler, (1968) Nuclear quasi-equilibrium during silicon burning, *Astrophys. J. Suppl.* No. 148, **16** 299-371

[17] Donald D. Clayton, Stirling A. Colgate and G. J. Fishman (1969) Gamma ray lines from young supernova remnants, *Astrophys. J.*. **155** 175

[18] D. D. Clayton; S.A. Colgate; G.J. Fishman (1969). "Gamma ray lines from young supernova remnants". *The Astrophysical Journal* **155**: 75–82. Bibcode:1969ApJ...155...75C. doi:10.1086/149849.

[19] D. D. Clayton; L. R.Nittler (2004). "Astrophysics with Presolar stardust". *Annual Reviews of Astronomy and Astrophysics* **42** (1): 39–78. Bibcode:2004ARA&A..42...39C. doi:10.1146/annurev.astro.42.053102.134022.

[20] Stromberg, Joseph. "All the Gold in the Universe Could Come From the Collisions of Neutron Stars". *Smithsonian*. Retrieved 27 April 2014.

11.9   Further reading

- E. M. Burbidge, G. R. Burbidge, W. A. Fowler, F. Hoyle, *Synthesis of the Elements in Stars*, Rev. Mod. Phys. 29 (1957) 547 (article at the Physical Review Online Archive (subscription required)).

- M. Meneguzzi, J. Audouze, H. Reeves, « The production of the elements Li, Be, B by galactic cosmic rays in space and its relation with stellar observations », Astronomy and Astrophysics, vol. 15, 1971, p. 337–359

- F. Hoyle, Monthly Notices Roy. Astron. Soc. 106, 366 (1946)

- F. Hoyle, Astrophys. J. Suppl. 1, 121 (1954)

- D. D. Clayton, "Principles of Stellar Evolution and Nucleosynthesis", McGraw-Hill, 1968; University of Chicago Press, 1983, ISBN 0-226-10952-6

- C. E. Rolfs, W. S. Rodney, *Cauldrons in the Cosmos*, Univ. of Chicago Press, 1988, ISBN 0-226-72457-3.

- D. D. Clayton, "Handbook of Isotopes in the Cosmos", Cambridge University Press, 2003, ISBN 0-521-82381-1.

- C. Iliadis, "Nuclear Physics of Stars", Wiley-VCH, 2007, ISBN 978-3-527-40602-9

Chapter 12

Chronology of the universe

See also: Timeline of the formation of the Universe
The **chronology of the universe** describes the history and

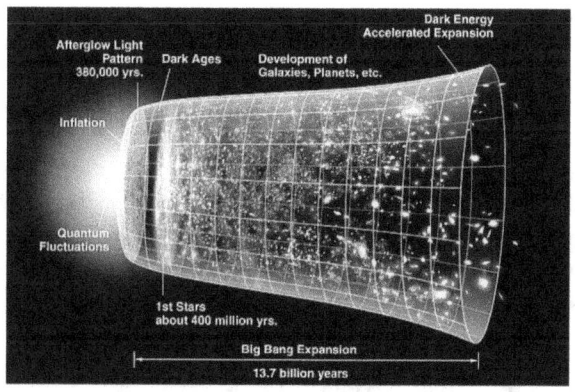

Diagram of Evolution of the universe from the Big Bang (left) - to the present.

future of the universe according to Big Bang cosmology, the prevailing scientific model of how the universe developed over time from the Planck epoch, using the cosmological time parameter of comoving coordinates. The model of the universe's expansion is known as the Big Bang. As of 2015, this expansion is estimated to have begun 13.799 ± 0.021 billion years ago.[1] It is convenient to divide the evolution of the universe so far into three phases.

12.1 Summary

In the first phase, the very earliest universe was so hot, or energetic, that initially no matter particles existed or could exist perhaps only fleetingly. According to prevailing scientific theories, at this time the distinct forces we see around us today were joined in one unified force. Space-time itself expanded during an inflationary epoch due to the immensity of the energies involved. Gradually the immense energies cooled – still to a temperature inconceivably hot compared to any we see around us now, but sufficiently to allow forces to gradually undergo symmetry breaking, a kind of repeated

condensation from one status quo to another, leading finally to the separation of the strong force from the electroweak force and the first particles.

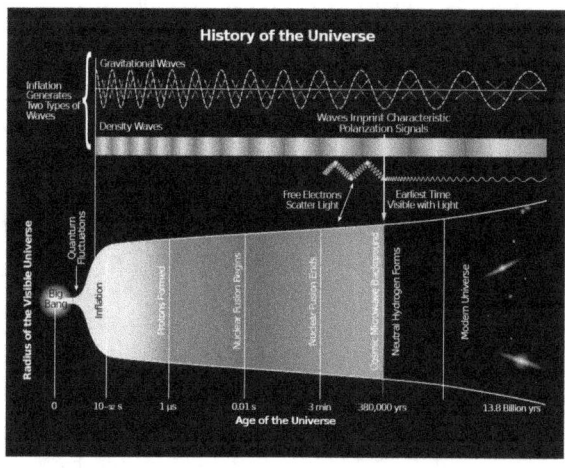

History of the Universe - gravitational waves are hypothesized to arise from cosmic inflation, a faster-than-light expansion just after the Big Bang (17 March 2014).[2][3][4]

In the second phase, the resulting quark–gluon plasma universe then cooled further, the current fundamental forces we know take their present forms through further symmetry breaking – notably the breaking of electroweak symmetry – and the full range of complex and composite particles we see around us today became possible, leading to a gravitationally dominated universe, the first neutral atoms (~ 80% hydrogen), and the cosmic microwave background radiation we can detect today. Modern high energy particle physics theories are satisfactory at these energy levels, and so physicists believe they have a good understanding of this and subsequent development of the fundamental universe around us. Because of these changes, space had also become largely transparent to light and other electromagnetic energy, rather than "foggy", by the end of this phase.

The third phase started after a short dark age with a universe whose fundamental particles and forces were as we know them, and witnessed the emergence of large scale stable structures, such as the earliest stars, quasars, galaxies, clusters of galaxies and superclusters, and the development of these to create the kind of universe we see today. Some researchers call the development of all this physical structure over billions of years "cosmic evolution". Other, more interdisciplinary, researchers refer to "cosmic evolution" as the entire scenario of growing complexity from big bang to humankind, thereby incorporating biology and culture into a unified view of all complex systems in the universe to date.[5]

Beyond the present day, scientists anticipate that the Earth will cease to be able to support life in about a billion years, and will be enveloped by a greatly-expanded Sun in about 5 billion years. On a far longer timescale, the Stelliferous Era will end as stars eventually die and fewer are born to replace them, leading to a darkening universe. Various theories suggest a number of subsequent possibilities. If particles such as protons are unstable then eventually matter may evaporate into low level energy in a kind of entropy related heat death. Alternatively the universe may collapse in a big crunch, although current data shows the rate of expansion is still increasing. If this is correct then it may end in a "big freeze" as matter and energy become very thinly spread and cool down. Alternative suggestions include a false vacuum catastrophe or a Big Rip as possible ends to the universe.

12.2   Very early universe

All ideas concerning the very early universe (cosmogony) are speculative. No accelerator experiments have yet probed energies of sufficient magnitude to provide any experimental insight into the behavior of matter at the energy levels that prevailed during this period. Proposed scenarios differ radically. Some examples are the Hartle–Hawking initial state, string landscape, brane inflation, string gas cosmology, and the ekpyrotic universe. Some of these are mutually compatible, while others are not.

12.2.1   Planck epoch

0 to 10^{-43} second after the Big Bang

Main article: Planck epoch

The Planck epoch is an era in traditional (non-inflationary) big bang cosmology wherein the temperature was so high that the four fundamental forces—electromagnetism, gravitation, weak nuclear interaction, and strong nuclear interaction—were one fundamental force. Little is understood about physics at this temperature; different hypotheses propose different scenarios. Traditional big bang cos-

mology predicts a gravitational singularity before this time, but this theory relies on general relativity and is expected to break down due to quantum effects.

In inflationary cosmology, times before the end of inflation (roughly 10^{-32} second after the Big Bang) do not follow the traditional big bang timeline.

12.2.2   Grand unification epoch

Between 10^{-43} second and 10^{-36} second after the Big Bang[6]

Main article: Grand unification epoch

As the universe expanded and cooled, it crossed transition temperatures at which forces separate from each other. These are phase transitions much like condensation and freezing. The grand unification epoch began when gravitation separated from the other forces of nature, which are collectively known as gauge forces. The non-gravitational physics in this epoch would be described by a so-called grand unified theory (GUT). The grand unification epoch ended when the GUT forces further separate into the strong and electroweak forces.

12.2.3   Electroweak epoch

Between 10^{-36} second (or the end of inflation) and 10^{-32} second after the Big Bang[6]

Main article: Electroweak epoch

According to traditional big bang cosmology, the Electroweak epoch began 10^{-36} second after the Big Bang, when the temperature of the universe was low enough (10^{28} K) to separate the strong force from the electroweak force (the name for the unified forces of electromagnetism and the weak interaction). In inflationary cosmology, the electroweak epoch ends when the inflationary epoch begins, at roughly 10^{-32} second.

Inflationary epoch

Unknown duration, ending 10^{-32}(?) second after the Big Bang

Main article: Inflationary epoch

Cosmic inflation was an era of accelerating expansion produced by a hypothesized field called the inflaton, which would have properties similar to the Higgs field and dark energy. While decelerating expansion would magnify deviations from homogeneity, making the universe more chaotic, accelerating expansion would make the universe more homogeneous. A sufficiently long period of inflationary expansion in the past could explain the high degree of homogeneity that is observed in the universe today at large scales, even if the state of the universe before inflation was highly disordered.

Inflation ended when the inflaton field decayed into ordinary particles in a process called "reheating", at which point ordinary Big Bang expansion began. The time of reheating is usually quoted as a time "after the Big Bang". This refers to the time that would have passed in traditional (non-inflationary) cosmology between the Big Bang singularity and the universe dropping to the same temperature that was produced by reheating, even though, in inflationary cosmology, the traditional Big Bang did not occur.

According to the simplest inflationary models, inflation ended at a temperature corresponding to roughly 10^{-32} second after the Big Bang. As explained above, this does not imply that the inflationary era lasted less than 10^{-32} second. In fact, in order to explain the observed homogeneity of the universe, the duration must be longer than 10^{-32} second. In inflationary cosmology, the earliest meaningful time "after the Big Bang" is the time of the end of inflation.

On March 17, 2014, astrophysicists of the BICEP2 collaboration announced the detection of inflationary gravitational waves in the B-mode power spectrum which was interpreted as clear experimental evidence for the theory of inflation.[2][3][4][7][8][9] However, on June 19, 2014, lowered confidence in confirming the cosmic inflation findings was reported [8][10][11] and finally, on February 2, 2015, a joint analysis of data from BICEP2/Keck and Planck satellite concluded that the statistical "significance [of the data] is too low to be interpreted as a detection of primordial B-modes" and can be attributed mainly to polarized dust in the Milky Way.[12][13][14][15]

Baryogenesis

Main article: Baryogenesis

There is currently insufficient observational evidence to explain why the universe contains far more baryons than antibaryons. A candidate explanation for this phenomenon must allow the Sakharov conditions to be satisfied at some time after the end of cosmological inflation. While particle physics suggests asymmetries under which these conditions are met, these asymmetries are too small empirically to account for the observed baryon-antibaryon asymmetry of the universe.

12.3 Early universe

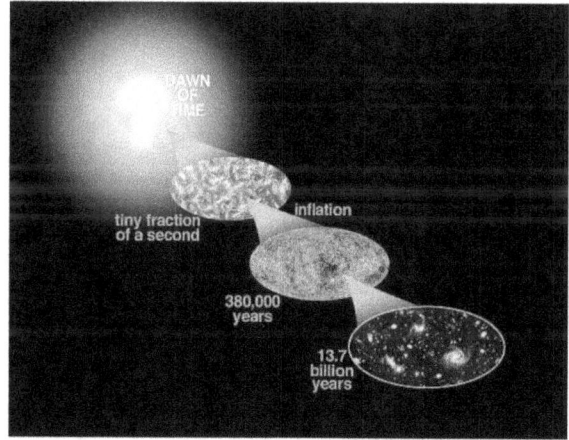

Cosmic History

After cosmic inflation ends, the universe is filled with a quark–gluon plasma. From this point onwards the physics of the early universe is better understood, and less speculative.

12.3.1 Supersymmetry breaking (speculative)

Main article: Supersymmetry breaking

If supersymmetry is a property of our universe, then it must be broken at an energy that is no lower than 1 TeV, the electroweak symmetry scale. The masses of particles and their superpartners would then no longer be equal, which could explain why no superpartners of known particles have ever been observed.

12.3.2 Electroweak symmetry breaking and the quark epoch

Between 10^{-12} second and 10^{-6} second after the Big Bang

Main articles: Electroweak symmetry breaking and Quark epoch

As the universe's temperature falls below a certain very high energy level, it is believed that the Higgs field spontaneously acquires a vacuum expectation value, which breaks electroweak gauge symmetry. This has two related effects:

1. The weak force and electromagnetic force, and their respective bosons (the W and Z bosons and photon)

manifest differently in the present universe, with different ranges;

2. Via the Higgs mechanism, all elementary particles interacting with the Higgs field become massive, having been massless at higher energy levels.

At the end of this epoch, the fundamental interactions of gravitation, electromagnetism, the strong interaction and the weak interaction have now taken their present forms, and fundamental particles have mass, but the temperature of the universe is still too high to allow quarks to bind together to form hadrons.

12.3.3 Hadron epoch

Between 10^{-6} second and 1 second after the Big Bang

Main article: Hadron epoch

The quark–gluon plasma that composes the universe cools until hadrons, including baryons such as protons and neutrons, can form. At approximately 1 second after the Big Bang neutrinos decouple and begin traveling freely through space. This cosmic neutrino background, while unlikely to ever be observed in detail since the neutrino energies are very low, is analogous to the cosmic microwave background that was emitted much later. (See above regarding the quark–gluon plasma, under the String Theory epoch.) However, there is strong indirect evidence that the cosmic neutrino background exists, both from Big Bang nucleosynthesis predictions of the helium abundance, and from anisotropies in the cosmic microwave background

12.3.4 Lepton epoch

Between 1 second and 10 seconds after the Big Bang

Main article: Lepton epoch

The majority of hadrons and anti-hadrons annihilate each other at the end of the hadron epoch, leaving leptons and anti-leptons dominating the mass of the universe. Approximately 10 seconds after the Big Bang the temperature of the universe falls to the point at which new lepton/anti-lepton pairs are no longer created and most leptons and anti-leptons are eliminated in annihilation reactions, leaving a small residue of leptons.[16]

12.3.5 Photon epoch

Between 10 seconds and 380,000 years after the Big Bang

Main article: Photon epoch

After most leptons and anti-leptons are annihilated at the end of the lepton epoch the energy of the universe is dominated by photons. These photons are still interacting frequently with charged protons, electrons and (eventually) nuclei, and continue to do so for the next 380,000 years.

Nucleosynthesis

Between 3 minutes and 20 minutes after the Big Bang[17]

Main article: Big Bang nucleosynthesis

During the photon epoch the temperature of the universe falls to the point where atomic nuclei can begin to form. Protons (hydrogen ions) and neutrons begin to combine into atomic nuclei in the process of nuclear fusion. Free neutrons combine with protons to form deuterium. Deuterium rapidly fuses into helium-4. Nucleosynthesis only lasts for about seventeen minutes, since the temperature and density of the universe has fallen to the point where nuclear fusion cannot continue. By this time, all neutrons have been incorporated into helium nuclei. This leaves about three times more hydrogen than helium-4 (by mass) and only trace quantities of other light nuclei.

Matter domination

70,000 years after the Big Bang

At this time, the densities of non-relativistic matter (atomic nuclei) and relativistic radiation (photons) are equal. The Jeans length, which determines the smallest structures that can form (due to competition between gravitational attraction and pressure effects), begins to fall and perturbations, instead of being wiped out by free-streaming radiation, can begin to grow in amplitude.

According to ΛCDM, at this stage, cold dark matter dominates, paving the way for gravitational collapse to amplify the tiny inhomogeneities left by cosmic inflation, making dense regions denser and rarefied regions more rarefied. However, because present theories as to the nature of dark matter are inconclusive, there is as yet no consensus as to its origin at earlier times, as currently exist for baryonic matter.

Recombination

ca. 377,000 years after the Big Bang

Main article: Recombination (cosmology)
Hydrogen and helium *atoms* begin to form as the density

9 year WMAP data (2012) shows the cosmic microwave background radiation variations throughout the universe from our perspective, though the actual variations are much smoother than the diagram suggests.[18][19]

of the universe falls. This is thought to have occurred about 377,000 years after the Big Bang.[20] Hydrogen and helium are at the beginning ionized, i.e., no electrons are bound to the nuclei, which (containing positively charged protons) are therefore electrically charged (+1 and +2 respectively). As the universe cools down, the electrons get captured by the ions, forming electrically neutral atoms. This process is relatively fast (and faster for the helium than for the hydrogen), and is known as recombination.[21] At the end of recombination, most of the protons in the universe are bound up in neutral atoms. Therefore, the photons' mean free path becomes effectively infinite and the photons can now travel freely (see Thomson scattering): the universe has become transparent. This cosmic event is usually referred to as *decoupling*.

The photons present at the time of decoupling are the same photons that we see in the cosmic microwave background (CMB) radiation, after being greatly cooled by the expansion of the universe. Around the same time, existing pressure waves within the electron-baryon plasma — known as baryon acoustic oscillations — became embedded in the distribution of matter as it condensed, giving rise to a very slight preference in distribution of large scale objects. Therefore the cosmic microwave background is a picture of the universe at the end of this epoch including the tiny fluctuations generated during inflation (see diagram), and the spread of objects such as galaxies in the universe is an indication of the scale and size of the universe as it developed over time.[22]

Habitable epoch

See also: Abiogenesis

The chemistry of life may have begun shortly after the Big Bang, 13.8 billion years ago, during a habitable epoch when the Universe was only 10-17 million years old.[23][24][25]

Dark Ages

See also: Hydrogen line

Before decoupling occurred, most of the photons in the universe were interacting with electrons and protons in the photon–baryon fluid. The universe was opaque or "foggy" as a result. There was light but not light we can now observe through telescopes. The baryonic matter in the universe consisted of ionized plasma, and it only became neutral when it gained free electrons during "recombination", thereby releasing the photons creating the CMB. When the photons were released (or decoupled) the universe became transparent. At this point the only radiation emitted was the 21 cm spin line of neutral hydrogen. There is currently an observational effort underway to detect this faint radiation, as it is in principle an even more powerful tool than the cosmic microwave background for studying the early universe. The Dark Ages are currently thought to have lasted between 150 million to 800 million years after the Big Bang. The October 2010 discovery of UDFy-38135539, the first observed galaxy to have existed during the following reionization epoch, gives us a window into these times. The galaxy earliest in this period observed and thus also the most distant galaxy ever observed is currently on the record of Leiden University's Richard J. Bouwens and Garth D. Illingsworth from UC Observatories/Lick Observatory. They found the galaxy UDFj-39546284 to be at a time some 480 million years after the Big Bang or about halfway through the Cosmic Dark Ages at a distance of about 13.2 billion light-years. More recently, the UDFj-39546284 galaxy was found to be around "380 million years" after the Big Bang and at a distance of 13.37 billion light-years.[26]

12.4   Structure formation

See also: Large-scale structure of the cosmos and Structure formation
Structure formation in the big bang model proceeds hierarchically, with smaller structures forming before larger ones. The first structures to form are quasars, which are thought to be bright, early active galaxies, and population III stars.

The Hubble Ultra Deep Fields often showcase galaxies from an ancient era that tell us what the early Stelliferous Age was like.

Another Hubble image shows an infant galaxy forming nearby, which means this happened very recently on the cosmological timescale. This shows that new galaxy formation in the universe is still occurring.

Before this epoch, the evolution of the universe could be understood through linear cosmological perturbation theory: that is, all structures could be understood as small deviations from a perfect homogeneous universe. This is computationally relatively easy to study. At this point non-linear structures begin to form, and the computational problem becomes much more difficult, involving, for example, N-body simulations with billions of particles.

12.4.1   Reionization

150 million to 1 billion years after the Big Bang

See also: Reionization and 21 centimeter radiation

The first stars and quasars form from gravitational collapse. The intense radiation they emit reionizes the surrounding universe. From this point on, most of the universe is composed of plasma.

12.4.2   Formation of stars

See also: Star formation

The first stars, most likely Population III stars, form and start the process of turning the light elements that were formed in the Big Bang (hydrogen, helium and lithium) into heavier elements. However, as yet there have been no observed Population III stars, and understanding of them is currently based on computational models of their formation and evolution. Fortunately observations of the Cosmic Microwave Background radiation can be used to date when star formation began in earnest. Analysis of such observations made by the European Space Agency's Planck telescope, as reported by BBC News in early February, 2015, concludes that the first generation of stars lit up 560 million years after the Big Bang. [27] [28]

12.4.3   Formation of galaxies

See also: Galaxy formation and evolution

Large volumes of matter collapse to form a galaxy. Population II stars are formed early on in this process, with Population I stars formed later.

Johannes Schedler's project has identified a quasar CFHQS 1641+3755 at 12.7 billion light-years away,[29] when the universe was just 7% of its present age.

On July 11, 2007, using the 10-metre Keck II telescope on Mauna Kea, Richard Ellis of the California Institute of Technology at Pasadena and his team found six star forming galaxies about 13.2 billion light years away and therefore created when the universe was only 500 million years old.[30] Only about 10 of these extremely early objects are currently known.[31] More recent observations have shown these ages to be shorter than previously indicated. The most distant galaxy observed as of October 2013 has been reported to be 13.1 billion light years away.[32]

The Hubble Ultra Deep Field shows a number of small galaxies merging to form larger ones, at 13 billion light years, when the universe was only 5% its current age.[33] This age estimate is now believed to be slightly shorter.[32]

Based upon the emerging science of nucleocosmochronology, the Galactic thin disk of the Milky Way is estimated to have been formed 8.8 ± 1.7 billion years ago.[34]

12.4.4 Formation of groups, clusters and superclusters

See also: Large-scale structure of the cosmos

Gravitational attraction pulls galaxies towards each other to form groups, clusters and superclusters.

12.4.5 Formation of the Solar System

9 billion years after the Big Bang

Main article: Formation and evolution of the Solar System

The Solar System began forming about 4.6 billion years ago, or about 9 billion years after the Big Bang. A fragment of a molecular cloud made mostly of hydrogen and traces of other elements began to collapse, forming a large sphere in the center which would become the Sun, as well as a surrounding disk. The surrounding accretion disk would coalesce into a multitude of smaller objects that would become planets, asteroids, and comets. The Sun is a late-generation star, and the Solar System incorporates matter created by previous generations of stars.

12.4.6 Today

13.8 billion years after the Big Bang

The Big Bang is estimated to have occurred about 13.8 billion years ago.[35] Since the expansion of the universe appears to be accelerating, its large-scale structure is likely to be the largest structure that will ever form in the universe. The present accelerated expansion prevents any more inflationary structures entering the horizon and prevents new gravitationally bound structures from forming.

12.5 Ultimate fate of the universe

Main article: Ultimate fate of the universe

As with interpretations of what happened in the very early universe, advances in fundamental physics are required before it will be possible to know the ultimate fate of the universe with any certainty. Below are some of the main possibilities.

12.5.1 Fate of the Solar System: 1 to 5 billion years

Main articles: Formation and evolution of the Solar System § Future, Stability of the Solar System, Future of the Earth § Solar evolution and Red giant § The Sun as a red giant

Over a timescale of a billion years or more, the Earth and

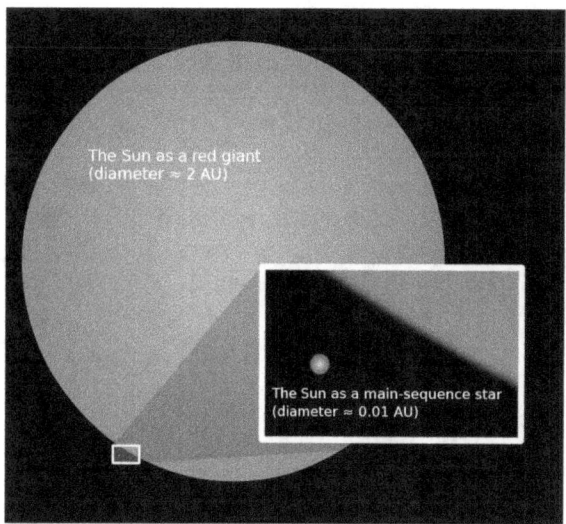

Relative size of our Sun as it is now (inset) compared to its estimated future size as a red giant

Solar System are unstable. Earth's existing biosphere is expected to vanish in about a billion years, as the Sun's heat production gradually increases to the point that liquid water and life are unlikely:[36] the Earth's magnetic fields, axial tilt and atmosphere are subject to long-term change; and the Solar System itself is chaotic over million- and billion-year timescales.[37] Eventually in around 5.4 billion years from now, the core of the Sun will become hot enough to trigger hydrogen fusion in its surrounding shell.[36] This will cause the outer layers of the star to expand greatly, and the star will enter a phase of its life in which it is called a red giant.[38][39] Within 7.5 billion years, the Sun will have expanded to a radius of 1.2 AU—256 times its current size, and studies announced in 2008 show that due to tidal interaction between Sun and Earth, Earth would actually fall back into a lower orbit, and get engulfed and incorporated inside the Sun before the Sun reaches its largest size, despite the Sun losing about 38% of its mass.[40] The Sun itself will continue to exist for many billions of years, passing through a number of phases, and eventually ending up as a long-lived white dwarf. Eventually, after billions more years, the Sun will finally cease to shine altogether, becoming a black dwarf.[41]

12.5.2 Big Rip: ≥20 billion years from now

See also: Big Rip

This scenario is possible only if the energy density of dark energy actually increases without limit over time. Such dark energy is called phantom energy and is unlike any known kind of energy. In this case, the expansion rate of the universe will increase without limit. Gravitationally bound systems, such as clusters of galaxies, galaxies, and ultimately the Solar System will be torn apart. Eventually the expansion will be so rapid as to overcome the electromagnetic forces holding molecules and atoms together. Finally even atomic nuclei will be torn apart and the universe as we know it will end in an unusual kind of gravitational singularity. At the time of this singularity, the expansion rate of the universe will reach infinity, so that any and all forces (no matter how strong) that hold composite objects together (no matter how closely) will be overcome by this expansion, literally tearing everything apart.

12.5.3 Big Crunch: ≥10^2 billion years from now

See also: Big Crunch

If the energy density of dark energy were negative or the universe were closed, then it would be possible that the expansion of the universe would reverse and the universe would contract towards a hot, dense state. This is a required element of oscillatory universe scenarios, such as the cyclic model, although a Big Crunch does not necessarily imply an oscillatory universe. Current observations suggest that this model of the universe is unlikely to be correct, and the expansion will continue or even accelerate.

12.5.4 Big Freeze: ≥10^5 billion years from now

Main articles: Future of an expanding universe and Heat death of the universe

This scenario is generally considered to be the most likely, as it occurs if the universe continues expanding as it has been. Over a time scale on the order of 10^{14} years or less, existing stars burn out, stars cease to be created, and the universe goes dark.[42], §IID. Over a much longer time scale in the eras following this, the galaxy evaporates as the stellar remnants comprising it escape into space, and black holes evaporate via Hawking radiation.[42], §III, §IVG. In some grand unified theories, proton decay after at least 10^{34} years will convert the remaining interstellar gas and stellar remnants into leptons (such as positrons and electrons) and photons. Some positrons and electrons will then recombine into photons.[42], §IV, §VF. In this case, the universe has reached a high-entropy state consisting of a bath of particles and low-energy radiation. It is not known however whether it eventually achieves thermodynamic equilibrium.[42], §VIB, VID.

12.5.5 Heat Death: 10^{1000} years from now

See also: Heat death of the universe

The heat death is a possible final state of the universe, estimated at after 10^{1000} years, in which it has "run down" to a state of no thermodynamic free energy to sustain motion or life. In physical terms, it has reached maximum entropy (because of this, the term "entropy" has often been confused with heat death, to the point of entropy being labelled as the "force killing the universe"). The hypothesis of a universal heat death stems from the 1850s ideas of William Thomson (Lord Kelvin)[43] who extrapolated the theory of heat views of mechanical energy loss in nature, as embodied in the first two laws of thermodynamics, to universal operation.

12.5.6 Vacuum metastability event

See also: False vacuum

If our universe is in a very long-lived false vacuum, it is possible that a small region of the universe will tunnel into a lower energy state (see Bubble nucleation). If this happens, all structures within will be destroyed instantaneously and the region will expand at near light speed, bringing destruction without any forewarning.

12.6   See also

- Cosmic Calendar (age of universe scaled to a single year)
- Cyclic model
- Dark-energy-dominated era
- Dyson's eternal intelligence

- Entropy (arrow of time)

- Graphical timeline from Big Bang to Heat Death

- Graphical timeline of the Big Bang

- Graphical timeline of the Stelliferous Era

- Illustris project

- Matter-dominated era

- Radiation-dominated era

- Timeline of the far future

- Ultimate fate of the universe

12.7 References

[1] Planck Collaboration (2015). "Planck 2015 results. XIII. Cosmological parameters (See Table 4 on page 31 of pfd).". arXiv:1502.01589.

[2] Staff (17 March 2014). "BICEP2 2014 Results Release". *National Science Foundation*. Retrieved 18 March 2014.

[3] Clavin, Whitney (17 March 2014). "NASA Technology Views Birth of the Universe". *NASA*. Retrieved 17 March 2014.

[4] Overbye, Dennis (17 March 2014). "Detection of Waves in Space Buttresses Landmark Theory of Big Bang". *The New York Times*. Retrieved 17 March 2014.

[5] Chaisson, E., (2001). *Cosmic Evolution: The Rise of Complexity in Nature*, Harvard University Press, ISBN 0-674-00987-8; see also Cosmic Evolution

[6] Ryden B: "Introduction to Cosmology", pg. 196 Addison-Wesley 2003

[7] Overbye, Dennis (March 24, 2014). "Ripples From the Big Bang". *New York Times*. Retrieved March 24, 2014.

[8] Ade, P.A.R. (BICEP2 Collaboration) et al. (June 19, 2014). "Detection of B-Mode Polarization at Degree Angular Scales by BICEP2" (PDF). *Physical Review Letters* **112**: 241101. arXiv:1403.3985. Bibcode:2014PhRvL.112x1101A. doi:10.1103/PhysRevLett.112.241101. PMID 24996078. Retrieved June 20, 2014.

[9] http://www.math.columbia.edu/~{}woit/wordpress/?p=6865

[10] Overbye, Dennis (June 19, 2014). "Astronomers Hedge on Big Bang Detection Claim". *New York Times*. Retrieved June 20, 2014.

[11] Amos, Jonathan (June 19, 2014). "Cosmic inflation: Confidence lowered for Big Bang signal". *BBC News*. Retrieved June 20, 2014.

[12] BICEP2/Keck, Planck Collaborations (2015). "A Joint Analysis of BICEP2/Keck Array and Planck Data (Provisionally accepted by PRL)". *arXiv*. arXiv:1502.00612v1. Retrieved 13 February 2015.

[13] Clavin, Whitney (30 January 2015). "Gravitational Waves from Early Universe Remain Elusive". *NASA*. Retrieved 30 January 2015.

[14] Overbye, Dennis (30 January 2015). "Speck of Interstellar Dust Obscures Glimpse of Big Bang". *New York Times*. Retrieved 31 January 2015.

[15] "Gravitational waves from early universe remain elusive". *Science Daily*. 31 January 2015. Retrieved 3 February 2015.

[16] The Timescale of Creation

[17] Detailed timeline of Big Bang nucleosynthesis processes

[18] Gannon, Megan (December 21, 2012). "New 'Baby Picture' of Universe Unveiled". Space.com. Retrieved December 21, 2012.

[19] Bennett, C.L.; Larson, L.; Weiland, J.L.; Jarosk, N.; Hinshaw, N.; Odegard, N.; Smith, K.M.; Hill, R.S.; Gold, B.; Halpern, M.; Komatsu, E.; Nolta, M.R.; Page, L.; Spergel, D.N.; Wollack, E.; Dunkley, J.; Kogut, A.; Limon, M.; Meyer, S.S.; Tucker, G.S.; Wright, E.L. (December 20, 2012). "Nine-Year Wilkinson Microwave Anisotropy Probe (WMAP) Observations: Final Maps and Results". *The Astrophysical Journal Supplement Series* **208**: 20. arXiv:1212.5225. Bibcode:2013ApJS..208...20B. doi:10.1088/0067-0049/208/2/20. Retrieved December 22, 2012.

[20] Hinshaw, G. et al. (2009). "Five-Year Wilkinson Microwave Anisotropy Probe (WMAP) Observations: Data Processing, Sky Maps, and Basic Results" (PDF). *Astrophysical Journal Supplement* **180** (2): 225–245. arXiv:0803.0732. Bibcode:2009ApJS..180..225H. doi:10.1088/0067-0049/180/2/225.

[21] Mukhanov, V: "Physical foundations of Cosmology", pg. 120, Cambridge 2005

[22] Amos, Jonathan (2012-11-13). "Quasars illustrate dark energy's roller coaster ride". *BBC News*. Retrieved 13 November 2012.

[23] Loeb, Abraham (October 2014). "The Habitable Epoch of the Early Universe". *International Journal of Astrobiology* **13** (04): 337–339. arXiv:1312.0613. Bibcode:2014IJAsB..13..337L. doi:10.1017/S1473550414000196. Retrieved 15 December 2014.

[24] Loeb, Abraham (2 December 2013). "The Habitable Epoch of the Early Universe" (PDF). *Arxiv*. arXiv:1312.0613v3. Retrieved 15 December 2014.

[25] Dreifus, Claudia (2 December 2014). "Much-Discussed Views That Go Way Back - Avi Loeb Ponders the Early Universe, Nature and Life". *New York Times*. Retrieved 3 December 2014.

[26] Wall, Mike (December 12, 2012). "Ancient Galaxy May Be Most Distant Ever Seen". Space.com. Retrieved December 12, 2012.

[27] *Ferreting Out The First Stars*; physorg.com

[28]

[29] APOD: 2007 September 6 - Time Tunnel

[30] "New Scientist" 14 July 2007

[31] HET Helps Astronomers Learn Secrets of One of Universe's Most Distant Objects

[32] Scientists confirm most distant galaxy ever

[33] APOD: 2004 March 9 – The Hubble Ultra Deep Field

[34] Eduardo F. del Peloso a1a, Licio da Silva a1, Gustavo F. Porto de Mello and Lilia I. Arany-Prado (2005), "The age of the Galactic thin disk from Th/Eu nucleocosmochronology: extended sample" (Proceedings of the International Astronomical Union (2005), 1: 485-486 Cambridge University Press)

[35] "Cosmic Detectives". The European Space Agency (ESA). 2013-04-02. Retrieved 2013-04-15.

[36] K. P. Schroder, Robert Connon Smith (2008). "Distant future of the Sun and Earth revisited". *Monthly Notices of the Royal Astronomical Society* **386** (1): 155–163. arXiv:0801.4031. Bibcode:2008MNRAS.386..155S. doi:10.1111/j.1365-2966.2008.13022.x.

[37] J. Laskar (1994). "Large-scale chaos in the solar system". *Astronomy and Astrophysics* **287**: L9–L12. Bibcode:1994A&A...287L...9L.

[38] Zeilik & Gregory 1998, p. 320–321.

[39] "Introduction to Cataclysmic Variables (CVs)". *NASA Goddard Space Center*. 2006. Retrieved 2006-12-29.

[40] Palmer, Jason (22 February 2008). "Hope dims that Earth will survive Sun's death". *New Scientist*.

[41] G. Fontaine, P. Brassard, P. Bergeron (2001). "The Potential of White Dwarf Cosmochronology". *Publications of the Astronomical Society of the Pacific* **113** (782): 409–435. Bibcode:2001PASP..113..409F. doi:10.1086/319535. Retrieved 2008-05-11.

[42] A dying universe: the long-term fate and evolution of astrophysical objects, Fred C. Adams and Gregory Laughlin, *Reviews of Modern Physics* **69**, #2 (April 1997), pp. 337–372. Bibcode: 1997RvMP...69..337A. doi:10.1103/RevModPhys.69.337.

[43] Thomson, William. (1851). "On the Dynamical Theory of Heat, with numerical results deduced from Mr Joule's equivalent of a Thermal Unit, and M. Regnault's Observations on Steam." Excerpts. [§§1-14 & §§99-100], *Transactions of the Royal Society of Edinburgh*, March, 1851; and *Philosophical Magazine* IV. 1852, [from *Mathematical and Physical Papers*, vol. i, art. XLVIII, pp. 174]

12.8   External links

- PBS Online (2000). From the Big Bang to the End of the Universe – The Mysteries of Deep Space Timeline. Retrieved March 24, 2005.

- Schulman, Eric (1997). The History of the Universe in 200 Words or Less. Retrieved March 24, 2005.

- Space Telescope Science Institute Office of Public Outreach (2005). Home of the Hubble Space Telescope. Retrieved March 24, 2005.

- Fermilab graphics (see "Energy time line from the Big Bang to the present" and "History of the Universe Poster")

- Exploring Time from Planck time to the lifespan of the Universe

- Cosmic Evolution is a multi-media web site that explores the cosmic-evolutionary scenario from big bang to humankind.

- Astronomers' first detailed hint of what was going on less than a trillionth of a second after time began

- The Universe Adventure

- Cosmology FAQ, Professor Edward L. Wright, UCLA

- Sean Carroll on the arrow of time (Part 1), *The origin of the universe and the arrow of time*, Sean Carroll, video, CHAST 2009, Templeton, Faculty of science, University of Sydney, November 2009, TED.com

- A Universe From Nothing, video, Lawrence Krauss, AAI 2009, YouTube.com

- Once Upon A Universe - Story of the Universe told in 13 chapters. Science communication site supported by STFC.

- Cosmic Evolution through Time - an interactive timeline explains the main events in the history of our Universe

Chapter 13

Chemical bond

A **chemical bond** is an attraction between atoms that allows the formation of chemical substances that contain two or more atoms. The bond is caused by the electrostatic force of attraction between opposite charges, either between electrons and nuclei, or as the result of a dipole attraction. The strength of chemical bonds varies considerably; there are "strong bonds" such as covalent or ionic bonds and "weak bonds" such as Dipole-dipole interaction, the London dispersion force and hydrogen bonding.

Since opposite charges attract via a simple electromagnetic force, the negatively charged electrons that are orbiting the nucleus and the positively charged protons in the nucleus attract each other. An electron positioned between two nuclei will be attracted to both of them, and the nuclei will be attracted toward electrons in this position. This attraction constitutes the chemical bond. Due to the matter wave nature of electrons and their smaller mass, they must occupy a much larger amount of volume compared with the nuclei, and this volume occupied by the electrons keeps the atomic nuclei relatively far apart, as compared with the size of the nuclei themselves. This phenomenon limits the distance between nuclei and atoms in a bond.

In general, strong chemical bonding is associated with the sharing or transfer of electrons between the participating atoms. The atoms in molecules, crystals, metals and diatomic gases—indeed most of the physical environment around us—are held together by chemical bonds, which dictate the structure and the bulk properties of matter.

All bonds can be explained by quantum theory, but, in practice, simplification rules allow chemists to predict the strength, directionality, and polarity of bonds. The octet rule and VSEPR theory are two examples. More sophisticated theories are valence bond theory which includes orbital hybridization and resonance, and the linear combination of atomic orbitals molecular orbital method which includes ligand field theory. Electrostatics are used to describe bond polarities and the effects they have on chemical substances.

Examples of Lewis dot-style representations of chemical bonds between carbon (C), hydrogen (H), and oxygen (O). Lewis dot diagrams were an early attempt to describe chemical bonding and are still widely used today.

13.1 Overview of main types of chemical bonds

A chemical bond is an attraction between atoms. This attraction may be seen as the result of different behaviors of the outermost or valence electrons of atoms. Although all of these behaviors merge into each other seamlessly in various bonding situations so that there is no clear line to be drawn between them, the behaviors of atoms become so **qualitatively** different as the character of the bond changes **quantitatively**, that it remains useful and customary to differentiate between the bonds that cause these different properties of condensed matter.

In the simplest view of a so-called 'covalent' bond, one or more electrons (often a pair of electrons) are drawn into the space between the two atomic nuclei. Here the negatively charged electrons are attracted to the positive charges of *both* nuclei, instead of just their own. This overcomes the repulsion between the two positively charged nuclei of the two atoms, and so this overwhelming attraction holds the two nuclei in a fixed configuration of equilibrium, even though they will still vibrate at equilibrium position. Thus, covalent bonding involves sharing of electrons in which the positively charged nuclei of two or more atoms simultaneously attract the negatively charged electrons that are being shared between them. These bonds exist between two particular identifiable atoms and have a direction in space, allowing them to be shown as single connecting lines between atoms in drawings, or modeled as sticks between spheres in models. In a polar covalent bond, one or more electrons are unequally shared between two nuclei. Covalent bonds often result in the formation of small collections of better-connected atoms called molecules, which in solids and liquids are bound to other molecules by forces that are often much weaker than the covalent bonds that hold the molecules internally together. Such weak intermolecular bonds give organic molecular substances, such as waxes and oils, their soft bulk character, and their low melting points (in liquids, molecules must cease most structured or oriented contact with each other). When covalent bonds link long chains of atoms in large molecules, however (as in polymers such as nylon), or when covalent bonds extend in networks through solids that are not composed of discrete molecules (such as diamond or quartz or the silicate minerals in many types of rock) then the structures that result may be both strong and tough, at least in the direction oriented correctly with networks of covalent bonds. Also, the melting points of such covalent polymers and networks increase greatly.

In a simplified view of an *ionic* bond, the bonding electron is not shared at all, but transferred. In this type of bond, the outer atomic orbital of one atom has a vacancy which allows the addition of one or more electrons. These newly added electrons potentially occupy a lower energy-state (effectively closer to more nuclear charge) than they experience in a different atom. Thus, one nucleus offers a more tightly bound position to an electron than does another nucleus, with the result that one atom may transfer an electron to the other. This transfer causes one atom to assume a net positive charge, and the other to assume a net negative charge. The *bond* then results from electrostatic attraction between atoms and the atoms become positive or negatively charged ions. Ionic bonds may be seen as extreme examples of polarization in covalent bonds. Often, such bonds have no particular orientation in space, since they result from equal electrostatic attraction of each ion to all ions around them. Ionic bonds are strong (and thus ionic substances require high temperatures to melt) but also brittle, since the forces between ions are short-range and do not easily bridge cracks and fractures. This type of bond gives rise to the physical characteristics of crystals of classic mineral salts, such as table salt.

A less often mentioned type of bonding is *metallic* bonding. In this type of bonding, each atom in a metal donates one or more electrons to a "sea" of electrons that reside between many metal atoms. In this sea, each electron is free (by virtue of its wave nature) to be associated with great many atoms at once. The bond results because the metal atoms become somewhat positively charged due to loss of their electrons while the electrons remain attracted to many atoms, without being part of any given atom. Metallic bonding may be seen as an extreme example of delocalization of electrons over a large system of covalent bonds, in which every atom participates. This type of bonding is often very strong (resulting in the tensile strength of metals). However, metallic bonding is more collective in nature than other types, and so they allow metal crystals to more easily deform, because they are composed of atoms attracted to each other, but not in any particularly-oriented ways. This results in the malleability of metals. The sea of electrons in metallic bonding causes the characteristically good electrical and thermal conductivity of metals, and also their "shiny" reflection of most frequencies of white light.

13.2 History

Main articles: History of chemistry and History of the molecule

Early speculations into the nature of the **chemical bond**, from as early as the 12th century, supposed that certain types of chemical species were joined by a type of chemical affinity. In 1704, Sir Isaac Newton famously outlined his atomic bonding theory, in "Query 31" of his Opticks, whereby atoms attach to each other by some "force". Specifically, after acknowledging the various popular theories in vogue at the time, of how atoms were reasoned to attach to each other, i.e. "hooked atoms", "glued together by rest", or "stuck together by conspiring motions", Newton states that he would rather infer from their cohesion, that "particles attract one another by some force, which in immediate contact is exceedingly strong, at small distances performs the chemical operations, and reaches not far from the particles with any sensible effect."

In 1819, on the heels of the invention of the voltaic pile, Jöns Jakob Berzelius developed a theory of chemical combination stressing the electronegative and electropositive character of the combining atoms. By the mid 19th century, Edward Frankland, F.A. Kekulé, A.S. Couper, Alexander Butlerov, and Hermann Kolbe, building on the theory of radicals, developed the theory of valency, originally called "combining power", in which compounds were joined owing to an attraction of positive and negative poles. In 1916, chemist Gilbert N. Lewis developed the concept of the electron-pair bond, in which two atoms may share one to six electrons, thus forming the single electron bond, a single bond, a double bond, or a triple bond; in Lewis's own words, "An electron may form a part of the shell of two different atoms and cannot be said to belong to either one exclusively."[1]

That same year, Walther Kossel put forward a theory similar to Lewis' only his model assumed complete transfers of electrons between atoms, and was thus a model of ionic bonding. Both Lewis and Kossel structured their bonding models on that of Abegg's rule (1904).

In 1927, the first mathematically complete quantum description of a simple chemical bond, i.e. that produced by one electron in the hydrogen molecular ion, H_2^+, was derived by the Danish physicist Oyvind Burrau.[2] This work showed that the quantum approach to chemical bonds could be fundamentally and quantitatively correct, but the mathematical methods used could not be extended to molecules containing more than one electron. A more practical, albeit less quantitative, approach was put forward in the same year by Walter Heitler and Fritz London. The Heitler-London method forms the basis of what is now called valence bond theory. In 1929, the linear combination of atomic orbitals molecular orbital method (LCAO) approximation was introduced by Sir John Lennard-Jones, who also suggested methods to derive electronic structures of molecules of F_2 (fluorine) and O_2 (oxygen) molecules, from basic quantum principles. This molecular orbital theory represented a covalent bond as an orbital formed by combining the quantum mechanical Schrödinger atomic orbitals which had been hypothesized for electrons in single atoms. The equations for bonding electrons in multi-electron atoms could not be solved to mathematical perfection (i.e., *analytically*), but approximations for them still gave many good qualitative predictions and results. Most quantitative calculations in modern quantum chemistry use either valence bond or molecular orbital theory as a starting point, although a third approach, density functional theory, has become increasingly popular in recent years.

In 1933, H. H. James and A. S. Coolidge carried out a calculation on the dihydrogen molecule that, unlike all previous calculation which used functions only of the distance of the electron from the atomic nucleus, used functions which also explicitly added the distance between the two electrons.[3] With up to 13 adjustable parameters they obtained a result very close to the experimental result for the dissociation energy. Later extensions have used up to 54 parameters and gave excellent agreement with experiments. This calculation convinced the scientific community that quantum theory could give agreement with experiment. However this approach has none of the physical pictures of the valence bond and molecular orbital theories and is difficult to extend to larger molecules.

13.3 Bonds in chemical formulas

The fact that atoms and molecules are three-dimensional makes it difficult to use a single technique for indicating orbitals and bonds. In **molecular formulas** the chemical bonds (binding orbitals) between atoms are indicated by various methods according to the type of discussion. Sometimes, they are completely neglected. For example, in organic chemistry chemists are sometimes concerned only with the functional groups of the molecule. Thus, the molecular formula of ethanol may be written in a paper in conformational form, three-dimensional form, full two-dimensional form (indicating every bond with no three-dimensional directions), compressed two-dimensional form (CH_3–CH_2–OH), by separating the functional group from another part of the molecule (C_2H_5OH), or by its atomic constituents (C_2H_6O), according to what is discussed. Sometimes, even the non-bonding valence shell electrons (with the two-dimensional approximate directions) are marked, i.e. for elemental carbon $\dot{.}C\dot{.}$. Some chemists may also mark the respective orbitals, i.e. the hypothetical ethene^{-4} anion ($\backslash^/C{=}C/^\backslash\ ^{-4}$) indicating the possibility of bond formation.

13.4 Strong chemical bonds

Strong chemical bonds are the *intramolecular* forces which hold atoms together in molecules. A strong chemical bond is formed from the transfer or sharing of electrons between atomic centers and relies on the electrostatic attraction between the protons in nuclei and the electrons in the orbitals.

The types of strong bond differ due to the difference in electronegativity of the constituent elements. A large difference in electronegativity leads to more polar (ionic) character in the bond.

13.4.1 Ionic bond

Main article: Ionic bonding

Ionic bonding is a type of electrostatic interaction between atoms which have a large electronegativity difference. There is no precise value that distinguishes ionic from covalent bonding, but a difference of electronegativity of over 1.7 is likely to be ionic, and a difference of less than 1.7 is likely to be covalent.[5] Ionic bonding leads to separate positive and negative ions. Ionic charges are commonly between −3e to +3e. Ionic bonding commonly occurs in metal salts such as sodium chloride (table salt). A typical feature of ionic bonds is that the species form into ionic crystals, in which no ion is specifically paired with any single other ion, in a specific directional bond. Rather, each species of ion is surrounded by ions of the opposite charge, and the spacing between it and each of the oppositely charged ions near it, is the same for all surrounding atoms of the same type. It is thus no longer possible to associate an ion with any specific other single ionized atom near it. This is a situation unlike that in covalent crystals, where covalent bonds between specific atoms are still discernible from the shorter distances between them, as measured via such techniques as X-ray diffraction.

Ionic crystals may contain a mixture of covalent and ionic species, as for example salts of complex acids, such as sodium cyanide, NaCN. Many minerals are of this type. X-ray diffraction shows that in NaCN, for example, the bonds between sodium cations (Na^+) and the cyanide anions (CN^-) are *ionic*, with no sodium ion associated with any particular cyanide. However, the bonds between C and N atoms in cyanide are of the *covalent* type, making each of the carbon and nitrogen associated with *just one* of its opposite type, to which it is physically much closer than it is to other carbons or nitrogens in a sodium cyanide crystal.

When such crystals are melted into liquids, the ionic bonds are broken first because they are non-directional and allow the charged species to move freely. Similarly, when such salts dissolve into water, the ionic bonds are typically broken by the interaction with water, but the covalent bonds continue to hold. For example, in solution, the cyanide ions, still bound together as single CN^- ions, move independently through the solution, as do sodium ions, as Na^+. In water, charged ions move apart because each of them are more strongly attracted to a number of water molecules, than to each other. The attraction between ions and water molecules in such solutions is due to a type of weak dipole-dipole type chemical bond. In melted ionic compounds, the ions continue to be attracted to each other, but not in any ordered or crystalline way.

13.4.2   Covalent bond

Main article: Covalent bond

Covalent bonding is a common type of bonding, in which the electronegativity difference between the bonded atoms is small or nonexistent. Bonds within most organic compounds are described as covalent. See sigma bonds and pi bonds for LCAO-description of such bonding.

A polar covalent bond is a covalent bond with a significant ionic character. This means that the electrons are closer to one of the atoms than the other, creating an imbalance of charge. They occur as a bond between two atoms with moderately different electronegativities and give rise to dipole-

dipole interactions. The electronegativity of these bonds is 0.3 to 1.7.

A coordinate covalent bond is one where both bonding electrons are from one of the atoms involved in the bond. These bonds give rise to Lewis acids and bases. The electrons are shared roughly equally between the atoms in contrast to ionic bonding. Such bonding occurs in molecules such as the ammonium ion (NH_4^+) and are shown by an arrow pointing to the Lewis acid. Also known as non-polar covalent bond, the electronegativity of these bonds range from 0 to 0.3.

Molecules which are formed primarily from non-polar covalent bonds are often immiscible in water or other polar solvents, but much more soluble in non-polar solvents such as hexane.

Single and multiple bonds

A single bond between two atoms corresponds to the sharing of one pair of electrons. The electron density of these two bonding electrons is concentrated in the region between the two atoms, which is the defining quality of a sigma bond.

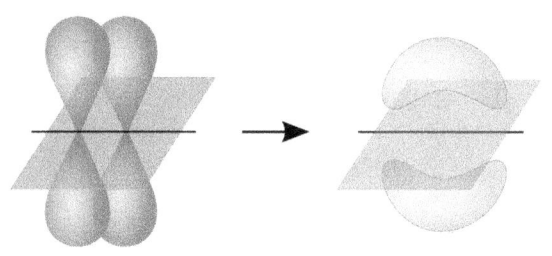

Two p-orbitals forming a pi-bond.

A double bond between two atoms is formed by the sharing of two pairs of electrons, one in a sigma bond and one in a pi bond, with electron density concentrated on two opposite sides of the internuclear axis. A triple bond consists of three shared electron pairs, forming one sigma and two pi bonds.

Quadruple and higher bonds are very rare and occur only between certain transition metal atoms.

13.4.3   Metallic bonding

Main article: Metallic bonding

In metallic bonding, bonding electrons are delocalized over a lattice of atoms. By contrast, in ionic compounds, the locations of the binding electrons and their charges are static.

The freely-moving or delocalization of bonding electrons leads to classical metallic properties such as luster (surface light reflectivity), electrical and thermal conductivity, ductility, and high tensile strength.

13.5　Intermolecular bonding

Main article: Intermolecular Force

There are four basic types of bonds that can be formed between two or more (otherwise non-associated) molecules, ions or atoms. Intermolecular forces cause molecules to be attracted or repulsed by each other. Often, these define some of the physical characteristics (such as the melting point) of a substance.

- A large difference in electronegativity between two bonded atoms will cause a permanent charge separation, or dipole, in a molecule or ion. Two or more molecules or ions with permanent dipoles can interact within **dipole-dipole interactions**. The bonding electrons in a molecule or ion will, on average, be closer to the more electronegative atom more frequently than the less electronegative one, giving rise to partial charges on each atom, and causing electrostatic forces between molecules or ions.

- A **hydrogen bond** is effectively a strong example of an interaction between two permanent dipoles. The large difference in electronegativities between hydrogen and any of fluorine, nitrogen and oxygen, coupled with their lone pairs of electrons cause strong electrostatic forces between molecules. Hydrogen bonds are responsible for the high boiling points of water and ammonia with respect to their heavier analogues.

- The **London dispersion force** arises due to instantaneous dipoles in neighbouring atoms. As the negative charge of the electron is not uniform around the whole atom, there is always a charge imbalance. This small charge will induce a corresponding dipole in a nearby molecule; causing an attraction between the two. The electron then moves to another part of the electron cloud and the attraction is broken.

- A **cation–pi interaction** occurs between a pi bond and a cation.

13.6　Theories of chemical bonding

In the (unrealistic) limit of "pure" ionic bonding, electrons are perfectly localized on one of the two atoms in the bond. Such bonds can be understood by classical physics. The forces between the atoms are characterized by isotropic continuum electrostatic potentials. Their magnitude is in simple proportion to the charge difference.

Covalent bonds are better understood by valence bond theory or molecular orbital theory. The properties of the atoms involved can be understood using concepts such as oxidation number. The electron density within a bond is not assigned to individual atoms, but is instead delocalized between atoms. In valence bond theory, the two electrons on the two atoms are coupled together with the bond strength depending on the overlap between them. In molecular orbital theory, the linear combination of atomic orbitals (LCAO) helps describe the delocalized molecular orbital structures and energies based on the atomic orbitals of the atoms they came from. Unlike pure ionic bonds, covalent bonds may have directed anisotropic properties. These may have their own names, such as sigma bond and pi bond.

In the general case, atoms form bonds that are intermediate between ionic and covalent, depending on the relative electronegativity of the atoms involved. This type of bond is sometimes called polar covalent.

13.7　References

[1] Lewis, Gilbert N. (1916). "The Atom and the Molecule". *Journal of the American Chemical Society* **38** (4): 772. doi:10.1021/ja02261a002. a copy

[2] Laidler, K. J. (1993). *The World of Physical Chemistry*. Oxford University Press. p. 346. ISBN 0-19-855919-4.

[3] James, H. H.; Coolidge, A. S. (1933). "The Ground State of the Hydrogen Molecule". *Journal of Chemical Physics* (American Institute of Physics) **1** (12): 825–835. doi:10.1063/1.1749252.

[4] "Bond Lengths and Energies". Science.uwaterloo.ca. Retrieved 2013-10-15.

[5] Atkins, Peter; Loretta Jones (1997). *Chemistry: Molecules, Matter and Change*. New York: W. H. Freeman & Co. pp. 294–295. ISBN 0-7167-3107-X.

13.8　External links

- W. Locke (1997). Introduction to Molecular Orbital Theory. Retrieved May 18, 2005.

- Carl R. Nave (2005). HyperPhysics. Retrieved May 18, 2005.

- Linus Pauling and the Nature of the Chemical Bond: A Documentary History. Retrieved February 29, 2008.

Chapter 14

Chemical compound

A **chemical compound** (or just **compound** if used in the context of chemistry) is an entity consisting of two or more different atoms which associate via chemical bonds. There are four types of compounds, depending on how the constituent atoms are held together: molecules held together by covalent bonds, salts held together by ionic bonds, intermetallic compounds held together by metallic bonds, and certain complexes held together by coordinate covalent bonds. Many chemical compounds have a unique numerical identifier assigned by the Chemical Abstracts Service (CAS): its CAS number.

A chemical formula is a way of expressing information about the proportions of atoms that constitute a particular chemical compound, using the standard abbreviations for the chemical elements, and subscripts to indicate the number of atoms involved. For example, water is composed of two hydrogen atoms bonded to one oxygen atom: the chemical formula is H_2O.

A compound can be converted to a different chemical composition by interaction with a second chemical compound via a chemical reaction. In this process, bonds between atoms are broken in both of the interacting compounds, and then bonds are reformed so that new associations are made between atoms. Schematically, this reaction could be described as AB + CD --> AC + BD, where A, B, C, and D are each unique atoms; and AB, CD, AC, and BD are each unique compounds.

A chemical element bonded to an identical chemical element is not a chemical compound since only one element, not two different elements, is involved. Examples are the diatomic molecule hydrogen (H_2) and the polyatomic molecule sulfur (S_8).

14.1   Definitions

Any substance consisting of two or more different types of atoms (chemical elements) in a fixed proportion of its atoms (i.e., stoichiometry) can be termed a *chemical compound*;

the concept is most readily understood when considering pure chemical substances.[1]:15 [2][3] It follows from their being composed of fixed proportions of two or more types of atoms is that chemical compounds can be converted, via chemical reaction, into compounds or substances each having fewer atoms.[4] In the case of non-stoichiometric compounds, the proportions may be reproducible with regard to their preparation, and give a fixed proportions of their component elements, but proportions that are not integral [e.g., for palladium hydride, PdHx $(0.02 < x < 0.58)$]. Chemical compounds have a unique and defined chemical structure held together in a defined spatial arrangement by chemical bonds. Chemical compounds can be molecular compounds held together by covalent bonds, salts held together by ionic bonds, intermetallic compounds held together by metallic bonds, or the subset of complexes that are held together by coordinate covalent bonds. Pure chemical elements are generally not considered chemical compounds, failing the two or more atom requirement, though they often consist of molecules composed of multiple atoms (such as in the diatomic molecule H_2, or the polyatomic molecule S_8, etc.).

There is varying and sometimes inconsistent nomenclature differentiating substances, which include truly non-stoichiometric examples, from chemical compounds, which require the fixed ratios. Many solid chemical substances—for example many silicate minerals—are chemical substances, but do not have simple formulae reflecting chemically bonding of elements to one another in fixed ratios; even so, these crystalline substances are sometimes called "non-stoichiometric compounds". It may be argued that they are related to, rather than being chemical compounds, insofar as the variability in their compositions is often due to either the presence of foreign elements trapped within the crystal structure of an otherwise known true *chemical compound*, or due to perturbations in structure relative to the known compound that arise because of an excess of deficit of the constituent elements at places in its structure; such non-stoichiometric substances form most of the crust and mantle of the Earth. Other compounds regarded as chemically identical may have varying amounts of heavy or light isotopes of the constituent elements, which changes the ratio of elements by mass slightly.

14.2   Elementary concepts

Characteristic properties of compounds include that *elements in a compound are present in a definite proportion*. For example, the molecule of the compound water is composed of hydrogen and oxygen in a ratio of 2:1.water. In addition, *compounds have a definite set of properties*, and the elements that comprise a compound do not retain their original properties. For example, hydrogen, which is combustible and non-supportive of combustion, combines with oxygen, which is non-combustible and supportive of combustion, to produce the compound water, which is non-combustible and non-supportive of combustion.

14.3   Comparison to mixtures

The physical and chemical properties of compounds differ from those of their constituent elements. This is one of the main criteria that distinguish a compound from a mixture of elements or other substances—in general, a mixture's properties are closely related to, and depend on, the properties of its constituents. Another criterion that distinguishes a compound from a mixture is that constituents of a mixture can usually be separated by simple mechanical means, such as filtering, evaporation, or magnetic force, but components of a compound can be separated only by a chemical reaction. However, mixtures can be created by mechanical means alone, but a compound can be created (either from elements or from other compounds, or a combination of the two) only by a chemical reaction.

Some mixtures are so intimately combined that they have some properties similar to compounds and may easily be mistaken for compounds. One example is alloys. Alloys are made mechanically, most commonly by heating the constituent metals to a liquid state, mixing them thoroughly, and then cooling the mixture quickly so that the constituents are trapped in the base metal. Other examples of compound-like mixtures include intermetallic compounds and solutions of alkali metals in a liquid form of ammonia.

14.4   Formula

Main article: Chemical formula

A chemical formula is a way of expressing information about the proportions of atoms that constitute a particular chemical compound, using a single line of chemical element symbols, numbers, and sometimes also other symbols, such as parentheses, dashes, brackets, commas and plus (+) and minus (−) signs.

Compounds may be described using formulas in various formats. For compounds that exist as molecules, the formula for the molecular unit is shown. For polymeric materials, such as minerals and many metal oxides, the empirical formula is normally given, e.g. NaCl for table salt.

The elements in a chemical formula are normally listed in a specific order, called the Hill system. In this system, the carbon atoms (if there are any) are usually listed first, any hydrogen atoms are listed next, and all other elements follow in alphabetical order. If the formula contains no carbon, then all of the elements, including hydrogen, are listed alphabetically. There are, however, several important exceptions to the normal rules. For ionic compounds, the positive ion is almost always listed first and the negative ion is listed second. For oxides, oxygen is usually listed last.

In general, organic acids follow the normal rules with C and H coming first in the formula. For example, the formula for trifluoroacetic acid is usually written as $C_2HF_3O_2$. More descriptive formulas can convey structural information, such as writing the formula for trifluoroacetic acid as CF_3CO_2H. On the other hand, the chemical formulas for most inorganic acids and bases are exceptions to the normal rules. They are written according to the rules for ionic compounds (positive first, negative second), but they also follow rules that emphasize their Arrhenius definitions. To be specific, the formula for most inorganic acids begins with hydrogen and the formula for most bases ends with the hydroxide ion (OH^-). Formulas for inorganic compounds do not often convey structural information, as illustrated by the common use of the formula H_2SO_4 for a molecule (sulfuric acid) that contains no H-S bonds. A more descriptive presentation would be $O_2S(OH)_2$, but it is almost never written this way.

14.5   Phases and thermal properties

Compounds may have several possible phases. All compounds can exist as solids, at least at low enough temperatures. Molecular compounds may also exist as liquids, gases, and, in some cases, even plasmas. All compounds decompose upon applying heat. The temperature at which such fragmentation occurs is often called the decomposition temperature. Decomposition temperatures are not sharp and depend on pressure, temperature, and the concentration of each species in the compound.

14.6   See also

- Chemical element

- Chemical Revolution

- Chemical structure

- Chemical name

- Dictionary of chemical formulas

- List of chemical compounds

- Addition to pi ligands

14.7   References

[1] Whitten, Kenneth W.; Davis, Raymond E.; Peck, M. Larry (2000), *General Chemistry* (6th ed.), Fort Worth, TX: Saunders College Publishing/Harcourt College Publishers, ISBN 978-0-03-072373-5

[2] Brown, Theodore L.; LeMay, H. Eugene; Bursten, Bruce E.; Murphy, Catherine J.; Woodward, Patrick (2009), *Chemistry: The Central Science, AP Edition* (11th ed.), Upper Saddle River, NJ: Pearson/Prentice Hall, pp. 5–6, ISBN 0-13-236489-1

[3] Hill, John W.; Petrucci, Ralph H.; McCreary, Terry W.; Perry, Scott S. (2005), *General Chemistry* (4th ed.), Upper Saddle River, NJ: Pearson/Prentice Hall, p. 6, ISBN 978-0-13-140283-6

[4] Wilbraham, Antony; Matta, Michael; Staley, Dennis; Waterman, Edward (2002), *Chemistry* (1st ed.), Upper Saddle River, NJ: Pearson/Prentice Hall, p. 36, ISBN 0-13-251210-6

14.8   Further reading

- Robert Siegfried (2002), *From elements to atoms: a history of chemical composition*, American Philosophical Society, ISBN 978-0-87169-924-4

14.9   External links

Chapter 15

Mineral

For other uses, see Mineral (disambiguation).

A **mineral** is a naturally occurring substance that is solid

Amethyst, a variety of quartz

and inorganic representable by a chemical formula, usually abiogenic, and has an ordered atomic structure. It is different from a rock, which can be an aggregate of minerals or non-minerals and does not have a specific chemical composition. The exact definition of a mineral is under debate, especially with respect to the requirement a valid species be abiogenic, and to a lesser extent with regard to it having an ordered atomic structure. The study of minerals is called mineralogy.

There are over 4,900 known mineral species; over 4,660 of these have been approved by the International Mineralogical Association (IMA). The silicate minerals compose over 90% of the Earth's crust. The diversity and abundance of mineral species is controlled by the Earth's chemistry. Silicon and oxygen constitute approximately 75% of the Earth's crust, which translates directly into the predominance of silicate minerals. Minerals are distinguished by various chemical and physical properties. Differences in chemical composition and crystal structure distinguish various species, and these properties in turn are influenced by the mineral's geological environment of formation. Changes in the temperature, pressure, or bulk composition of a rock mass cause changes in its minerals.

Minerals can be described by various physical properties which relate to their chemical structure and composition. Common distinguishing characteristics include crystal structure and habit, hardness, lustre, diaphaneity, colour, streak, tenacity, cleavage, fracture, parting, and specific gravity. More specific tests for minerals include magnetism, taste or smell, radioactivity and reaction to acid.

Minerals are classified by key chemical constituents; the two dominant systems are the Dana classification and the Strunz classification. The silicate class of minerals is subdivided into six subclasses by the degree of polymerization in the chemical structure. All silicate minerals have a base unit of a $[SiO_4]^{4-}$ silica tetrahedra—that is, a silicon cation coordinated by four oxygen anions, which gives the shape of a tetrahedron. These tetrahedra can be polymerized to give the subclasses: orthosilicates (no polymerization, thus single tetrahedra), disilicates (two tetrahedra bonded together), cyclosilicates (rings of tetrahedra), inosilicates (chains of tetrahedra), phyllosilicates (sheets of tetrahedra), and tectosilicates (three-dimensional network of tetrahedra). Other important mineral groups include the native elements, sulfides, oxides, halides, carbonates, sulfates, and phosphates.

15.1 Definition

15.1.1 Basic definition

The general definition of a mineral encompasses the following criteria:[1]

1. Naturally occurring

2. Stable at room temperature

3. Represented by a chemical formula

4. Usually abiogenic (not resulting from the activity of living organisms)

5. Ordered atomic arrangement

The first three general characteristics are less debated than the last two.[1] The first criterion means that a mineral has to form by a natural process, which excludes anthropogenic compounds. Stability at room temperature, in the simplest sense, is synonymous to the mineral being solid. More specifically, a compound has to be stable or metastable at 25 °C. Classical examples of exceptions to this rule include native mercury, which crystallizes at −39 °C, and water ice, which is solid only below 0 °C; as these two minerals were described prior to 1959, they were grandfathered by the International Mineralogical Association (IMA).[2][3] Modern advances have included extensive study of liquid crystals, which also extensively involve mineralogy. Minerals are chemical compounds, and as such they can be described by fixed or a variable formula. Many mineral groups and species are composed of a solid solution; pure substances are not usually found because of contamination or chemical substitution. For example, the olivine group is described by the variable formula $(Mg, Fe)_2SiO_4$, which is a solid solution of two end-member species, magnesium-rich forsterite and iron-rich fayalite, which are described by a fixed chemical formula. Mineral species themselves could have a variable compositions, such as the sulfide mackinawite, $(Fe, Ni)_9S_8$, which is mostly a ferrous sulfide, but has a very significant nickel impurity that is reflected in its formula.[1][4]

The requirement of a valid mineral species to be abiogenic has also been described as similar to have to be inorganic; however, this criterion is imprecise and organic compounds have been assigned a separate classification branch. Finally, the requirement of an ordered atomic arrangement is usually synonymous to being crystalline; however, crystals are periodic in addition to being ordered, so the broader criterion is used instead.[1] The presence of an ordered atomic arrangement translates to a variety of macroscopic physical properties, such as crystal form, hardness, and cleavage.[5] There have been several recent proposals to amend the definition to consider biogenic or amorphous substances as minerals. The formal definition of a mineral approved by the IMA in 1995:

"A mineral is an element or chemical compound that is normally crystalline and that has been formed as a result of geological processes."[6]

In addition, biogenic substances were explicitly excluded:

"Biogenic substances are chemical compounds produced entirely by biological processes without a geological component (e.g., urinary calculi, oxalate crystals in plant tissues, shells of marine molluscs, etc.) and are not regarded as minerals. However, if geological processes were involved in the genesis of the compound, then the product can be accepted as a mineral."[6]

15.1.2 Recent advances

Mineral classification schemes and their definitions are evolving to match recent advances in mineral science. Recent changes have included the addition of an organic class, in both the new Dana and the Strunz classification schemes.[7][8] The organic class includes a very rare group of minerals with hydrocarbons. The IMA Commission on New Minerals and Mineral Names adopted in 2009 a hierarchical scheme for the naming and classification of mineral groups and group names[9] and established seven commissions and four working groups to review and classify minerals into an official listing of their published names.[10] According to these new rules, "mineral species can be grouped in a number of different ways, on the basis of chemistry, crystal structure, occurrence, association, genetic history, or resource, for example, depending on the purpose to be served by the classification."[9]

The Nickel (1995) exclusion of biogenic substances was not universally adhered to. For example, Lowenstam (1981) stated that "organisms are capable of forming a diverse array of minerals, some of which cannot be formed inorganically in the biosphere."[11] The distinction is a matter of classification and less to do with the constituents of the minerals themselves. Skinner (2005) views all solids as potential minerals and includes biominerals in the mineral kingdom, which are those that are created by the metabolic activities of organisms. Skinner expanded the previous definition of a mineral to classify "element or compound, amorphous or crystalline, formed through *biogeochemical* processes," as a mineral.[12]

Recent advances in high-resolution genetic and x-ray absorption spectroscopy is opening new revelations on the biogeochemical relations between microorganisms and minerals that may make Nickel's (1995)[6] biogenic mineral exclusion obsolete and Skinner's (2005) biogenic mineral inclusion a necessity.[12] For example, the IMA commissioned 'Environmental Mineralogy and Geochemistry Working Group'[13] deals with minerals in the hydrosphere, atmosphere, and biosphere. Mineral forming microorganisms inhabit the areas that this working group deals with. These organisms exist on nearly every rock, soil, and particle surface spanning the globe reaching depths at 1600 metres below the sea floor (possibly further) and 70 kilometres into the stratosphere (possibly entering the mesosphere).[14][15][16] Biologists and geologists have started to research and appreciate the magnitude of mineral geoengineering that these creatures are capable of. Bacteria have contributed to the formation of minerals for billions of years and critically define the biogeochemical cycles on this planet. Microorganisms can precipitate metals from solution contributing to the formation of ore deposits in addition to their ability to catalyze mineral dissolution, to respire, precipitate, and form minerals.[17][18][19]

Prior to the International Mineralogical Association's listing, over 60 biominerals had been discovered, named, and published.[20] These minerals (a sub-set tabulated in Lowenstam (1981)[111]) are considered minerals proper according to the Skinner (2005) definition.[12] These biominerals are not listed in the International Mineral Association official list of mineral names,[21] however, many of these biomineral representatives are distributed amongst the 78 mineral classes listed in the Dana classification scheme.[12] Another rare class of minerals (primarily biological in origin) include the mineral liquid crystals that are crystalline and liquid at the same time. To date over 80,000 liquid crystalline compounds have been identified.[22][23]

> Concerning the use of the term "mineral" to name this family of liquid crystals, one can argue that the term inorganic would be more appropriate. However, inorganic liquid crystals have long been used for organometallic liquid crystals. Therefore, in order to avoid any confusion between these fairly chemically different families, and taking into account that a large number of these liquid crystals occur naturally in nature, we think that the use of the old fashioned but adequate "mineral" adjective taken sensus largo is more specific that an alternative such as "purely inorganic", to name this subclass of the inorganic liquid crystals family.[23]

The Skinner (2005) definition[12] of a mineral takes this matter into account by stating that a mineral can be crystalline or amorphous. Liquid mineral crystals are amorphous. Biominerals and liquid mineral crystals, however, are not the primary form of minerals, most are geological in origin,[24] but these groups do help to identify at the margins of what constitutes a mineral proper. The formal Nickel (1995) definition explicitly mentioned crystalline nature as a key to defining a substance as a mineral. A 2011 article defined icosahedrite, an aluminium-iron-copper alloy as mineral; named for its unique natural icosahedral symmetry, it is also a quasicrystal. Unlike a true crystal, quasicrystals are ordered but not periodic.[25][26]

15.1.3 Rocks, ores, and gems

Schist is a metamorphic rock characterized by an abundance of platy minerals. In this example, the rock has prominent sillimanite porphyroblasts as large as 3 cm (1.2 in).

Minerals are not equivalent to rocks. Whereas a mineral is a naturally occurring usually solid substance, stable at room temperature, representable by a chemical formula, usually abiogenic, and has an ordered atomic structure, a rock is either an aggregate of one or more minerals, or not composed of minerals at all.[27] Rocks like limestone or quartzite are composed primarily of one mineral—calcite or aragonite in the case of limestone, and quartz in the latter case.[28][29] Other rocks can be defined by relative abundances of key (essential) minerals; a granite is defined by proportions of quartz, alkali feldspar, and plagioclase feldspar.[30] The other minerals in the rock are termed accessory, and do not greatly affect the bulk composition of the rock. Rocks can also be composed entirely of non-mineral material; coal is a sedimentary rock composed primarily of organically derived carbon.[27][31]

In rocks, some mineral species and groups are much more abundant than others; these are termed the rock-forming minerals. The major examples of these are quartz, the feldspars, the micas, the amphiboles, the pyroxenes, the olivines, and calcite; except the last one, all of the minerals are silicates.[32] Overall, around 150 minerals are considered particularly important, whether in terms of their abundance or aesthetic value in terms of collecting.[33]

Commercially valuable minerals and rocks are referred to as industrial minerals. For example, muscovite, a white mica, can be used for windows (sometimes referred to as isinglass), as a filler, or as an insulator.[34] Ores are minerals that have a high concentration of a certain element, typically a metal. Examples are cinnabar (HgS), an ore of mercury, sphalerite (ZnS), an ore of zinc, or cassiterite (SnO_2), an ore of tin. Gems are minerals with an ornamental value, and are distinguished from non-gems by their beauty, durability, and usually, rarity. There are about 20 mineral species that qualify as gem minerals, which constitute about 35 of the most common gemstones. Gem minerals are often present in several varieties, and so one mineral can account for several different gemstones; for example, ruby and sapphire are both corundum, Al_2O_3.[35]

15.1.4 Nomenclature and classification

In general, a mineral is defined as naturally occurring solid, that is stable at room temperature, representable by a chemical formula, usually abiogenic, and has an ordered atomic structure. However, a mineral can be also narrowed down in terms of a mineral group, series, species, or variety, in order from most broad to least broad. The basic level of definition is that of mineral species, which is distinguished from other species by specific and unique chemical and physical properties. For example, quartz is defined by its formula, SiO_2, and a specific crystalline structure that distinguishes it from other minerals with the same chemical formula (termed polymorphs). When there exists a range of composition between two minerals species, a mineral series is defined. For example, the biotite series is represented by variable amounts of the endmembers phlogopite, siderophyllite, annite, and eastonite. In contrast, a mineral group is a grouping of mineral species with some common chemical properties that share a crystal structure. The pyroxene group has a common formula of $XY(Si,Al)_2O_6$, where X and Y are both cations, with X typically bigger than Y; the pyroxenes are single-chain silicates that crystallize in

either the orthorhombic or monoclinic crystal systems. Finally, a mineral variety is a specific type of mineral species that differs by some physical characteristic, such as colour or crystal habit. An example is amethyst, which is a purple variety of quartz.[36]

Two common classifications are used for minerals; both the Dana and Strunz classifications rely on the composition of the mineral, specifically with regards to important chemical groups, and its structure. The Dana *System of Mineralogy* was first published in 1837 by James Dwight Dana, a leading geologist of his time; it is in its eighth edition (1997 ed.). The Dana classification, assigns a four-part number to a mineral species. First is its class, based on important compositional groups; next, the type gives the ratio of cations to anions in the mineral; finally, the last two numbers group minerals by structural similarity with a given type or class. The less commonly used Strunz classification, named for German mineralogist Karl Hugo Strunz, is based on the Dana system, but combines both chemical and structural criteria, the latter with regards to distribution of chemical bonds.[37]

There are over 4,660 approved mineral species.[38] They are most commonly named after a person (45%), followed by discovery location (23%); names based on chemical composition (14%) and physical properties (8%) are the two other major groups of mineral name etymologies.[36][39] The common suffix *-ite* of mineral names descends from the ancient Greek suffix - ί τ η ς (-ites), meaning "connected with or belonging to".[40]

15.2 Mineral chemistry

The abundance and diversity of minerals is controlled directly by their chemistry, in turn dependent on elemental abundances in the Earth. The majority of minerals observed are derived from the Earth's crust. Eight elements account for most of the key components of minerals, due to their abundance in the crust. These eight elements, summing to over 98% of the crust by weight, are, in order of decreasing abundance: oxygen, silicon, aluminium, iron, magnesium, calcium, sodium and potassium. Oxygen and silicon are by far the two most important — oxygen composes 46.6% of the crust by weight, and silicon accounts for 27.7%.[41]

The minerals that form are directly controlled by the bulk chemistry of the parent body. For example, a magma rich in iron and magnesium will form mafic minerals, such as olivine and the pyroxenes; in contrast, a more silica-rich magma will crystallize to form minerals than incorporate more SiO_2, such as the feldspars and quartz. In a limestone, calcite or aragonite (both $CaCO_3$) form because the rock

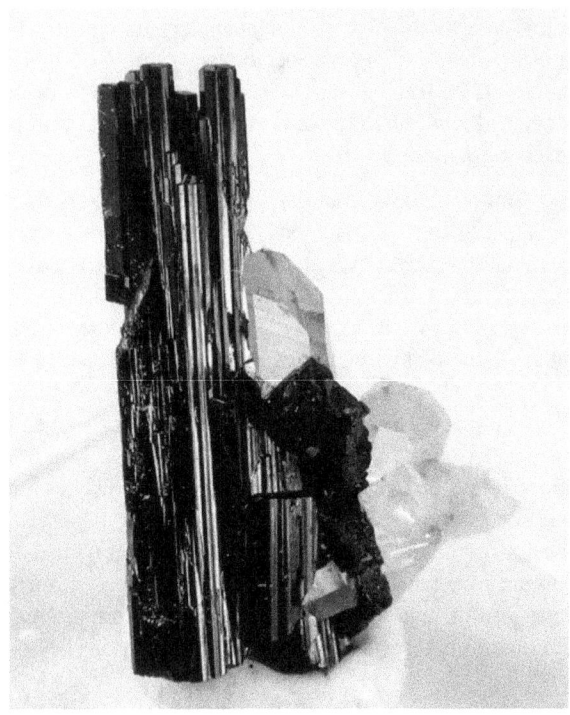

Hübnerite, the manganese-rich end-member of the wolframite series, with minor quartz in the background

is rich in calcium and carbonate. A corollary is that a mineral will not be found in a rock whose bulk chemistry does not resemble the bulk chemistry of a given mineral with the exception of trace minerals. For example, kyanite, Al_2SiO_5 forms from the metamorphism of aluminium-rich shales; it would not likely occur in aluminium-poor rock, such quartzite.

The chemical composition may vary between end member species of a mineral series. For example, the plagioclase feldspars comprise a continuous series from sodium-rich end member albite ($NaAlSi_3O_8$) to calcium-rich anorthite ($CaAl_2Si_2O_8$) with four recognized intermediate varieties between them (given in order from sodium- to calcium-rich): oligoclase, andesine, labradorite, and bytownite.[42] Other examples of series include the olivine series of magnesium-rich forsterite and iron-rich fayalite, and the wolframite series of manganese-rich hübnerite and iron-rich ferberite.

Chemical substitution and coordination polyhedra explain this common feature of minerals. In nature, minerals are not pure substances, and are contaminated by whatever other elements are present in the given chemical system. As a result, it is possible for one element to be substituted for another.[43] Chemical substitution will occur between ions of a similar size and charge; for example, K^+ will not substitute for Si^{4+} because of chemical and structural incompatibilities caused by a big difference in size and charge. A

common example of chemical substitution is that of Si^{4+} by Al^{3+}, which are close in charge, size, and abundance in the crust. In the example of plagioclase, there are three cases of substitution. Feldspars are all framework silicates, which have a silicon-oxygen ratio of 2:1, and the space for other elements is given by the substitution of Si^{4+} by Al^{3+} to give a base unit of $[AlSi_3O_8]^-$; without the substitution, the formula would be charge-balanced as SiO_2, giving quartz.[44] The significance of this structural property will be explained further by coordination polyhedra. The second substitution occurs between Na^+ and Ca^{2+}; however, the difference in charge has to accounted for by making a second substitution of Si^{4+} by Al^{3+}.[45]

Coordination polyhedra are geometric representation of how a cation is surrounded by an anion. In mineralogy, due its abundance in the crust, coordination polyhedra are usually considered in terms of oxygen. The base unit of silicate minerals is the silica tetrahedron — one Si^{4+} surrounded by four O^{2-}. An alternate way of describing the coordination of the silicate is by a number: in the case of the silica tetrahedron, the silicon is said to have a coordination number of 4. Various cations have a specific range of possible coordination numbers; for silicon, it is almost always 4, except for very high-pressure minerals where compound is compressed such that silicon is in six-fold (octahedral) coordination by oxygen. Bigger cations have a bigger coordination number because of the increase in relative size as compared to oxygen (the last orbital subshell of heavier atoms is different too). Changes in coordination numbers between leads to physical and mineralogical differences; for example, at high pressure such as in the mantle, many minerals, especially silicates such as olivine and garnet will change to a perovskite structure, where silicon is in octahedral coordination. Another example are the aluminosilicates kyanite, andalusite, and sillimanite (polymorphs, as they share the formula Al_2SiO_5), which differ by the coordination number of the Al^{3+}; these minerals transition from one another as a response to changes in pressure and temperature.[41] In the case of silicate materials, the substitution of Si^{4+} by Al^{3+} allows for a variety of minerals because of the need to balance charges.[46]

Changes in temperature and pressure, and composition alter the mineralogy of a rock sample. Changes in composition can be caused by processes such as weathering or metasomatism (hydrothermal alteration). Changes in temperature and pressure occur when the host rock undergoes tectonic or magmatic movement into differing physical regimes. Changes in thermodynamic conditions make it favourable for mineral assemblages to react with each other to produce new minerals; as such, it is possible for two rocks to have an identical or a very similar bulk rock chemistry without having a similar mineralogy. This process of mineralogical alteration is related to the rock cycle. An example

When minerals react, the products will sometimes assume the shape of the reagent; the product mineral is termed to be a pseudomorph of (or after) the reagent. Illustrated here is a pseudomorph of kaolinite after orthoclase. Here, the pseudomorph preserved the Carlsbad twinning common in orthoclase.

of a series of mineral reactions is illustrated as follows.[47]

Orthoclase feldspar ($KAlSi_3O_8$) is a mineral commonly found in granite, a plutonic igneous rock. When exposed to weathering, it reacts to form kaolinite ($Al_2Si_2O_5(OH)_4$, a sedimentary mineral, and silicic acid):

$$2\,KAlSi_3O_8 + 5\,H_2O + 2\,H^+ \rightarrow Al_2Si_2O_5(OH)_4 + 4\,H_2SiO_3 + 2\,K^+$$

Under low-grade metamorphic conditions, kaolinite reacts with quartz to form pyrophyllite ($Al_2Si_4O_{10}(OH)_2$):

$$Al_2Si_2O_5(OH)_4 + SiO_2 \rightarrow Al_2Si_4O_{10}(OH)_2 + H_2O$$

As metamorphic grade increases, the pyrophyllite reacts to form kyanite and quartz:

$$Al_2Si_4O_{10}(OH)_2 \rightarrow Al_2SiO_5 + 3\,SiO_2 + H_2O$$

Alternatively, a mineral may change its crystal structure as a consequence of changes in temperature and pressure without reacting. For example, quartz will change into a variety of its SiO_2 polymorphs, such as tridymite and cristobalite at high temperatures, and coesite at high pressures.[48]

15.3 Physical properties of minerals

Classifying minerals ranges from simple to difficult. A mineral can be identified by several physical properties, some of them being sufficient for full identification without equivocation. In other cases, minerals can only be classified by more complex optical, chemical or X-ray diffraction analysis; these methods, however, can be costly and

time-consuming. Physical properties applied for classification include crystal structure and habit, hardness, lustre, diaphaneity, colour, streak, cleavage and fracture, and specific gravity. Other less general tests include fluorescence, phosphorescence, magnetism, radioactivity, tenacity (response to mechanical induced changes of shape or form), piezoelectricity and reactivity to dilute acids.[49]

15.3.1 Crystal structure and habit

Main articles: Crystal system and Crystal habit
See also: Crystal twinning
Crystal structure results from the orderly geometric spatial

Topaz has a characteristic orthorhombic elongated crystal shape.

arrangement of atoms in the internal structure of a mineral. This crystal structure is based on regular internal atomic or ionic arrangement that is often expressed in the geometric form that the crystal takes. Even when the mineral grains are too small to see or are irregularly shaped, the underlying crystal structure is always periodic and can be determined by X-ray diffraction.[1] Minerals are typically described by their symmetry content. Crystals are restricted to 32 point groups, which differ by their symmetry. These groups are classified in turn into more broad categories, the most encompassing of these being the six crystal families.[50]

These families can be described by the relative lengths of the three crystallographic axes, and the angles between them; these relationships correspond to the symmetry operations that define the narrower point groups. They are summarized below; a, b, and c represent the axes, and α, β, γ represent the angle opposite the respective crystallographic axis (e.g. α is the angle opposite the a-axis, viz. the angle between the b and c axes):[50]

The hexagonal crystal family is also split into two crystal *systems* — the trigonal, which has a three-fold axis of symmetry, and the hexagonal, which has a six-fold axis of symmetry.

Chemistry and crystal structure together define a mineral. With a restriction to 32 point groups, minerals of different chemistry may have identical crystal structure. For example, halite (NaCl), galena (PbS), and periclase (MgO) all belong to the hexaoctahedral point group (isometric family), as they have a similar stoichiometry between their different constituent elements. In contrast, polymorphs are groupings of minerals that share a chemical formula but have a different structure. For example, pyrite and marcasite, both iron sulfides, have the formula FeS_2; however, the former is isometric while the latter is orthorhombic. This polymorphism extends to other sulfides with the generic AX_2 formula; these two groups are collectively known as the pyrite and marcasite groups.[51]

Polymorphism can extend beyond pure symmetry content. The aluminosilicates are a group of three minerals — kyanite, andalusite, and sillimanite — which share the chemical formula Al_2SiO_5. Kyanite is triclinic, while andalusite and sillimanite are both orthorhombic and belong to the dipyramidal point group. These difference arise correspond to how aluminium is coordinated within the crystal structure. In all minerals, one aluminium ion is always in six-fold coordination by oxygen; the silicon, as a general rule is in four-fold coordination in all minerals; an exception is a case like stishovite (SiO_2, an ultra-high pressure quartz polymorph with rutile structure).[52] In kyanite, the second aluminium is in six-fold coordination; its chemical formula can be expressed as $Al^{[6]}Al^{[6]}SiO_5$, to reflect its crystal structure. Andalusite has the second aluminium in five-fold coordination ($Al^{[6]}Al^{[5]}SiO_5$) and sillimanite has it in four-fold coordination ($Al^{[6]}Al^{[4]}SiO_5$).[53]

Differences in crystal structure and chemistry greatly influence other physical properties of the mineral. The carbon allotropes diamond and graphite have vastly different properties; diamond is the hardest natural substance, has an adamantine lustre, and belongs to the isometric crystal family, whereas as graphite is very soft, has a greasy lustre, and crystallises in the hexagonal family. This difference is accounted by differences in bonding. In diamond, the carbons are in sp^3 hybrid orbitals, which means they form a framework where each carbon is covalently bonded to three neighbours in a tetrahedral fashion; on the other hand, graphite is composed of sheets of carbons in sp^2 hybrid orbitals, where each carbon is bonded covalently to only two others. These sheets are held together by much weaker van der Waals forces, and this discrepancy translates to big macroscopic differences.[54]

Contact twins, as seen in spinel

Twinning is the intergrowth of two or more crystal of a single mineral species. The geometry of the twinning is controlled by the mineral's symmetry. As a result, there are several types of twins, including contact twins, reticulated twins, geniculated twins, penetration twins, cyclic twins, and polysynthetic twins. Contact, or simple twins, consist of two crystals joined at a plane; this type of twinning is common in spinel. Reticulated twins, common in rutile, are interlocking crystals resembling netting. Geniculated twins have a bend in the middle that is caused by start of the twin. Penetration twins consist of two single crystals that have grown into each other; examples of this twinning include cross-shaped staurolite twins and Carlsbad twinning in orthoclase. Cyclic twins are caused by repeated twinning around a rotation axis. It occurs around three, four, five, six, or eight-fold axes, and the corresponding patterns are called threelings, fourlings, fivelings, sixlings, and eightlings. Sixlings are common in aragonite. Polysynthetic twins are similar to cyclic twinning by the presence of repetitive twinning; however, instead of occurring around a rotational axis, it occurs along parallel planes, usually on a microscopic scale.[55][56]

Crystal habit refers to the overall shape of crystal. Several terms are used to describe this property. Common habits include acicular, which described needlelike crystals like in natrolite, bladed, dendritic (tree-pattern, common in native copper), equant, which is typical of garnet, prismatic (elongated in one direction), and tabular, which differs from bladed habit in that the former is platy whereas the latter has a defined elongation. Related to crystal form, the quality of crystal faces is diagnostic of some minerals, especially with a petrographic microscope. Euhedral crystals have a defined external shape, while anhedral crystals do not; those intermediate forms are termed subhedral.[57][58]

15.3.2 Hardness

Main article: Mohs scale of mineral hardness
The hardness of a mineral defines how much it can re-

Diamond is the hardest natural material, and has a Mohs hardness of 10.

sist scratching. This physical property is controlled by the chemical composition and crystalline structure of a mineral. A mineral's hardness is not necessarily constant for all sides, which is a function of its structure; crystallographic weakness renders some directions softer than others.[59] An example of this property exists in kyanite, which has a Mohs hardness of 5½ parallel to [001] but 7 parallel to [100].[60]

The most common scale of measurement is the ordinal Mohs hardness scale. Defined by ten indicators, a mineral with a higher index scratches those below it. The scale ranges from talc, a phyllosilicate, to diamond, a carbon polymorph that is the hardest natural material. The scale is provided below:[59]

15.3.3 Lustre and diaphaneity

Main article: Lustre (mineralogy)
 Lustre indicates how light reflects from the mineral's surface, with regards to its quality and intensity. There are numerous qualitative terms used to describe this property, which are split into metallic and non-metallic categories. Metallic and sub-metallic minerals have high reflectivity like metal; examples of minerals with this lustre are

Pyrite has a metallic lustre.

galena and pyrite. Non-metallic lustres include: adamantine, such as in diamond; vitreous, which is a glassy lustre very common in silicate minerals; pearly, such as in talc and apophyllite, resinous, such as members of the garnet group, silky which common in fibrous minerals such as asbestiform chrysotile.[61]

The diaphaneity of a mineral describes the ability of light to pass through it. Transparent minerals do not diminish the intensity of light passing through it. An example of such a mineral is muscovite (potassium mica); some varieties are sufficiently clear to have been used for windows. Translucent minerals allow some light to pass, but less than those that are transparent. Jadeite and nephrite (mineral forms of jade are examples of minerals with this property). Minerals that do not allow light to pass are called opaque.[62][63]

The diaphaneity of a mineral depends on thickness of the sample. When a mineral is sufficiently thin (e.g., in a thin section for petrography), it may become transparent even if that property is not seen in hand sample. In contrast, some minerals, such as hematite or pyrite are opaque even in thin-section.[63]

15.3.4 Colour and streak

Main article: Streak (mineralogy)

Colour is the most obvious property of a mineral, but it is often non-diagnostic.[64] It is caused by electromagnetic radiation interacting with electrons (except in the case of incandescence, which does not apply to minerals).[65] Two broad classes of elements are defined with regards to their contribution to a mineral's colour. Idiochromatic elements are essential to a mineral's composition; their contribution to a mineral's colour is diagnostic.[62][66] Examples of such minerals are malachite (green) and azurite (blue). In contrast, allochromatic elements in minerals are present in trace amounts as impurities. An example of such a mineral would be the ruby and sapphire varieties of the mineral corundum.[66] The colours of pseudochromatic minerals are the result of interference of light waves. Examples include labradorite and bornite.

In addition to simple body colour, minerals can have various other distinctive optical properties, such as play of colours, asterism, chatoyancy, iridescence, tarnish, and pleochroism. Several of these properties involve variability in colour. Play of colour, such as in opal, results in the sample reflecting different colours as it is turned, while pleochroism describes the change in colour as light passes through a mineral in a different orientation. Iridescence is a variety of the play of colours where light scatters off a coating on the surface of crystal, cleavage planes, or off layers having minor gradations in chemistry.[67] In contrast, the play of colours in opal is caused by light refracting from ordered microscopic silica spheres within its physical structure.[68] Chatoyancy ("cat's eye") is the wavy banding of colour that is observed as the sample is rotated; asterism, a variety of chatoyancy, gives the appearance of a star on the mineral grain. The latter property is particularly common in gem-quality corundum.[67][68]

The streak of a mineral refers to the colour of a mineral in powdered form, which may or may not be identical to its body colour.[66] The most common way of testing this property is done with a streak plate, which is made out of porcelain and coloured either white or black. The streak of a mineral is independent of trace elements[62] or any weathering surface.[66] A common example of this property is illustrated with hematite, which is coloured black, silver, or red in hand sample, but has a cherry-red[62] to reddish-brown streak.[66] Streak is more often distinctive for metallic minerals, in contrast to non-metallic minerals whose body colour is created by allochromatic elements.[62] Streak testing is constrained by the hardness of the mineral, as those harder than 7 powder the *streak plate* instead.[66]

15.3.5 Cleavage, parting, fracture, and tenacity

Main articles: Cleavage (crystal) and Fracture (mineralogy)
By definition, minerals have a characteristic atomic arrangement. Weakness in this crystalline structure causes planes of weakness, and the breakage of a mineral along such planes is termed cleavage. The quality of cleavage can be described based on how cleanly and easily the mineral breaks; common descriptors, in order of decreasing quality, are "perfect", "good", "distinct", and "poor". In particularly transparent mineral, or in thin-section, cleavage can be seen a series of parallel lines marking the planar surfaces when viewed at a side. Cleavage is not a universal prop-

Perfect basal cleavage as seen in biotite (black), and good cleavage seen in the matrix (pink orthoclase).

erty among minerals; for example, quartz, consisting of extensively interconnected silica tetrahedra, does not have a crystallographic weakness which would allow it to cleave. In contrast, micas, which have perfect basal cleavage, consist of sheets of silica tetrahedra which are very weakly held together.[69][70]

As cleavage is a function of crystallography, there are a variety of cleavage types. Cleavage occurs typically in either one, two, three, four, or six directions. Basal cleavage in one direction is a distinctive property of the micas. Two-directional cleavage is described as prismatic, and occurs in minerals such as the amphiboles and pyroxenes. Minerals such as galena or halite have cubic (or isometric) cleavage in three directions, at 90°; when three directions of cleavage are present, but not at 90°, such as in calcite or rhodochrosite, it is termed rhombohedral cleavage. Octahedral cleavage (four directions) is present in fluorite and diamond, and sphalerite has six-directional dodecahedral cleavage.[69][70]

Minerals with many cleavages might not break equally well in all of the directions; for example, calcite has good cleavage in three direction, but gypsum has perfect cleavage in one direction, and poor cleavage in two other directions. Angles between cleavage planes vary between minerals. For example, as the amphiboles are double-chain silicates and the pyroxenes are single-chain silicates, the angle between their cleavage planes is different. The pyroxenes cleave in two directions at approximately 90°, whereas the amphiboles distinctively cleave in two directions separated by approximately 120° and 60°. The cleavage angles can be measured with a contact goniometer, which is similar to a protractor.[69][70]

Parting, sometimes called "false cleavage", is similar in appearance to cleavage but is instead produced by structural defects in the mineral as opposed to systematic weakness. Parting varies from crystal to crystal of a mineral, whereas all crystals of a given mineral will cleave if the atomic structure allows for that property. In general, parting is caused by some stress applied to a crystal. The sources of the stresses include deformation (e.g. an increase in pressure), exsolution, or twinning. Minerals that often display parting include the pyroxenes, hematite, magnetite, and corundum.[69][71]

When a mineral is broken in a direction that does not correspond to a plane of cleavage, it is termed to have been fractured. There are several types of uneven fracture. The classic example is conchoidal fracture, like that of quartz; rounded surfaces are created, which are marked by smooth curved lines. This type of fracture occurs only in very homogeneous minerals. Other types of fracture are fibrous, splintery, and hackly. The latter describes a break along a rough, jagged surface; an example of this property is found in native copper.[72]

Tenacity is related to both cleavage and fracture. Whereas fracture and cleavage describes the surfaces that are created when a mineral is broken, tenacity describes how resistant a mineral is to such breaking. Minerals can be described as brittle, ductile, malleable, sectile, flexible, or elastic.[73]

Galena, PbS, is a mineral with a high specific gravity.

mass. A generalization is that minerals with metallic or adamantine lustre tend to have higher specific gravities than those having a non-metallic to dull lustre. For example, hematite, Fe_2O_3, has a specific gravity of 5.26[75] while galena, PbS, has a specific gravity of 7.2–7.6,[76] which is a result of their high iron and lead content, respectively. A very high specific gravity becomes very pronounced in native metals; kamacite, an iron-nickel alloy common in iron meteorites has a specific gravity of 7.9,[77] and gold has an observed specific gravity between 15 and 19.3.[74][78]

15.3.6 Specific gravity

Specific gravity numerically describes the density of a mineral. The dimensions of density are mass divided by volume with units: kg/m^3 or g/cm^3. Specific gravity measures how much water a mineral sample displaces. Defined as the quotient of the mass of the sample and difference between the weight of the sample in air and its corresponding weight in water, specific gravity is a unitless ratio. Among most minerals, this property is not diagnostic. Rock forming minerals — typically silicates or occasionally carbonates — have a specific gravity of 2.5–3.5.[74]

High specific gravity is a diagnostic property of a mineral. A variation in chemistry (and consequently, mineral class) correlates to a change in specific gravity. Among more common minerals, oxides and sulfides tend to have a higher specific gravity as they include elements with higher atomic

15.3.7 Other properties

Other properties can be used to diagnose minerals. These are less general, and apply to specific minerals.

Dropping dilute acid (often 10% HCl) aids in distinguishing carbonates from other mineral classes. The acid reacts with the carbonate ($[CO_3]^{2-}$) group, which causes the affected area to effervesce, giving off carbon dioxide gas. This test can be further expanded to test the mineral in its original crystal form or powdered. An example of this test is done when distinguish calcite from dolomite, especially within rocks (limestone and dolostone respectively). Calcite immediately effervesces in acid, whereas acid must be applied to powdered dolomite (often to a scratched surface in a rock), for it to effervesce.[79] Zeolite minerals will not effervesce in acid; instead, they become frosted after 5–10 minutes, and if left in acid for a day, they dissolve or be-

Carnotite (yellow) is a radioactive uranium-bearing mineral.

come a silica gel.[80]

When tested, magnetism is a very conspicuous property of minerals. Among common minerals, magnetite exhibits this property strongly, and it is also present, albeit not as strongly, in pyrrhotite and ilmenite.[79]

Minerals can also be tested for taste or smell. Halite, NaCl, is table salt; its potassium-bearing counterpart, sylvite, has a pronounced bitter taste. Sulfides have a characteristic smell, especially as samples are fractured, reacting, or powdered.[79]

Radioactivity is a rare property; minerals may be composed of radioactive elements. They could be a defining constituent, such as uranium in uraninite, autunite, and carnotite, or as trace impurities. In the latter case, the decay of a radioactive element damages the mineral crystal; the result, termed a *radioactive halo* or *pleochroic halo*, is observable by various techniques, such as thin-section petrography.[79]

15.4   Mineral classes

As the composition of the Earth's crust is dominated by silicon and oxygen, silicate elements are by far the most important class of minerals in terms of rock formation and diversity. However, non-silicate minerals are of great economic importance, especially as ores.[81][82]

Non-silicate minerals are subdivided into several other classes by their dominant chemistry, which included native elements, sulfides, halides, oxides and hydroxides, carbonates and nitrates, borates, sulfates, phosphates, and organic compounds. The majority of non-silicate mineral species are extremely rare (constituting in total 8% of the Earth's crust), although some are relative common, such as calcite, pyrite, magnetite, and hematite. There are two major struc-

tural styles observed in non-silicates: close-packing and silicate-like linked tetrahedra. The close-packed structures, which is a way to densely pack atoms while minimizing interstitial space. Hexagonal close-packing involves stacking layers where every other layer is the same ("ababab"), whereas cubic close-packing involves stacking groups of three layers ("abcabcabc"). Analogues to linked silica tetrahedra include SO_4 (sulfate), PO_4 (phosphate), AsO_4 (arsenate), and VO_4 (vanadate). The non-silicates have great economic importance, as they concentrate elements more than the silicate minerals do.[83]

The largest grouping of minerals by far are the silicates; most rocks are composed of greater than 95% silicate minerals, and over 90% of the Earth's crust is composed of these minerals.[84] The two main constituents of silicates are silicon and oxygen, which are the two most abundant elements in the Earth's crust. Other common elements in silicate minerals correspond to other common elements in the Earth's crust, such aluminium, magnesium, iron, calcium, sodium, and potassium.[85] Some important rock-forming silicates include the feldspars, quartz, olivines, pyroxenes, amphiboles, garnets, and micas.

15.4.1   Silicates

Main article: Silicate minerals

The base of unit of a silicate mineral is the $[SiO_4]^{4-}$ tetrahedron. In the vast majority of cases, silicon is in four-fold or tetrahedral coordination with oxygen. In very high-pressure situations, silicon will be six-fold or octahedral coordination, such as in the perovskite structure or the quartz polymorph stishovite (SiO_2). In the latter case, the mineral no longer has a silicate structure, but that of rutile (TiO_2), and its associated group, which are simple oxides. These silica tetrahedra are then polymerized to some degree to create various structures, such as one-dimensional chains, two-dimensional sheets, and three-dimensional frameworks. The basic silicate mineral where no polymerization of the tetrahedra has occurred requires other elements to balance out the base 4- charge. In other silicate structures, different combinations of elements are required to balance out the resultant negative charge. It is common for the Si^{4+} to be substituted by Al^{3+} because of similarity in ionic radius and charge; in those case, the $[AlO_4]^{5-}$ tetrahedra form the same structures as do the unsubstituted tetrahedra, but their charge-balancing requirements are different.[86]

The degree of polymerization can be described by both the structure formed and how many tetrahedral corners (or coordinating oxygens) are shared (for aluminium and silicon in tetrahedral sites).[87] Orthosilicates (or nesosilicates) have no linking of polyhedra, thus tetrahedra share no cor-

Natrolite is a mineral series in the zeolite group; this sample has a very prominent acicular crystal habit.

Aegirine, an iron-sodium clinopyroxene, is part of the inosilicate subclass.

ners. Disilicates (or sorosilicates) have two tetrahedra sharing one oxygen atom. Inosilicates are chain silicates; single-chain silicates have two shared corners, whereas double-chain silicates have two or three shared corners. In phyllosilicates, a sheet structure is formed which requires three shared oxygens; in the case of double-chain silicates, some tetrahedra must share two corners instead of three as otherwise a sheet structure would result. Framework silicates, or tectosilicates, have tetrahedra that share all four corners. The ring silicates, or cyclosilicates, only need tetrahedra to share two corners to form the cyclical structure.[88]

The silicate subclasses are described below in order of decreasing polymerization.

Tectosilicates

Main category: Tectosilicates

Tectosilicates, also known as framework silicates, have the highest degree of polymerization. With all corners of a tetrahedra shared, the silicon:oxygen ratio becomes 1:2. Examples are quartz, the feldspars, feldspathoids, and the zeolites. Framework silicates tend to be particularly chem-

ically stable as a result of strong covalent bonds.[89]

Forming 12% of the Earth's crust, quartz (SiO_2) is the most abundant mineral species. It is characterized by its high chemical and physical resistivity. Quartz has several polymorphs, including tridymite and cristobalite at high temperatures, high-pressure coesite, and ultra-high pressure stishovite. The latter mineral can only be formed on Earth by meteorite impacts, and its structure has been composed so much that it had changed from a silicate structure to that of rutile (TiO_2). The silica polymorph that is most stable at the Earth's surface is α-quartz. Its counterpart, β-quartz, is present only at high temperatures and pressures (changes to α-quartz below 573 °C at 1 bar). These two polymorphs differ by a "kinking" of bonds; this change in structure gives β-quartz greater symmetry than α-quartz, and they are thus also called high quartz (β) and low quartz (α).[84][90]

Feldspars are the most abundant group in the Earth's crust, at about 50%. In the feldspars, Al^{3+} substitutes for Si^{4+}, which creates a charge imbalance that must be accounted for by the addition of cations. The base structure becomes either $[AlSi_3O_8]^-$ or $[Al_2Si_2O_8]^{2-}$ There are 22 mineral species of feldspars, subdivided into two major subgroups—alkali and plagioclase—and two less common groups—celsian and banalsite. The alkali feldspars are most commonly in a series between potassium-rich orthoclase and sodium-rich albite; in the case of plagioclase, the most common series ranges from albite to calcium-rich anorthite. Crystal twinning is common in feldspars, especially polysynthetic twins in plagioclase and Carlsbad twins in alkali feldspars. If the latter subgroup cools slowly from a melt, it forms exsolution lamellae because the two components—orthoclase and albite—are unstable in solid solution. Exsolution can be on a scale from microscopic to readily observable in hand-sample; perthitic texture forms when Na-rich feldspar exsolve in a K-rich host. The opposite texture (antiperthitic), where K-rich feldspar exsolves in a Na-rich host, is very rare.[91]

Feldsapthoids are structurally similar to feldspar, but differ in that they form in Si-deficient conditions which allows for further substitution by Al^{3+}. As a result, feldsapthoids cannot be associated with quartz. A common example of a feldsapthoid is nepheline (($Na, K)AlSiO_4$); compared to alkali feldspar, nepheline has an Al_2O_3:SiO_2 ratio of 1:2, as opposed to 1:6 in the feldspar.[92] Zeolites often have distinctive crystal habits, occurring in needles, plates, or blocky masses. They form in the presence of water at low temperatures and pressures, and have channels and voids in their structure. Zeolites have several industrial applications, especially in waste water treatment.[93]

Phyllosilicates

Main category: Phyllosilicates
 Phyllosilicates consist of sheets of polymerized tetrahe-

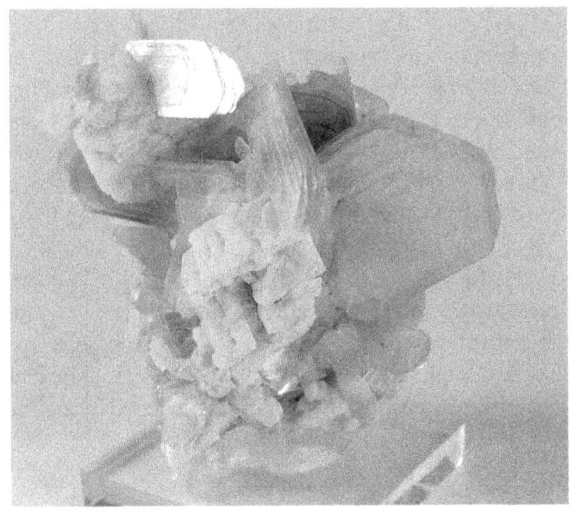

Muscovite, a mineral species in the mica group, within the phyllosilicate subclass

dra. They are bound at three oxygen sites, which gives a characteristic silicon:oxygen ratio of 2:5. Important examples include the mica, chlorite, and the kaolinite-serpentine groups. The sheets are weakly bound by van der Waals forces or hydrogen bonds, which causes a crystallographic weakness, in turn leading to a prominent basal cleavage among the phyllosilicates.[94] In addition to the tetrahedra, phyllosilicates have a sheet of octahedra (elements in sixfold coordination by oxygen) that balanced out the basic tetrahedra, which have a negative charge (e.g. $[Si_4O_{10}]^{4-}$) These tetrahedra (T) and octahedra (O) sheets are stacked in a variety of combinations to create phyllosilicate groups. Within an octahedral sheet, there are three octahedral sites

in a unit structure; however, not all of the sites may be occupied. In that case, the mineral is termed dioctahedral, whereas in other case it is termed trioctahedral.[95]

The kaolinite-serpentine group consists of T-O stacks (the 1:1 clay minerals); their hardness ranges from 2 to 4, as the sheets are held by hydrogen bonds. The 2:1 clay minerals (pyrophyllite-talc) consist of T-O-T stacks, but they are softer (hardness from 1 to 2), as they are instead held together by van der Waals forces. These two groups of minerals are subgrouped by octahedral occupation; specifically, kaolinite and pyrophyllite are dioctahedral whereas serpentine and talc trioctahedral.[96]

Micas are also T-O-T-stacked phyllosilicates, but differ from the other T-O-T and T-O-stacked subclass members in that they incorporate aluminium into the tetrahedral sheets (clay minerals have Al^{3+} in octahedral sites). Common examples of micas are muscovite, and the biotite series. The chlorite group is related to mica group, but a brucite-like ($Mg(OH)_2$) layer between the T-O-T stacks.[97]

Because of their chemical structure, phyllosilicates typically have flexible, elastic, transparent layers that are electrical insulators and can be split into very thin flakes. Micas can be used in electronics as insulators, in construction, as optical filler, or even cosmetics. Chrysotile, a species of serpentine, is the most common mineral species in industrial asbestos, as it is less dangerous in terms of health than the amphibole asbestos.[98]

Inosilicates

Main category: Inosilicates
 Inosilicates consist of tetrahedra repeatedly bonded in chains. These chains can be single, where a tetrahedron is bound to two others to form a continuous chain; alternatively, two chains can be merged to create double-chain silicates. Single-chain silicates have a silicon:oxygen ratio of 1:3 (e.g. $[Si_2O_6]^{4-}$), whereas the double-chain variety has a ratio of 4:11, e.g. $[Si_8O_{22}]^{12-}$. Inosilicates contain two important rock-forming mineral groups; single-chain silicates are most commonly pyroxenes, while double-chain silicates are often amphiboles.[99] Higher-order chains exist (e.g. three-member, four-member, five-member chains, etc.) but they are rare.[100]

The pyroxene group consists of 21 mineral species.[101] Pyroxenes have a general structure formula of $XY(Si_2O_6)$, where X is an octahedral site, while Y can vary in coordination number from six to eight. Most varieties of pyroxene consist of permutations of Ca^{2+}, Fe^{2+} and Mg^{2+} to balance the negative charge on the backbone. Pyroxenes are common in the Earth's crust (about 10%) and are a key constituent of mafic igneous rocks.[102]

Asbestiform tremolite, part of the amphibole group in the inosilicate subclass

An example of elbaite, a species of tourmaline, with distinctive colour banding.

Amphiboles have great variability in chemistry, described variously as a "mineralogical garbage can" or a "mineralogical shark swimming a sea of elements". The backbone of the amphiboles is the $[Si_8O_{22}]^{12-}$; it is balanced by cations in three possible positions, although the third position is not always used, and one element can occupy both remaining ones. Finally, the amphiboles are usually hydrated, that is, they have a hydroxyl group ($[OH]^-$), although it can be replaced by a fluoride, a chloride, or an oxide ion.[103] Because of the variable chemistry, there are over 80 species of amphibole, although variations, as in the pyroxenes, most commonly involve mixtures of Ca^{2+}, Fe^{2+} and Mg^{2+}.[101] Several amphibole mineral species can have an asbestiform crystal habit. These asbestos minerals form long, thin, flexible, and strong fibres, which are electrical insulators, chemically inert and heat-resistant; as such, they have several applications, especially in construction materials. However, asbestos are known carcinogens, and cause various other illnesses, such as asbestosis; amphibole asbestos (anthophyllite, tremolite, actinolite, grunerite, and riebeckite) are considered more dangerous than chrysotile serpentine asbestos.[104]

Cyclosilicates

Main category: Cyclosilicates

Cyclosilicates, or ring silicates, have a ratio of silicon to oxygen of 1:3. Six-member rings are most common,

with a base structure of $[Si_6O_{18}]^{12-}$; examples include the tourmaline group and beryl. Other ring structures exist, with 3, 4, 8, 9, 12 having been described.[105] Cyclosilicates tend to be strong, with elongated, striated crystals.[106]

Tourmalines have a very complex chemistry that can be described by a general formula $XY_3Z_6(BO_3)_3T_6O_{18}V_3W$. The T_6O_{18} is the basic ring structure, where T is usually Si^{4+}, but substitutable by Al_{3+} or B^{3+}. Tourmalines can be subgrouped by the occupancy of the X site, and from there further subdivided by the chemistry of the W site. The Y and Z sites can accommodate a variety of cations, especially various transition metals; this variability in structural transition metal content gives the tourmaline group greater variability in colour. Other cyclosilicates include beryl, $Al_2Be_3Si_6O_{18}$, whose varieties include the gemstones emerald (green) and aquamarine (bluish). Cordierite is structurally similar to beryl, and is a common metamorphic mineral.[107]

Sorosilicates

Main category: Sorosilicates

Sorosilicates, also termed disilicates, have tetrahedron-tetrahedron bonding at one oxygen, which results in a 2:7 ratio of silicon to oxygen. The resultant common structural element is the $[Si_2O_7]^{6-}$ group. The most common

Epidote often has a distinctive pistachio-green colour.

disilicates by far are members of the epidote group. Epidotes are found in variety of geologic settings, ranging from mid-ocean ridge to granites to metapelites. Epidotes are built around the structure $[(SiO_4)(Si_2O_7)]^{10-}$ structure; for example, the mineral *species* epidote has calcium, aluminium, and ferric iron to charge balance: $Ca_2Al_2(Fe^{3+}, Al)(SiO_4)(Si_2O_7)O(OH)$. The presence of iron as Fe^{3+} and Fe^{2+} helps understand oxygen fugacity, which in turn is a significant factor in petrogenesis.[108]

Other examples of sorosilicates include lawsonite, a metamorphic mineral forming in the blueschist facies (subduction zone setting with low temperature and high pressure), vesuvianite, which takes up a significant amount of calcium in its chemical structure.[108][109]

Orthosilicates

Main category: Nesosilicates

Orthosilicates consist of isolated tetrahedra that are charge-balanced by other cations.[110] Also termed nesosilicates, this type of silicate has a silicon:oxygen ratio of 1:4 (e.g. SiO_4). Typical orthosilicates tend to form blocky equant crystals, and are fairly hard.[111] Several rock-forming minerals are part of this subclass, such as the aluminosilicates, the olivine group, and the garnet group.

The aluminosilicates—kyanite, andalusite, and sillimanite, all Al_2SiO_5—are structurally composed of one $[SiO_4]^{4-}$ tetrahedron, and one Al^{3+} in octahedral coordination. The remaining Al^{3+} can be in six-fold coordination (kyanite), five-fold (andalusite) or four-fold (sillimanite); which mineral forms in a given environment is depend on pressure and temperature conditions. In the olivine structure, the main olivine series of $(Mg, Fe)_2SiO_4$ consist of magnesium-rich forsterite and iron-rich fayalite. Both iron and magnesium are in octahedral by oxygen. Other mineral species having this structure exist, such as tephroite, Mn_2SiO_4.[112] The garnet group has a general

Black andradite, an end-member of the orthosilicate garnet group.

formula of $X_3Y_2(SiO_4)_3$, where X is a large eight-fold coordinated cation, and Y is a smaller six-fold coordinated cation. There are six ideal endmembers of garnet, split into two group. The pyralspite garnets have Al^{3+} in the Y position: pyrope ($Mg_3Al_2(SiO_4)_3$), almandine ($Fe_3Al_2(SiO_4)_3$), and spessartine ($Mn_3Al_2(SiO_4)_3$). The ugrandite garnets have Ca^{2+} in the X position: uvarovite ($Ca_3Cr_2(SiO_4)_3$), grossular ($Ca_3Al_2(SiO_4)_3$) and andradite ($Ca_3Fe_2(SiO_4)_3$). While there are two subgroups of garnet, solid solutions exist between all six end-members.[110]

Other orthosilicates include zircon, staurolite, and topaz. Zircon ($ZrSiO_4$) is useful in geochronology as the Zr^{4+} can be substituted by U^{6+}; furthermore, because of its very resistant structure, it is difficult to reset it as a chronometer. Staurolite is a common metamorphic intermediate-grade index mineral. It has a particularly complicated crystal structure that was only fully described in 1986. Topaz ($Al_2SiO_4(F, OH)_2$, often found in granitic pegmatites associated with tourmaline, is a common gemstone mineral.[113]

15.4.2 Non-silicates

Native elements

Main article: Native element minerals

Native elements are those that are not chemically bonded to other elements. This mineral group includes native metals,

Native gold. Rare specimen of stout crystals growing off of a central stalk, size 3.7 x 1.1 x 0.4 cm, from Venezuela.

semi-metals, and non-metals, and various alloys and solid solutions. The metals are held together by metallic bonding, which confers distinctive physical properties such as their shiny metallic lustre, ductility and malleability, and electrical conductivity. Native elements are subdivided into groups by their structure or chemical attributes.

The gold group, with a cubic close-packed structure, includes metals such as gold, silver, and copper. The platinum group is similar in structure to the gold group. The iron-nickel group is characterized by several iron-nickel alloy species. Two examples are kamacite and taenite, which are found in iron meteorites; these species differ by the amount of Ni in the alloy; kamacite has less than 5–7% nickel and is a variety of native iron, whereas the nickel content of taenite ranges from 7–37%. Arsenic group minerals consist of semi-metals, which have only some metallic; for example, they lack the malleability of metals. Native carbon occurs in two allotropes, graphite and diamond; the latter forms at very high pressure in the mantle, which gives it a much stronger structure than graphite.[114]

Sulfides

Main article: Sulfide minerals

The sulfide minerals are chemical compounds of one or more metals or semimetals with a sulfur; tellurium, arsenic, or selenium can substitute for the sulfur. Sulfides tend to be soft, brittle minerals with a high specific gravity. Many powdered sulfides, such as pyrite, have a sulfurous smell

Red cinnabar (HgS), a mercury ore, on dolomite

when powdered. Sulfides are susceptible to weathering, and many readily dissolve in water; these dissolved minerals can be later redeposited, which creates enriched secondary ore deposits.[115] Sulfides are classified by the ratio of the metal or semimetal to the sulfur, such as M:S equal to 2:1, or 1:1.[116] Many sulfide minerals are economically important as metal ores; examples include sphalerite (ZnS), an ore of zinc, galena (PbS), an ore of lead, cinnabar (HgS), an ore of mercury, and molybdenite (MoS_2, an ore of molybdenum.[117] Pyrite (FeS_2), is the most commonly occurring sulfide, and can be found in most geological environments. It is not, however, an ore of iron, but can be instead oxidized to produce sulfuric acid.[118] Related to the sulfides are the rare sulfosalts, in which a metallic element is bonded to sulfur and a semimetal such as antimony, arsenic, or bismuth. Like the sulfides, sulfosalts are typically soft, heavy, and brittle minerals.[119]

Oxides

Main article: Oxide minerals

Oxide minerals are divided into three categories: simple oxides, hydroxides, and multiple oxides. Simple oxides are characterized by O^{2-} as the main anion and primarily ionic bonding. They can be further subdivided by the ratio of oxygen to the cations. The periclase group consists of minerals with a 1:1 ratio. Oxides with a 2:1 ratio include cuprite (Cu_2O) and water ice. Corundum group minerals have a 2:3 ratio, and includes minerals such as corundum (Al_2O_3), and hematite (Fe_2O_3). Rutile group minerals have a ratio of 1:2; the eponymous species, rutile (TiO_2) is the chief ore of titanium; other examples include cassiterite (SnO_2; ore of tin), and pyrolusite (MnO_2; ore of manganese).[120][121] In hydroxides, the dominant anion is the hydroxyl ion, OH^-. Bauxites are the chief aluminium ore, and are a heterogeneous mixture of the hydroxide minerals diaspore, gibbsite, and bohmite; they form in areas with a very high rate of chemical weathering (mainly tropical conditions).[122] Finally, multiple oxides are compounds of two metals with oxygen. A major group within this class are the spinels, with a general formula of $X^{2+}Y^{3+}_2O_4$. Examples of species include spinel ($MgAl_2O_4$), chromite ($FeCr_2O_4$), and magnetite (Fe_3O_4). The latter is readily distinguishable by its strong magnetism, which occurs as it has iron in two oxidation states ($Fe^{2+}Fe^{3+}_2O_4$), which makes it a multiple oxide instead of a single oxide.[123]

Halides

Main article: Halide minerals
 The halide minerals are compounds where a halogen (flu-

Pink cubic halite (NaCl; halide class) crystals on a nahcolite matrix (NaHCO$_3$; a carbonate, and mineral form of sodium bicarbonate, used as baking soda).

orine, chlorine, iodine, and bromine) is the main anion. These minerals tend to be soft, weak, brittle, and water-soluble. Common examples of halides include halite ($NaCl$, table salt), sylvite (KCl), fluorite (CaF_2). Halite and sylvite commonly form as evaporites, and can be dominant minerals in chemical sedimentary rocks. Cryolite, Na_3AlF_6, is a key mineral in the extraction of aluminium from bauxites; however, as the only significant occurrence at Ivittuut, Greenland, in a granitic pegmatite, was depleted, synthetic cryolite can be made from fluorite.[124]

Carbonates

Main article: Carbonate minerals

The carbonate minerals are those were the main anionic group is carbonate, $[CO_3]^{2-}$. Carbonates tend to be brittle, many have rhombohedral cleavage, and all react with acid.[125] Due to the last characteristic, field geologists often carry dilute hydrochloric acid to distinguish carbonates from non-carbonates. The reaction of acid with carbonates, most commonly found as the polymorph calcite and aragonite ($CaCO_3$), relates to the dissolution and precipitation of the mineral, which is a key in the formation of limestone caves, features within them such as stalactite and stalagmites, and karst landforms. Carbonates are most often formed as biogenic or chemical sediments in marine environments. The carbonate group is structurally a triangle, where a central C^{4+} cation is surrounded by three O^{2-} anions; different groups of minerals form from different arrangements of these triangles.[126] The most common carbonate mineral is calcite, and is the primary constituent of sedimentary limestone and metamorphic marble. Calcite, $CaCO_3$, can have a high magnesium impurity; under high-Mg conditions, its polymorph aragonite will form instead; the marine geochemistry in this regard can be described as an aragonite or calcite sea, depending on which mineral preferentially forms. Dolomite is a double carbonate, with the formula $CaMg(CO_3)_2$. Secondary dolomitization of limestone is common, where calcite or aragonite are converted to dolomite; this reaction increases pore space (the unit cell volume of dolomite is 88% that of calcite), which can create a reservoir for oil and gas. These two minerals species are members of eponymous mineral groups: the calcite group includes carbonates with the general formula XCO_3, and the dolomite group constitutes minerals with general formula $XY(CO_3)_2$.[127]

Sulfates

Main article: Sulfate minerals
 The sulfate minerals all contain the sulfate anion, $[SO_4]^{2-}$. They tend to be transparent to translucent, soft, and many are fragile.[128] Sulfate minerals commonly form as evaporites, where they precipitate out of evaporating saline waters; alternative, sulfates can also be found in hydrothermal vein systems associated with sulfides,[129] or as oxidation products of sulfides.[130] Sulfates can be subdivided into anhydrous and hydrous minerals. The most common hydrous sulfate by far is gypsum, $CaSO_4 \cdot 2H_2O$. It forms as an evaporite, and is associated with other evaporites such as calcite and halite; if it incorporates sand grains as it crystallizes, gypsum can form desert roses. Gypsum has very low thermal conductivity and maintains a low temperature when

Gypsum desert rose

heated as it loses that heat by dehydrating; as such, gypsum is used as an insulator in materials such as plaster and drywall. The anhydrous equivalent of gypsum is anhydrite; it can form directly from seawater in highly arid conditions. The barite group has the general formula XSO_4, where the X is a large 12-coordinated cation. Examples include barite ($BaSO_4$), celestine ($SrSO_4$), and anglesite ($PbSO_4$); anhydrite is not part of the barite group, as the smaller Ca^{2+} is only in eight-fold coordination.[131]

Phosphates

Main article: Phosphate minerals

The phosphate minerals are characterized by the tetrahedral $[PO_4]^{3-}$ unit, although the structure can be generalized, and phosphorus is replaced by antimony, arsenic, or vanadium. The most common phosphate is the apatite group; common species within this group are fluorapatite ($Ca_5(PO_4)_3F$), chlorapatite ($Ca_5(PO_4)_3Cl$) and hydroxylapatite ($Ca_5(PO_4)_3(OH)$). Minerals in this group are the main crystalline constituents of teeth and bones in vertebrates. The relatively abundant monazite group has a general structure of ATO_4, where T is phosphorus or arsenic, and A is often a rare-earth element (REE). Monazite is important in two ways: first, as a REE "sink", it can sufficiently concentrate these elements to become an ore; secondly, monazite group elements can incorporate relatively large amounts of uranium and thorium, which can be used to date the rock based on the decay of the U and Th to lead.[132]

Organic minerals

Main article: Organic minerals

The Strunz classification includes a class for organic minerals. These rare compounds contain organic carbon, but can be formed by a geologic process. For example, whewellite, $CaC_2O_4 \cdot H_2O$ is an oxalate that can be deposited in hydrothermal ore veins. While hydrated calcium oxalate can be found in coal seams and other sedimentary deposits involving organic matter, the hydrothermal occurrence is not considered to be related to biological activity.[82]

15.5   Astrobiology

It has been suggested that biominerals could be important indicators of extraterrestrial life and thus could play an important role in the search for past or present life on the planet Mars. Furthermore, organic components (biosignatures) that are often associated with biominerals are believed to play crucial roles in both pre-biotic and biotic reactions.[133]

On January 24, 2014, NASA reported that current studies by the *Curiosity* and *Opportunity* rovers on Mars will now be searching for evidence of ancient life, including a biosphere based on autotrophic, chemotrophic and/or chemolithoautotrophic microorganisms, as well as ancient water, including fluvio-lacustrine environments (plains related to ancient rivers or lakes) that may have been habitable.[134][135][136][137] The search for evidence of habitability, taphonomy (related to fossils), and organic carbon on the planet Mars is now a primary NASA objective.[134][135]

15.6   See also

- Amateur geology
- Asterism
- Dietary mineral
- Isomorphism (crystallography)
- List of minerals
- List of minerals (complete)
- Mineral collecting
- Polymorphism (materials science)

15.7   Bibliography

- Busbey, A.B.; Coenraads, R.E.; Roots, D.; Willis, P. (2007). *Rocks and Fossils*. San Francisco: Fog City Press. ISBN 978-1-74089-632-0.

- Chesterman, C.W.; Lowe, K.E. (2008). *Field guide to North American rocks and minerals.* Toronto: Random House of Canada. ISBN 0-394-50269-8.

- Dyar, M.D.; Gunter, M.E. (2008). *Mineralogy and Optical Mineralogy.* Chantilly, Virginia: Mineralogical Society of America. ISBN 978-0-939950-81-2.

15.8   References

[1] Dyar, Gunter, and Tasa (2007). *Mineralogy and Optical Mineralogy.* Mineralogical Society of America. pp. 2–4. ISBN 978-0939950812.

[2] "Mercury". Mindat.org. Retrieved 2012-08-13.

[3] "Ice". Mindat.org. Retrieved 2012-08-13.

[4] "Mackinawite". Mindat.org. Retrieved 2012-08-13.

[5] Chesterman and Lowe, pp. 13–14

[6] Nickel, Ernest H. (1995). "The definition of a mineral". *The Canadian Mineralogist* **33** (3): 689–690. alt version

[7] Dana Classification 8th edition – Organic Compounds. Mindat.org. Retrieved on 2011-10-20.

[8] Strunz Classification – Organic Compounds. Mindat.org. Retrieved on 2011-10-20.

[9] Mills, J. S.; Hatert, F.; Nickel, E. H.; Ferraris, G. (2009). "The standardisation of mineral group hierarchies: application to recent nomenclature proposals" (PDF). *European Journal of Mineralogy* **21** (5): 1073–1080. doi:10.1127/0935-1221/2009/0021-1994.

[10] IMA divisions. Ima-mineralogy.org (2011-01-12). Retrieved on 2011-10-20.

[11] H. A., Lowenstam (1981). "Minerals formed by organisms". *Science* **211** (4487): 1126–1131. Bibcode:1981Sci...211.1126L. doi:10.1126/science.7008198. JSTOR 1685216. PMID 7008198.

[12] Skinner, H. C. W. (2005). "Biominerals". *Mineralogical Magazine* **69** (5): 621–641. doi:10.1180/0026461056950275.

[13] Working Group On Environmental Mineralogy (Wgem). Ima-mineralogy.org. Retrieved on 2011-10-20.

[14] Takai, K. (2010). "Limits of life and the biosphere: Lessons from the detection of microorganisms in the deep sea and deep subsurface of the Earth.". In Gargaud, M.; Lopez-Garcia, P.; Martin, H. *Origins and Evolution of Life: An Astrobiological Perspective.* Cambridge, UK: Cambridge University Press. pp. 469–486.

[15] Roussel, E. G.; Cambon Bonavita, M.; Querellou, J.; Cragg, B. A.; Prieur, D.; Parkes, R. J.; Parkes, R. J. (2008). "Extending the Sub-Sea-Floor Biosphere". *Science* **320** (5879): 1046–1046. Bibcode:2008Sci...320.1046R. doi:10.1126/science.1154545.

[16] Pearce, D. A.; Bridge, P. D.; Hughes, K. A.; Sattler, B.; Psenner, R.; Russel, N. J. (2009). "Microorganisms in the atmosphere over Antarctica". *FEMS Microbiology Ecology* **69** (2): 143–157. doi:10.1111/j.1574-6941.2009.00706.x. PMID 19527292.

[17] Newman, D. K.; Banfield, J. F. (2002). "Geomicrobiology: How Molecular-Scale Interactions Underpin Biogeochemical Systems". *Science* **296** (5570): 1071–1077. doi:10.1126/science.1010716. PMID 12004119.

[18] Warren, L. A.; Kauffman, M. E. (2003). "Microbial geoengineers". *Science* **299** (5609): 1027–1029. doi:10.1126/science.1072076. JSTOR 3833546. PMID 12586932.

[19] González-Muñoz, M. T.; Rodriguez-Navarro, C.; Martínez-Ruiz, F.; Arias, J. M.; Merroun, M. L.; Rodriguez-Gallego, M. "Bacterial biomineralization: new insights from Myxococcus-induced mineral precipitation". *Geological Society, London, Special Publications* **336** (1): 31–50. Bibcode:2010GSLSP.336...31G. doi:10.1144/SP336.3.

[20] Veis, A. (1990). "Biomineralization. Cell Biology and Mineral Deposition. by Kenneth Simkiss; Karl M. Wilbur On Biomineralization. by Heinz A. Lowenstam; Stephen Weiner". *Science* **247** (4946): 1129–1130. Bibcode:1990Sci...247.1129S. doi:10.1126/science.247.4946.1129. JSTOR 2874281. PMID 17800080.

[21] Official IMA list of mineral names (updated from March 2009 list). uws.edu.au

[22] Bouligand, Y. (2006). "Liquid crystals and morphogenesis.". In Bourgine, P.; Lesne, A. *Morphogenesis: Origins of Patterns and Shape.* Cambridge, UK: Springer Verlag. pp. 49–.

[23] Gabriel, C. P.; Davidson, P. (2003). "Mineral Liquid Crystals from Self-Assembly of Anisotropic Nanosystems" (PDF). *Topics in Current Chemistry* **226**: 119–172. doi:10.1007/b10827.

[24] K., Hefferan; J., O'Brien (2010). *Earth Materials.* Wiley-Blackwell. ISBN 978-1-4443-3460-9.

[25] Bindi, L.; Paul J. Steinhardt; Nan Yao; Peter J. Lu (2011). "Icosahedrite, $Al_{63}Cu_{24}Fe_{13}$, the first natural quasicrystal" (PDF). *American Mineralogist* **96**: 928–931. doi:10.2138/am.2011.3758.

[26] Commission on New Minerals and Mineral Names, Approved as new mineral

[27] Chesterman and Lowe, pp. 15–16

[28] Chesterman and Lowe, p. 719–721

[29] Chesterman and Lowe, p. 747–748

[30] Chesterman and Lowe, p. 694–696

[31] Chesterman and Lowe, pp. 728–730

[32] Dyar and Gunter, p. 15

[33] Chesterman and Lowe, p. 14

[34] Chesterman and Cole, pp. 531–532

[35] Chesterman and Lowe, pp. 14–15

[36] Dyar and Gunter, pp. 20–22

[37] Dyar and Gunter, pp 558–559

[38] "IMA Mineral List with Database of Mineral Properties".

[39] Dyar and Gunter, p. 556

[40] Online Etymology Dictionary

[41] Dyar and Gunter, pp. 4–7

[42] Dyar and Gunter, p. 586

[43] Dyar and Gunter, p. 141

[44] Dyar and Gunter, p. 14

[45] Dyar and Gunter, p. 585

[46] Dyar and Gunter, pp. 12–17

[47] Dyar and Gunter, p. 549

[48] Dyar and Gunter, p. 579

[49] Dyar and Gunter, p. 22–23

[50] Dyar and Gunter, pp. 69–80

[51] Dyar and Gunter, pp. 654–655

[52] Dyar and Gunter, p. 581

[53] Dyar and Gunter, pp. 631–632

[54] Dyar and Gunter, p. 166

[55] Dyar and Gunter, pp. 41–43

[56] Chesterman and Lowe, p. 39

[57] Dyar and Gunter, pp. 32–39

[58] Chesterman and Lowe, p. 38

[59] Dyar and Gunter, pp. 28–29

[60] "Kyanite". Mindat.org. Retrieved 2012-08-01.

[61] Dyar and Darby, pp. 26–28

[62] Busbey et al., p. 72

[63] Dyar and Gunter, p. 25

[64] Dyar and Gunter, p. 23

[65] Dyar and Gunter, pp. 131–144

[66] Dyar and Gunter, p. 24

[67] Dyar and Gunter, pp. 24–26

[68] Busbey et al., p. 73

[69] Dyar and Gunter, pp. 39–40

[70] Chesterman and Lowe, pp. 29–30

[71] Chesterman and Lowe, pp. 30–31

[72] Dyar and Gunter, pp. 31–33

[73] Dyar and Gunter, pp. 30–31

[74] Dyar and Gunter, pp. 43–44

[75] "Hematite". Mindat.org. Retrieved 2012-08-02.

[76] "Galena". Mindat.org. Retrieved 2012-08-02.

[77] "Kamacite". Webmineral.com. Retrieved 2012-08-02.

[78] "Gold". Mindat.org. Retrieved 2012-08-02.

[79] Dyar and Gunter, pp. 44–45

[80] "Mineral Identification Key: Radioactivity, Magnetism, Acid Reactions". Mineralogical Society of America. Retrieved 2012-08-15.

[81] Dyar and Gunter, p. 641

[82] Dyar and Gunter, p. 681

[83] Dyar and Gunter, pp. 641–643

[84] Dyar and Gunter, p. 104

[85] Dyar and Gunter, p. 5

[86] Dyar and Gunter, pp. 104–120

[87] Dyar and Gunter, p. 105

[88] Dyar and Gunter, pp. 104–117

[89] Chesterman and Cole, p. 502

[90] Dyar and Gunter, pp. 578–583

[91] Dyar and Gunter, pp. 583–588

[92] Dyar and Gunter, p. 588

[93] Dyar and Gunter, pp. 589–593

[94] Chesterman and Lowe, p. 525

[95] Dyar and Gunter, p. 110

[96] Dyar and Gunter, pp. 110–113

[97] Dyar and Gunter, pp. 602–605

[98] Dyar and Gunter, p. 593–595

[99] Chesterman and Lowe, p. 537

[100] "09.D Inosilicates". Webmineral.com. Retrieved 2012-08-20.

[101] Dyar and Gunter, pp. 112

[102] Dyar and Gunter pp. 612–613

[103] Dyar and Gunter, pp. 606–612

[104] Dyar and Gunter, pp. 611–612

[105] Dyar and Gunter, pp. 113–115

[106] Chesterman and Lowe,p. 558

[107] Dyar and Gunter, pp. 617–621

[108] Dyar and Gunter, pp. 612–627

[109] Chesterman and Lowe, pp. 565–573

[110] Dyar and Gunter, pp. 116–117

[111] Chesterman and Lowe, p. 573

[112] Chesterman and Lowe, pp. 574–575

[113] Dyar and Gunter, pp. 627–634

[114] Dyar and Gunter, pp. 644–648

[115] Chesterman and Lowe, p. 357

[116] Dyar and Gunter, p. 649

[117] Dyar and Gunter, pp. 651–654

[118] Dyar and Gunter, p. 654

[119] Chesterman and Lowe, p. 383

[120] Chesterman and Lowe, pp. 400–403

[121] Dyar and Gunter, pp. 657–660

[122] Dyar and Gunter, pp. 663–664

[123] Dyar and Gunter, pp. 660–663

[124] Chesterman and Lowe, pp. 425–430

[125] Chesterman and Lowe, p. 431

[126] Dyar and Gunter, pp. 667

[127] Dyar and Gunter, pp. 668–669

[128] Chesterman and Lowe, p. 453

[129] Chesterman and Lowe, pp. 456–457

[130] Dyar and Gunter, p. 674

[131] Dyar and Gunter, pp. 672–673

[132] Dyar and Gunter, pp. 675–680

[133] Steele, Andrew; Beaty, David, eds. (September 26, 2006). "Final report of the MEPAG Astrobiology Field Laboratory Science Steering Group (AFL-SSG)". *The Astrobiology Field Laboratory* (.doc). U.S.A.: Mars Exploration Program Analysis Group (MEPAG) - NASA. p. 72. Retrieved 2009-07-22.

[134] Grotzinger, John P. (January 24, 2014). "Introduction to Special Issue - Habitability, Taphonomy, and the Search for Organic Carbon on Mars". *Science* **343** (6169): 386–387. doi:10.1126/science.1249944. Retrieved January 24, 2014.

[135] Various (January 24, 2014). "Special Issue - Table of Contents - Exploring Martian Habitability". *Science* **343** (6169): 345–452. Retrieved January 24, 2014.

[136] Various (January 24, 2014). "Special Collection - Curiosity - Exploring Martian Habitability". *Science*. Retrieved January 24, 2014.

[137] Grotzinger, J.P. et al. (January 24, 2014). "A Habitable Fluvio-Lacustrine Environment at Yellowknife Bay, Gale Crater, Mars". *Science* **343** (6169). doi:10.1126/science.1242777. Retrieved January 24, 2014.

15.9 External links

- Mindat mineralogical database, largest mineral database on the Internet
- "Mineralogy Database" by David Barthelmy (2009)
- "Mineral Identification Key II" Mineralogical Society of America
- "American Mineralogist Crystal Structure Database"
- Minerals and the Origins of Life (Robert Hazen, NASA) (video, 60m, April 2014).

Chapter 16

Noble gas

The **noble gases** make a group of chemical elements with similar properties. Under standard conditions, they are all odorless, colorless, monatomic gases with very low chemical reactivity. The six noble gases that occur naturally are helium (He), neon (Ne), argon (Ar), krypton (Kr), xenon (Xe), and the radioactive radon (Rn).

For the first six periods of the periodic table, the noble gases are exactly the members of **group 18** of the periodic table. It is possible that due to relativistic effects, the group 14 element flerovium exhibits some noble-gas-like properties,[1] instead of the group 18 element ununoctium.[2] Noble gases are typically highly unreactive except when under particular extreme conditions. The inertness of noble gases makes them very suitable in applications where reactions are not wanted. For example: argon is used in lightbulbs to prevent the hot tungsten filament from oxidizing; also, helium is breathed by deep-sea divers to prevent oxygen and nitrogen toxicity.

The properties of the noble gases can be well explained by modern theories of atomic structure: their outer shell of valence electrons is considered to be "full", giving them little tendency to participate in chemical reactions, and it has been possible to prepare only a few hundred noble gas compounds. The melting and boiling points for a given noble gas are close together, differing by less than 10 °C (18 °F); that is, they are liquids over only a small temperature range.

Neon, argon, krypton, and xenon are obtained from air in an air separation unit using the methods of liquefaction of gases and fractional distillation. Helium is sourced from natural gas fields which have high concentrations of helium in the natural gas, using cryogenic gas separation techniques, and radon is usually isolated from the radioactive decay of dissolved radium, thorium, or uranium compounds (since those compounds give off alpha particles). Noble gases have several important applications in industries such as lighting, welding, and space exploration. A helium-oxygen breathing gas is often used by deep-sea divers at depths of seawater over 55 m (180 ft) to keep the diver from experiencing oxygen toxemia, the lethal effect of

high-pressure oxygen, and nitrogen narcosis, the distracting narcotic effect of the nitrogen in air beyond this partial-pressure threshold. After the risks caused by the flammability of hydrogen became apparent, it was replaced with helium in blimps and balloons.

16.1 History

Noble gas is translated from the German noun *Edelgas*, first used in 1898 by Hugo Erdmann[3] to indicate their extremely low level of reactivity. The name makes an analogy to the term "noble metals", which also have low reactivity. The noble gases have also been referred to as *inert gases*, but this label is deprecated as many noble gas compounds are now known.[4] *Rare gases* is another term that was used,[5] but this is also inaccurate because argon forms a fairly considerable part (0.94% by volume, 1.3% by mass) of the Earth's atmosphere due to decay of radioactive potassium-40.[6]

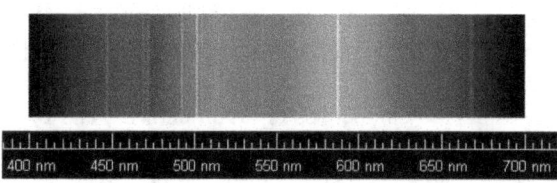

Helium was first detected in the Sun due to its characteristic spectral lines.

Pierre Janssen and Joseph Norman Lockyer discovered a new element on August 18, 1868 while looking at the chromosphere of the Sun, and named it helium after the Greek word for the Sun, ἥλιος (*ilios* or *helios*).[7] No chemical analysis was possible at the time, but helium was later found to be a noble gas. Before them, in 1784, the English chemist and physicist Henry Cavendish had discovered that air contains a small proportion of a substance less reactive than nitrogen.[8] A century later, in 1895, Lord Rayleigh discovered that samples of nitrogen from the air were of a different density than nitrogen resulting from chemical reactions. Along with Scottish scientist William Ramsay at University College, London, Lord Rayleigh theorized that the nitrogen extracted from air was mixed with another gas, leading to an experiment that successfully isolated a new element, argon, from the Greek word αργός (*argós*, "inactive").[8] With this discovery, they realized an entire class of gases was missing from the periodic table. During his search for argon, Ramsay also managed to isolate helium for the first time while heating cleveite, a mineral. In 1902, having accepted the evidence for the elements helium and argon, Dmitri Mendeleev included these noble gases as group 0 in his arrangement of the elements, which would later become the periodic table.[9]

Ramsay continued to search for these gases using the method of fractional distillation to separate liquid air into several components. In 1898, he discovered the elements krypton, neon, and xenon, and named them after the Greek words κρυπτός (*kryptós*, "hidden"), νέος (*néos*, "new"), and ξένος (*xénos*, "stranger"), respectively. Radon was first identified in 1898 by Friedrich Ernst Dorn,[10] and was named *radium emanation*, but was not considered a noble gas until 1904 when its characteristics were found to be similar to those of other noble gases.[11] Rayleigh and Ramsay received the 1904 Nobel Prizes in Physics and in Chemistry, respectively, for their discovery of the noble gases;[12][13] in the words of J. E. Cederblom, then president of the Royal Swedish Academy of Sciences, "the discovery of an entirely new group of elements, of which no single representative had been known with any certainty, is something utterly unique in the history of chemistry, being intrinsically an advance in science of peculiar significance".[13]

The discovery of the noble gases aided in the development of a general understanding of atomic structure. In 1895, French chemist Henri Moissan attempted to form a reaction between fluorine, the most electronegative element, and argon, one of the noble gases, but failed. Scientists were unable to prepare compounds of argon until the end of the 20th century, but these attempts helped to develop new theories of atomic structure. Learning from these experiments, Danish physicist Niels Bohr proposed in 1913 that the electrons in atoms are arranged in shells surrounding the nucleus, and that for all noble gases except helium the outermost shell always contains eight electrons.[11] In 1916, Gilbert N. Lewis formulated the *octet rule*, which concluded an octet of electrons in the outer shell was the most stable arrangement for any atom; this arrangement caused them to be unreactive with other elements since they did not require any more electrons to complete their outer shell.[14]

In 1962, Neil Bartlett discovered the first chemical compound of a noble gas, xenon hexafluoroplatinate.[15] Compounds of other noble gases were discovered soon after: in 1962 for radon, radon difluoride,[16] and in 1963 for kryp-

ton, krypton difluoride (KrF 2).[17] The first stable compound of argon was reported in 2000 when argon fluorohydride (HArF) was formed at a temperature of 40 K (−233.2 °C; −387.7 °F).[18]

In December 1998, scientists at the Joint Institute for Nuclear Research working in Dubna, Russia bombarded plutonium (Pu) with calcium (Ca) to produce a single atom of element 114,[19] flerovium (Fl).[20] Preliminary chemistry experiments have indicated this element may be the first superheavy element to show abnormal noble-gas-like properties, even though it is a member of group 14 on the periodic table.[21] In October 2006, scientists from the Joint Institute for Nuclear Research and Lawrence Livermore National Laboratory successfully created synthetically ununoctium (Uuo), the seventh element in group 18,[22] by bombarding californium (Cf) with calcium (Ca).[23]

16.2   Physical and atomic properties

The noble gases have weak interatomic force, and consequently have very low melting and boiling points. They are all monatomic gases under standard conditions, including the elements with larger atomic masses than many normally solid elements.[11] Helium has several unique qualities when compared with other elements: its boiling and melting points are lower than those of any other known substance; it is the only element known to exhibit superfluidity; it is the only element that cannot be solidified by cooling under standard conditions—a pressure of 25 standard atmospheres (2,500 kPa; 370 psi) must be applied at a temperature of 0.95 K (−272.200 °C; −457.960 °F) to convert it to a solid.[26] The noble gases up to xenon have multiple stable isotopes. Radon has no stable isotopes; its longest-lived isotope, ^{222}Rn, has a half-life of 3.8 days and decays to form helium and polonium, which ultimately decays to lead.[11] Melting and boiling points generally increase going down the group.

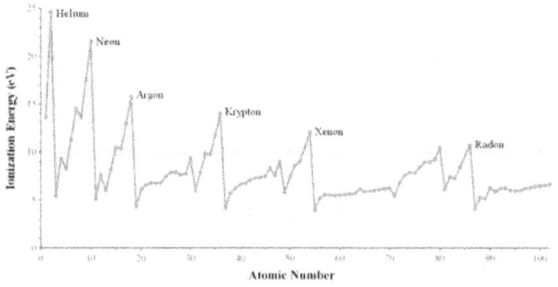

This is a plot of ionization potential versus atomic number. The noble gases, which are labeled, have the largest ionization potential for each period.

The noble gas atoms, like atoms in most groups, increase steadily in atomic radius from one period to the next due to the increasing number of electrons. The size of the atom is related to several properties. For example, the ionization potential decreases with an increasing radius because the valence electrons in the larger noble gases are farther away from the nucleus and are therefore not held as tightly together by the atom. Noble gases have the largest ionization potential among the elements of each period, which reflects the stability of their electron configuration and is related to their relative lack of chemical reactivity.[24] Some of the heavier noble gases, however, have ionization potentials small enough to be comparable to those of other elements and molecules. It was the insight that xenon has an ionization potential similar to that of the oxygen molecule that led Bartlett to attempt oxidizing xenon using platinum hexafluoride, an oxidizing agent known to be strong enough to react with oxygen.[15] Noble gases cannot accept an electron to form stable anions; that is, they have a negative electron affinity.[27]

The macroscopic physical properties of the noble gases are dominated by the weak van der Waals forces between the atoms. The attractive force increases with the size of the atom as a result of the increase in polarizability and the decrease in ionization potential. This results in systematic group trends: as one goes down group 18, the atomic radius, and with it the interatomic forces, increases, resulting in an increasing melting point, boiling point, enthalpy of vaporization, and solubility. The increase in density is due to the increase in atomic mass.[24]

The noble gases are nearly ideal gases under standard conditions, but their deviations from the ideal gas law provided important clues for the study of intermolecular interactions. The Lennard-Jones potential, often used to model intermolecular interactions, was deduced in 1924 by John Lennard-Jones from experimental data on argon before the development of quantum mechanics provided the tools for understanding intermolecular forces from first principles.[28] The theoretical analysis of these interactions became tractable because the noble gases are monatomic and the atoms spherical, which means that the interaction between the atoms is independent of direction, or isotropic.

16.3 Chemical properties

The noble gases are colorless, odorless, tasteless, and non-flammable under standard conditions. They were once labeled *group 0* in the periodic table because it was believed they had a valence of zero, meaning their atoms cannot combine with those of other elements to form compounds. However, it was later discovered some do indeed form compounds, causing this label to fall into disuse.[11]

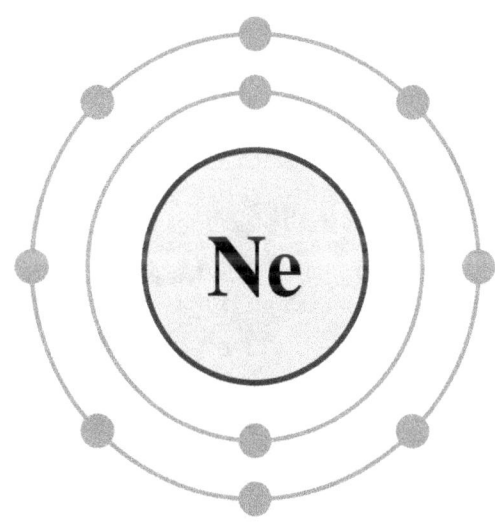

Neon, like all noble gases, has a full valence shell. Noble gases have eight electrons in their outermost shell, except in the case of helium, which has two.

16.3.1 Configuration

Main article: Noble gas configuration

Like other groups, the members of this family show patterns in its electron configuration, especially the outermost shells resulting in trends in chemical behavior:

The noble gases have full valence electron shells. Valence electrons are the outermost electrons of an atom and are normally the only electrons that participate in chemical bonding. Atoms with full valence electron shells are extremely stable and therefore do not tend to form chemical bonds and have little tendency to gain or lose electrons.[29] However, heavier noble gases such as radon are held less firmly together by electromagnetic force than lighter noble gases such as helium, making it easier to remove outer electrons from heavy noble gases.

As a result of a full shell, the noble gases can be used in conjunction with the electron configuration notation to form the *noble gas notation*. To do this, the nearest noble gas that precedes the element in question is written first, and then the electron configuration is continued from that point forward. For example, the electron notation of phosphorus is $1s^2\,2s^2\,2p^6\,3s^2\,3p^3$, while the noble gas notation is [Ne] $3s^2\,3p^3$. This more compact notation makes it easier to identify elements, and is shorter than writing out the full notation of atomic orbitals.[30]

16.3.2 Compounds

Main article: Noble gas compound
 The noble gases show extremely low chemical reactivity;

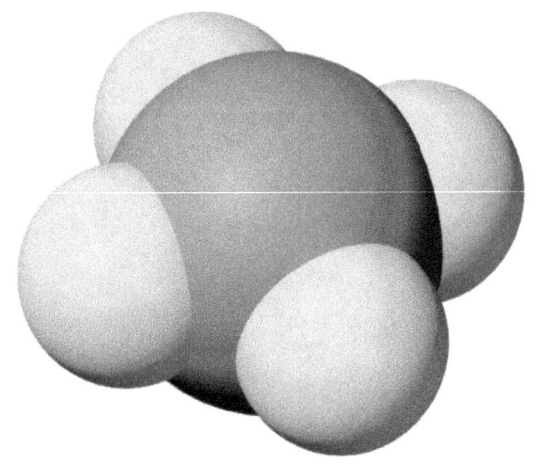

Structure of XeF
4, one of the first noble gas compounds to be discovered

consequently, only a few hundred noble gas compounds have been formed. Neutral compounds in which helium and neon are involved in chemical bonds have not been formed (although there is some theoretical evidence for a few helium compounds), while xenon, krypton, and argon have shown only minor reactivity.[31] The reactivity follows the order Ne < He < Ar < Kr < Xe < Rn.

In 1933, Linus Pauling predicted that the heavier noble gases could form compounds with fluorine and oxygen. He predicted the existence of krypton hexafluoride (KrF
6) and xenon hexafluoride (XeF
6), speculated that XeF
8 might exist as an unstable compound, and suggested xenic acid could form perxenate salts.[32][33] These predictions were shown to be generally accurate, except that XeF
8 is now thought to be both thermodynamically and kinetically unstable.[34]

Xenon compounds are the most numerous of the noble gas compounds that have been formed.[35] Most of them have the xenon atom in the oxidation state of +2, +4, +6, or +8 bonded to highly electronegative atoms such as fluorine or oxygen, as in xenon difluoride (XeF
2), xenon tetrafluoride (XeF
4), xenon hexafluoride (XeF
6), xenon tetroxide (XeO
4), and sodium perxenate (Na
4XeO
6). Xenon reacts with fluorine to form numerous xenon fluorides according to the following equations:

$$Xe + F_2 \rightarrow XeF_2$$
$$Xe + 2F_2 \rightarrow XeF_4$$
$$Xe + 3F_2 \rightarrow XeF_6$$

Some of these compounds have found use in chemical synthesis as oxidizing agents; XeF
2, in particular, is commercially available and can be used as a fluorinating agent.[36] As of 2007, about five hundred compounds of xenon bonded to other elements have been identified, including organoxenon compounds (containing xenon bonded to carbon), and xenon bonded to nitrogen, chlorine, gold, mercury, and xenon itself.[31][37] Compounds of xenon bound to boron, hydrogen, bromine, iodine, beryllium, sulphur, titanium, copper, and silver have also been observed but only at low temperatures in noble gas matrices, or in supersonic noble gas jets.[31]

In theory, radon is more reactive than xenon, and therefore should form chemical bonds more easily than xenon does. However, due to the high radioactivity and short half-life of radon isotopes, only a few fluorides and oxides of radon have been formed in practice.[38]

Krypton is less reactive than xenon, but several compounds have been reported with krypton in the oxidation state of +2.[31] Krypton difluoride is the most notable and easily characterized. Under extreme conditions, krypton reacts with fluorine to form KrF_2 according to the following equation:

$$Kr + F_2 \rightarrow KrF_2$$

Compounds in which krypton forms a single bond to nitrogen and oxygen have also been characterized,[39] but are only stable below −60 °C (−76 °F) and −90 °C (−130 °F) respectively.[31]

Krypton atoms chemically bound to other nonmetals (hydrogen, chlorine, carbon) as well as some late transition metals (copper, silver, gold) have also been observed, but only either at low temperatures in noble gas matrices, or in supersonic noble gas jets.[31] Similar conditions were used to obtain the first few compounds of argon in 2000, such as argon fluorohydride (HArF), and some bound to the late transition metals copper, silver, and gold.[31] As of 2007, no stable neutral molecules involving covalently bound helium or neon are known.[31]

The noble gases—including helium—can form stable molecular ions in the gas phase. The simplest is the helium hydride molecular ion, HeH^+, discovered in 1925.[40] Because it is composed of the two most abundant elements in the universe, hydrogen and helium, it is believed to occur naturally in the interstellar medium, although it has not been detected yet.[41] In addition to these ions, there are many known neutral excimers of the noble gases. These are compounds such as ArF and KrF that are stable only when in an excited electronic state; some of them find application in excimer lasers.

In addition to the compounds where a noble gas atom is involved in a covalent bond, noble gases also form non-covalent compounds. The clathrates, first described in 1949,[42] consist of a noble gas atom trapped within cavities of crystal lattices of certain organic and inorganic substances. The essential condition for their formation is that the guest (noble gas) atoms must be of appropriate size to fit in the cavities of the host crystal lattice. For instance, argon, krypton, and xenon form clathrates with hydroquinone, but helium and neon do not because they are too small or insufficiently polarizable to be retained.[43] Neon, argon, krypton, and xenon also form clathrate hydrates, where the noble gas is trapped in ice.[44]

An endohedral fullerene compound containing a noble gas atom

Noble gases can form endohedral fullerene compounds, in which the noble gas atom is trapped inside a fullerene molecule. In 1993, it was discovered that when C
60, a spherical molecule consisting of 60 carbon atoms, is exposed to noble gases at high pressure, complexes such as He@C
60 can be formed (the @ notation indicates He is contained inside C
60 but not covalently bound to it).[45] As of 2008, endohedral complexes with helium, neon, argon, krypton, and xenon have been obtained.[46] These compounds have found use in the study of the structure and reactivity of fullerenes by means of the nuclear magnetic resonance of the noble gas atom.[47]

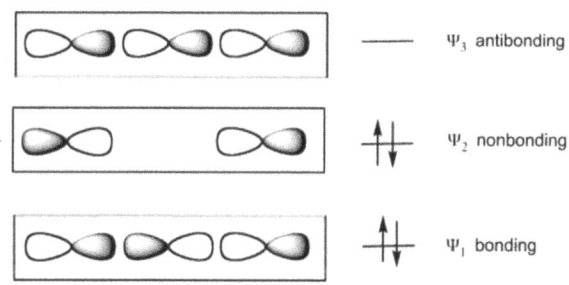

*Bonding in XeF
2 according to the 3-center-4-electron bond model*

Noble gas compounds such as xenon difluoride (XeF
2) are considered to be hypervalent because they violate the octet rule. Bonding in such compounds can be explained using a three-center four-electron bond model.[48][49] This model, first proposed in 1951, considers bonding of three collinear atoms. For example, bonding in XeF
2 is described by a set of three molecular orbitals (MOs) derived from p-orbitals on each atom. Bonding results from the combination of a filled p-orbital from Xe with one half-filled p-orbital from each F atom, resulting in a filled bonding orbital, a filled non-bonding orbital, and an empty antibonding orbital. The highest occupied molecular orbital is localized on the two terminal atoms. This represents a localization of charge which is facilitated by the high electronegativity of fluorine.[50]

The chemistry of heavier noble gases, krypton and xenon, are well established. The chemistry of the lighter ones, argon and helium, is still at an early stage, while a neon compound is yet to be identified.

16.4 Occurrence and production

The abundances of the noble gases in the universe decrease as their atomic numbers increase. Helium is the most common element in the universe after hydrogen, with a mass fraction of about 24%. Most of the helium in the universe was formed during Big Bang nucleosynthesis, but the amount of helium is steadily increasing due to the fusion of hydrogen in stellar nucleosynthesis (and, to a very slight degree, the alpha decay of heavy elements).[51][52] Abundances on Earth follow different trends; for example, helium is only the third most abundant noble gas in the atmosphere. The reason is that there is no primordial helium in the atmosphere; due to the small mass of the atom, helium cannot be retained by the Earth's gravitational field.[53] Helium on Earth comes from the alpha decay of heavy elements such as uranium and thorium found in the Earth's crust, and tends to accumulate in natural gas deposits.[53] The abundance of argon, on the other hand, is increased as a result of the beta decay of potassium-40, also found in the Earth's crust, to form argon-40, which is the most abundant isotope of argon on Earth despite being relatively rare in the Solar System. This process is the base for the potassium-argon dating method.[54] Xenon has an unexpectedly low abundance in the atmosphere, in what has been called the *missing xenon problem*; one theory is that the missing xenon may be trapped in minerals inside the Earth's crust.[55] After the discovery of xenon dioxide, a research showed that Xe can substitute for Si in the quartz.[56] Radon is formed in the lithosphere as from the alpha decay of radium. It can seep into buildings through cracks in their foundation and accumulate in areas that are not well ventilated. Due to its high radioactivity, radon presents a significant health hazard; it is implicated in an estimated 21,000 lung cancer deaths per year in the United States alone.[57]

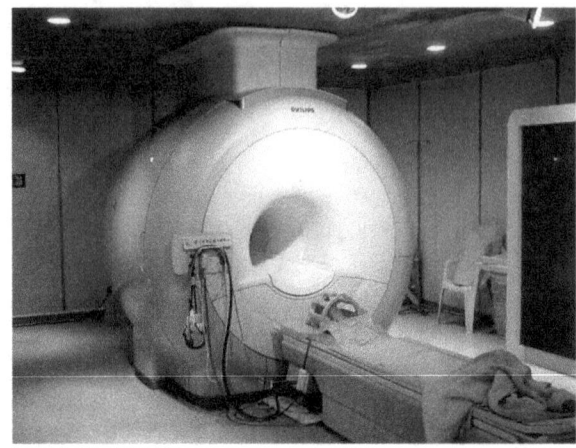

Liquid helium is used to cool the superconducting magnets in modern MRI scanners

For large-scale use, helium is extracted by fractional distillation from natural gas, which can contain up to 7% helium.[62]

Neon, argon, krypton, and xenon are obtained from air using the methods of liquefaction of gases, to convert elements to a liquid state, and fractional distillation, to separate mixtures into component parts. Helium is typically produced by separating it from natural gas, and radon is isolated from the radioactive decay of radium compounds.[11] The prices of the noble gases are influenced by their natural abundance, with argon being the cheapest and xenon the most expensive. As an example, the table to the right lists the 2004 prices in the United States for laboratory quantities of each gas.

16.5 Applications

Noble gases have very low boiling and melting points, which makes them useful as cryogenic refrigerants.[63] In particular, liquid helium, which boils at 4.2 K (−268.95 °C; −452.11 °F), is used for superconducting magnets, such as those needed in nuclear magnetic resonance imaging and nuclear magnetic resonance.[64] Liquid neon, although it does not reach temperatures as low as liquid helium, also finds use in cryogenics because it has over 40 times more refrigerating capacity than liquid helium and over three times more than liquid hydrogen.[60]

Helium is used as a component of breathing gases to replace nitrogen, due its low solubility in fluids, especially in lipids. Gases are absorbed by the blood and body tissues when under pressure like in scuba diving, which causes an anesthetic effect known as nitrogen narcosis.[65] Due to its reduced solubility, little helium is taken into cell membranes, and when helium is used to replace part of the breathing mixtures, such as in trimix or heliox, a decrease in the narcotic effect of the gas at depth is obtained.[66] Helium's reduced solubility offers further advantages for the condition known as decompression sickness, or *the bends*.[11][67] The reduced amount of dissolved gas in the body means that fewer gas bubbles form during the decrease in pressure of the ascent. Another noble gas, argon, is considered the best option for use as a drysuit inflation gas for scuba diving.[68] Helium is also used as filling gas in nuclear fuel rods for nuclear reactors.[69]

Goodyear Blimp

Since the *Hindenburg* disaster in 1937,[70] helium has replaced hydrogen as a lifting gas in blimps and balloons due to its lightness and incombustibility, despite an 8.6%[71] decrease in buoyancy.[11]

In many applications, the noble gases are used to provide an inert atmosphere. Argon is used in the synthesis of air-sensitive compounds that are sensitive to nitrogen. Solid argon is also used for the study of very unstable compounds, such as reactive intermediates, by trapping them in an inert matrix at very low temperatures.[72] Helium is used as the carrier medium in gas chromatography, as a filler gas for thermometers, and in devices for measuring radiation, such as the Geiger counter and the bubble chamber.[61] Helium and argon are both commonly used to shield welding arcs and the surrounding base metal from the atmosphere during welding and cutting, as well as in other metallurgical processes and in the production of silicon for the semiconductor industry.[60]

15,000-watt xenon short-arc lamp used in IMAX projectors

Noble gases are commonly used in lighting because of their lack of chemical reactivity. Argon, mixed with nitrogen, is used as a filler gas for incandescent light bulbs.[60] Krypton is used in high-performance light bulbs, which have higher color temperatures and greater efficiency, because it reduces the rate of evaporation of the filament more than argon; halogen lamps, in particular, use krypton mixed with small amounts of compounds of iodine or bromine.[60] The noble gases glow in distinctive colors when used inside gas-discharge lamps, such as "neon lights". These lights are called after neon but often contain other gases and phosphors, which add various hues to the orange-red color of neon. Xenon is commonly used in xenon arc lamps which, due to their nearly continuous spectrum that resembles daylight, find application in film projectors and as automobile headlamps.[60]

The noble gases are used in excimer lasers, which are based on short-lived electronically excited molecules known as excimers. The excimers used for lasers may be noble gas dimers such as Ar_2, Kr_2 or Xe_2, or more commonly, the noble gas is combined with a halogen in excimers such as ArF, KrF, XeF, or XeCl. These lasers produce ultraviolet light which, due to its short wavelength (193 nm for ArF and 248 nm for KrF), allows for high-precision imaging. Excimer lasers have many industrial, medical, and scientific applications. They are used for microlithography and microfabrication, which are essential for integrated circuit manufacture, and for laser surgery, including laser angioplasty and eye surgery.[73]

Some noble gases have direct application in medicine. Helium is sometimes used to improve the ease of breathing of asthma sufferers.[60] Xenon is used as an anesthetic because of its high solubility in lipids, which makes it more potent than the usual nitrous oxide, and because it is readily eliminated from the body, resulting in faster recovery.[74] Xenon finds application in medical imaging of the lungs through hyperpolarized MRI.[75] Radon, which is highly radioactive and is only available in minute amounts, is used in radiotherapy.[11]

16.6 Discharge color

The color of gas discharge emission depends on several factors, including the following:[76]

- discharge parameters (local value of current density and electric field, temperature, etc. – note the color variation along the discharge in the top row);

- gas purity (even small fraction of certain gases can affect color);

- material of the discharge tube envelope – note suppression of the UV and blue components in the bottom-row tubes made of thick household glass.

16.7 See also

- Noble gas (data page), for extended tables of physical properties.

- Noble metal, for metals that are resistant to corrosion or oxidation.

- Inert gas, for any gas that is not reactive under normal circumstances.

- Industrial gas

- Neutronium

- Noble gas configuration

16.8 Notes

[1] "Flerov laboratory of nuclear reactions" (PDF). JINR. Retrieved 2009-08-08.

[2] Nash, Clinton S. (2005). "Atomic and Molecular Properties of Elements 112, 114, and 118". *J. Phys. Chem. A* **109** (15): 3493–3500. doi:10.1021/jp050736o. PMID 16833687.

[3] Renouf, Edward (1901). "Noble gases". *Science* **13** (320): 268–270. Bibcode:1901Sci....13..268R. doi:10.1126/science.13.320.268.

[4] Ojima 2002, p. 30

[5] Ojima 2002, p. 4

[6] "argon". *Encyclopædia Britannica*. 2008.

[7] *Oxford English Dictionary* (1989), s.v. "helium". Retrieved December 16, 2006, from Oxford English Dictionary Online. Also, from quotation there: Thomson, W. (1872). *Rep. Brit. Assoc.* xcix: "Frankland and Lockyer find the yellow prominences to give a very decided bright line not far from D, but hitherto not identified with any terrestrial flame. It seems to indicate a new substance, which they propose to call Helium."

[8] Ojima 2002, p. 1

[9] Mendeleev 1903, p. 497

[10] Partington, J. R. (1957). "Discovery of Radon". *Nature* **179** (4566): 912. Bibcode:1957Natur.179..912P. doi:10.1038/179912a0.

[11] "Noble Gas". *Encyclopædia Britannica*. 2008.

[12] Cederblom, J. E. (1904). "The Nobel Prize in Physics 1904 Presentation Speech".

[13] Cederblom, J. E. (1904). "The Nobel Prize in Chemistry 1904 Presentation Speech".

[14] Gillespie, R. J.; Robinson, E. A. (2007). "Gilbert N. Lewis and the chemical bond: the electron pair and the octet rule from 1916 to the present day". *J Comput Chem* **28** (1): 87–97. doi:10.1002/jcc.20545. PMID 17109437.

[15] Bartlett, N. (1962). "Xenon hexafluoroplatinate $Xe^+[PtF_6]^-$". *Proceedings of the Chemical Society* (6): 218. doi:10.1039/PS9620000197.

[16] Fields, Paul R.; Stein, Lawrence; Zirin, Moshe H. (1962). "Radon Fluoride". *Journal of the American Chemical Society* **84** (21): 4164–4165. doi:10.1021/ja00880a048.

[17] Grosse, A. V.; Kirschenbaum, A. D.; Streng, A. G.; Streng, L. V. (1963). "Krypton Tetrafluoride: Preparation and Some Properties". *Science* **139** (3559): 1047–1048. Bibcode:1963Sci...139.1047G. doi:10.1126/science.139.3559.1047. PMID 17812982.

[18] Khriachtchev, Leonid; Pettersson, Mika; Runeberg, Nino; Lundell, Jan; Räsänen, Markku (2000). "A stable argon compound". *Nature* **406** (6798): 874–876. doi:10.1038/35022551. PMID 10972285.

[19] Oganessian, Yu. Ts.; Utyonkov, V.; Lobanov, Yu.; Abdullin, F.; Polyakov, A. et al. (1999). "Synthesis of Superheavy Nuclei in the ^{48}Ca + ^{244}Pu Reaction". *Physical Review Letters* (American Physical Society) **83** (16): 3154–3157. Bibcode:1999PhRvL..83.3154O. doi:10.1103/PhysRevLett.83.3154.

[20] Woods, Michael (2003-05-06). "Chemical element No. 110 finally gets a name—darmstadtium". *Pittsburgh Post-Gazette*. Retrieved 2008-06-26.

[21] "Gas Phase Chemistry of Superheavy Elements" (PDF). Texas A&M University. Retrieved 2008-05-31.

[22] Robert C. Barber, Paul J. Karol, Hiromichi Nakahara, Emanuele Vardaci, and Erich W. Vogt (2011). "Discovery of the elements with atomic numbers greater than or equal to 113 (IUPAC Technical Report)*" (PDF). *Pure Appl. Chem.* (IUPAC) **83** (7). doi:10.1515/ci.2011.33.5.25b. Retrieved 2014-05-30.

[23] Oganessian, Yu. Ts.; Utyonkov, V.; Lobanov, Yu.; Abdullin, F.; Polyakov, A., et al. (2006). "Synthesis of the isotopes of elements 118 and 116 in the 249Cf and 245Cm + 48Ca fusion reactions". *Physical Review C* **74** (4): 44602. Bibcode:2006PhRvC..74d4602O. doi:10.1103/PhysRevC.74.044602.

[24] Greenwood 1997, p. 891

[25] Allen, Leland C. (1989). "Electronegativity is the average one-electron energy of the valence-shell electrons in ground-state free atoms". *Journal of the American Chemical Society* **111** (25): 9003–9014. doi:10.1021/ja00207a003.

[26] "Solid Helium". University of Alberta. Retrieved 2008-06-22.

[27] Wheeler, John C. (1997). "Electron Affinities of the Alkaline Earth Metals and the Sign Convention for Electron Affinity". *Journal of Chemical Education* **74**: 123–127. Bibcode:1997JChEd..74..123W. doi:10.1021/ed074p123.; Kalcher, Josef; Sax, Alexander F. (1994). "Gas Phase Stabilities of Small Anions: Theory and Experiment in Cooperation". *Chemical Reviews* **94** (8): 2291–2318. doi:10.1021/cr00032a004.

[28] Mott, N. F. (1955). "John Edward Lennard-Jones. 1894–1954". *Biographical Memoirs of Fellows of the Royal Society* **1**: 175–184. doi:10.1098/rsbm.1955.0013.

[29] Ojima 2002, p. 35

[30] CliffsNotes 2007, p. 15

[31] Grochala, Wojciech (2007). "Atypical compounds of gases, which have been called noble". *Chemical Society Reviews* **36** (10): 1632–1655. doi:10.1039/b702109g. PMID 17721587.

[32] Pauling, Linus (1933). "The Formulas of Antimonic Acid and the Antimonates". *Journal of the American Chemical Society* **55** (5): 1895–1900. doi:10.1021/ja01332a016.

[33] Holloway 1968

[34] Seppelt, Konrad (1979). "Recent developments in the Chemistry of Some Electronegative Elements". *Accounts of Chemical Research* **12** (6): 211–216. doi:10.1021/ar50138a004.

[35] Moody, G. J. (1974). "A Decade of Xenon Chemistry". *Journal of Chemical Education* **51** (10): 628–630. Bibcode:1974JChEd..51..628M. doi:10.1021/ed051p628. Retrieved 2007-10-16.

[36] Zupan, Marko; Iskra, Jernej; Stavber, Stojan (1998). "Fluorination with XeF$_2$. 44. Effect of Geometry and Heteroatom on the Regioselectivity of Fluorine Introduction into an Aromatic Ring". *J. Org. Chem* **63** (3): 878–880. doi:10.1021/jo971496e. PMID 11672087.

[37] Harding 2002, pp. 90–99

[38] .Avrorin, V. V.; Krasikova, R. N.; Nefedov, V. D.; Toropova, M. A. (1982). "The Chemistry of Radon". *Russian Chemical Review* **51** (1): 12–20. Bibcode:1982RuCRv..51...12A. doi:10.1070/RC1982v051n01ABEH002787.

[39] Lehmann, J (2002). "The chemistry of krypton". *Coordination Chemistry Reviews*. 233–234: 1–39. doi:10.1016/S0010-8545(02)00202-3.

[40] Hogness, T. R.; Lunn, E. G. (1925). "The Ionization of Hydrogen by Electron Impact as Interpreted by Positive Ray Analysis". *Physical Review* **26**: 44–55. Bibcode:1925PhRv...26...44H. doi:10.1103/PhysRev.26.44.

[41] Fernandez, J.; Martin, F. (2007). "Photoionization of the HeH$_2^+$ molecular ion". *J. Phys. B: At. Mol. Opt. Phys* **40** (12): 2471–2480. Bibcode:2007JPhB...40.2471F. doi:10.1088/0953-4075/40/12/020.

[42] H. M. Powell and M. Guter (1949). "An Inert Gas Compound". *Nature* **164** (4162): 240–241. Bibcode:1949Natur.164..240P. doi:10.1038/164240b0.

[43] Greenwood 1997, p. 893

[44] Dyadin, Yuri A. et al. (1999). "Clathrate hydrates of hydrogen and neon". *Mendeleev Communications* **9** (5): 209–210. doi:10.1070/MC1999v009n05ABEH001104.

[45] Saunders, M.; Jiménez-Vázquez, H. A.; Cross, R. J.; Poreda, R. J. (1993). "Stable compounds of helium and neon. He@C60 and Ne@C60". *Science* **259** (5100): 1428–1430. Bibcode:1993Sci...259.1428S. doi:10.1126/science.259.5100.1428. PMID 17801275.

[46] Saunders, Martin; Jimenez-Vazquez, Hugo A.; Cross, R. James; Mroczkowski, Stanley; Gross, Michael L.; Giblin, Daryl E.; Poreda, Robert J. (1994). "Incorporation of helium, neon, argon, krypton, and xenon into fullerenes using high pressure". *J. Am. Chem. Soc.* **116** (5): 2193–2194. doi:10.1021/ja00084a089.

[47] Frunzi, Michael; Cross, R. Jame; Saunders, Martin (2007). "Effect of Xenon on Fullerene Reactions". *Journal of the American Chemical Society* **129** (43): 13343–6. doi:10.1021/ja075568n. PMID 17924634.

[48] Greenwood 1997, p. 897

[49] Weinhold 2005, pp. 275–306

[50] Pimentel, G. C. (1951). "The Bonding of Trihalide and Bifluoride Ions by the Molecular Orbital Method". *The Journal of Chemical Physics* **19** (4): 446–448. Bibcode:1951JChPh..19..446P. doi:10.1063/1.1748245.

[51] Weiss, Achim. "Elements of the past: Big Bang Nucleosynthesis and observation". Max Planck Institute for Gravitational Physics. Retrieved 2008-06-23.

[52] Coc, A. et al. (2004). "Updated Big Bang Nucleosynthesis confronted to WMAP observations and to the Abundance of Light Elements". *Astrophysical Journal* **600** (2): 544–552. arXiv:astro-ph/0309480. Bibcode:2004ApJ...600..544C. doi:10.1086/380121.

[53] Morrison, P.; Pine, J. (1955). "Radiogenic Origin of the Helium Isotopes in Rock". *Annals of the New York Academy of Sciences* **62** (3): 71–92. Bibcode:1955NYASA..62...71M. doi:10.1111/j.1749-6632.1955.tb35366.x.

[54] Scherer, Alexandra (2007-01-16). "^{40}Ar/^{39}Ar dating and errors". Technische Universität Bergakademie Freiberg. Archived from the original on 2007-10-14. Retrieved 2008-06-26.

[55] Sanloup, Chrystèle; Schmidt, Burkhard C. et al. (2005). "Retention of Xenon in Quartz and Earth's Missing Xenon". *Science* **310** (5751): 1174–1177. Bibcode:2005Sci...310.1174S. doi:10.1126/science.1119070. PMID 16293758.

[56] Tyler Irving (May 2011). "Xenon Dioxide May Solve One of Earth's Mysteries". L'Actualité chimique canadienne (Canadian Chemical News). Retrieved 2012-05-18.

[57] "A Citizen's Guide to Radon". U.S. Environmental Protection Agency. 2007-11-26. Retrieved 2008-06-26.

[58] Lodders, Katharina (July 10, 2003). "Solar System Abundances and Condensation Temperatures of the Elements" (PDF). *The Astrophysical Journal* (The American Astronomical Society) **591** (2): 1220–1247. Bibcode:2003ApJ...591.1220L. doi:10.1086/375492.

[59] "The Atmosphere". National Weather Service. Retrieved 2008-06-01.

[60] Häussinger, Peter; Glatthaar, Reinhard; Rhode, Wilhelm; Kick, Helmut; Benkmann, Christian; Weber, Josef; Wunschel, Hans-Jörg; Stenke, Viktor; Leicht, Edith; Stenger, Hermann (2002). "Noble gases". *Ullmann's Encyclopedia of Industrial Chemistry*. Wiley. doi:10.1002/14356007.a17_485.

[61] Hwang, Shuen-Chen; Lein, Robert D.; Morgan, Daniel A. (2005). "Noble Gases". *Kirk Othmer Encyclopedia of Chemical Technology*. Wiley. pp. 343–383. doi:10.1002/0471238961.0701190508230114.a01.

[62] Winter, Mark (2008). "Helium: the essentials". University of Sheffield. Retrieved 2008-07-14.

[63] "Neon". *Encarta*. 2008.

[64] Zhang, C. J.; Zhou, X. T.; Yang, L. (1992). "Demountable coaxial gas-cooled current leads for MRI superconducting magnets". *Magnetics, IEEE Transactions on* (IEEE) **28** (1): 957–959. Bibcode:1992ITM....28..957Z. doi:10.1109/20.120038.

[65] Fowler, B; Ackles, K. N.; Porlier, G. (1985). "Effects of inert gas narcosis on behavior—a critical review". *Undersea Biomed. Res.* **12** (4): 369–402. ISSN 0093-5387. OCLC 2068005. PMID 4082343. Retrieved 2008-04-08.

[66] Bennett 1998, p. 176

[67] Vann, R. D. (ed) (1989). "The Physiological Basis of Decompression". *38th Undersea and Hyperbaric Medical Society Workshop.* 75(Phys)6-1-89: 437. Retrieved 2008-05-31.

[68] Maiken, Eric (2004-08-01). "Why Argon?". Decompression. Retrieved 2008-06-26.

[69] Horhoianu, G; Ionescu, D.V; Olteanu, G (1999). "Thermal behaviour of CANDU type fuel rods during steady state and transient operating conditions". *Annals of Nuclear Energy* **26** (16): 1437–1445. doi:10.1016/S0306-4549(99)00022-5.

[70] "Disaster Ascribed to Gas by Experts". *The New York Times.* 1937-05-07. p. 1.

[71] Freudenrich, Craig (2008). "How Blimps Work". HowStuffWorks. Retrieved 2008-07-03.

[72] Dunkin, I. R. (1980). "The matrix isolation technique and its application to organic chemistry". *Chem. Soc. Rev.* **9**: 1–23. doi:10.1039/CS9800900001.

[73] Basting, Dirk; Marowsky, Gerd (2005). *Excimer Laser Technology.* Springer. ISBN 3-540-20056-8.

[74] Sanders, Robert D.; Ma, Daqing; Maze, Mervyn (2005). "Xenon: elemental anaesthesia in clinical practice". *British Medical Bulletin* **71** (1): 115–135. doi:10.1093/bmb/ldh034. PMID 15728132.

[75] Albert, M. S.; Balamore, D. (1998). "Development of hyperpolarized noble gas MRI". *Nuclear Instruments and Methods in Physics Research A* **402** (2–3): 441–453. Bibcode:1998NIMPA.402..441A. doi:10.1016/S0168-9002(97)00888-7. PMID 11543065.

[76] Ray, Sidney F. (1999). *Scientific photography and applied imaging.* Focal Press. pp. 383–384. ISBN 0-240-51323-1.

16.9   References

- Bennett, Peter B.; Elliott, David H. (1998). *The Physiology and Medicine of Diving.* SPCK Publishing. ISBN 0-7020-2410-4.

- Bobrow Test Preparation Services (2007-12-05). *CliffsAP Chemistry.* CliffsNotes. ISBN 0-470-13500-X.

- Greenwood, N. N.; Earnshaw, A. (1997). *Chemistry of the Elements* (2nd ed.). Oxford: Butterworth-Heinemann. ISBN 0-7506-3365-4.

- Harding, Charlie J.; Janes, Rob (2002). *Elements of the P Block.* Royal Society of Chemistry. ISBN 0-85404-690-9.

- Holloway, John H. (1968). *Noble-Gas Chemistry.* London: Methuen Publishing. ISBN 0-412-21100-9.

- Mendeleev, D. (1902–1903). *Osnovy Khimii (The Principles of Chemistry)* (in Russian) (7th ed.).

- Ojima, Minoru; Podosek, Frank A. (2002). *Noble Gas Geochemistry.* Cambridge University Press. ISBN 0-521-80366-7.

- Weinhold, F.; Landis, C. (2005). *Valency and bonding.* Cambridge University Press. ISBN 0-521-83128-8.

- Scerri, Eric R. (2007). *The Periodic Table, Its Story and Its Significance.* Oxford University Press. ISBN 0-19-530573-6.

Chapter 17

Noble metal

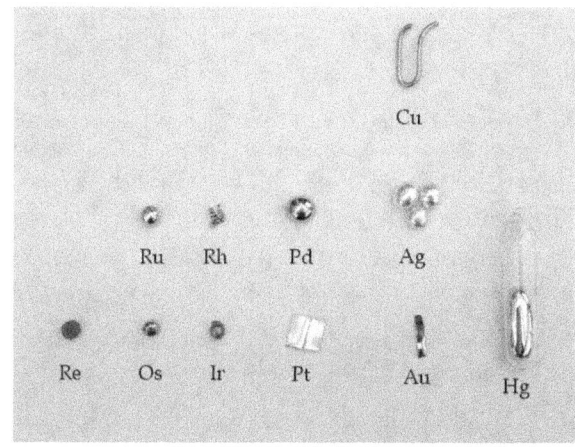

A collection of the noble metals, including copper, rhenium and mercury, which are included by some definitions. These are arranged according to their position in the periodic table.

In chemistry, the **noble metals** are metals that are resistant to corrosion and oxidation in moist air (unlike most base metals). The short list of chemically noble metals (those elements upon which almost all chemists agree) comprises ruthenium, rhodium, palladium, silver, osmium, iridium, platinum, and gold.[1]

More inclusive lists include one or more of mercury,[2][3][4] rhenium[5] or copper as noble metals. On the other hand, titanium, niobium, and tantalum are not included as noble metals although they are very resistant to corrosion.

While the noble metals tend to be valuable – due to both their rarity in the Earth's crust and their usefulness in areas like metallurgy, high technology, and ornamentation (jewelry, art, sacred objects, etc.) – the terms "noble metal" and "precious metal" are not synonymous.

The term *noble metal* can be traced back to at least the late 14th century[6] and has slightly different meanings in different fields of study and application. Only in atomic physics is there a strict definition. For this reason there are many quite different lists of "noble metals".

In addition to this term's function as a compound noun, there are circumstances where "noble" is used as an adjective for the noun "metal". A "galvanic series" is a hierarchy of metals (or other electrically conductive materials, including composites and semimetals) that runs from *noble* to *active*, and allows one to predict how materials will interact in the environment used to generate the series. In this sense of the word, graphite is more noble than silver and the relative nobility of many materials is highly dependent upon context, as for aluminium and stainless steel in conditions of varying pH.[7]

17.1   Properties

Palladium, platinum, gold and mercury can be dissolved in aqua regia, a highly concentrated mixture of hydrochloric acid and nitric acid, but iridium and silver cannot. Silver is, however, soluble in pure nitric acid. Ruthenium can be dissolved in aqua regia only when in the presence of oxygen, while rhodium must be in a fine pulverized form. Niobium and tantalum are resistant to all acids, including aqua regia. [8]

17.2   Physics

In physics, the definition of a noble metal is most strict. It requires that the d-bands of the electronic structure are filled. From this perspective, only copper, silver and gold are noble metals, as all d-like bands are filled and do not cross the Fermi level.[9] However, d-hybridized bands do cross the Fermi level to a minimal extent. For platinum, two d-bands cross the Fermi level, changing its chemical behaviour such that it can function as a catalyst. The difference in reactivity can easily be seen during the preparation of clean metal surfaces in an ultra-high vacuum: surfaces of "physically defined" noble metals (e.g., gold) are easy to clean and keep clean for a long time, while those of platinum or palladium, for example, are covered by carbon monoxide very quickly.[10]

17.3 Electrochemistry

Metallic elements, including noble and several non-noble metals (noble metals bolded):[11]

The columns *group* and *period* denote its position in the periodic table, hence electronic configuration. The simplified *reaction*s, listed in the next column, can also be read in detail from the Pourbaix diagrams of the considered element in water. Finally the column *potential* indicates the electric potential of the element measured against a Standard hydrogen electrode. All missing elements in this table are either not metals or have a negative standard potential.

Antimony is considered to be a metalloid and thus cannot be a noble metal. Also chemists and metallurgists consider copper and bismuth not noble metals because they easily oxidize due to the reaction O
2 + 2 H
2O + 4 e
- ⇌ 4 OH−
(aq) + 0.40 V which is possible in moist air.

The film of silver is due to its high sensitivity to hydrogen sulfide. Chemically patina is caused by an attack of oxygen in wet air and by CO
2 afterward.[8] On the other hand, rhenium coated mirrors are said to be very durable,[8] although rhenium and technetium are said to tarnish slowly in moist atmosphere.[13]

17.4 See also

- Base metal
- Minor metals
- Precious metal

17.5 References

[1] A. Holleman, N. Wiberg, "Lehrbuch der Anorganischen Chemie", de Gruyter, 1985, 33. edition, p. 1486

[2] Die Adresse für Ausbildung, Studium und Beruf

[3] "Dictionary of Mining, Mineral, and Related Terms", Compiled by the American Geological Institute, 2nd edition, 1997

[4] Scoullos, M.J., Vonkeman, G.H., Thornton, I., Makuch, Z., "Mercury - Cadmium - Lead: Handbook for Sustainable Heavy Metals Policy and Regulation",Series: Environment & Policy, Vol. 31, Springer-Verlag, 2002

[5] The New Encyclopædia Britannica, 15th edition, Vol. VII, 1976

[6] http://dictionary.reference.com/browse/noble+metal

[7] Everett Collier, "The Boatowner's Guide to Corrosion", International Marine Publishing, 2001, p. 21

[8] A. Holleman, N. Wiberg, "Inorganic Chemistry", Academic Press, 2001

[9] Hüger, E.; Osuch, K. (2005). "Making a noble metal of Pd". *EPL (Europhysics Letters)* **71** (2): 276. Bibcode:2005EL.....71..276H. doi:10.1209/epl/i2005-10075-5.

[10] S. Fuchs, T.Hahn, H.G. Lintz, "The oxidation of carbon monoxide by oxygen over platinum, palladium and rhodium catalysts from 10^{-10} to 1 bar", Chemical engineering and processing, 1994, V 33(5), pp. 363-369

[11] D. R. Lidle editor, "CRC Handbook of Chemistry and Physics", 86th edition, 2005

[12] A. J. Bard, "Encyclopedia of the Electrochemistry of the Elements", Vol. IV, Marcel Dekker Inc., 1975

[13] R. D. Peack, "The Chemistry of Technetium and Rhenium", Elsevier, 1966

Notes

- R. R. Brooks, "Noble metals and biological systems: their role in Medicine, Mineral Exploration, and the Environment", CRC Press, 1992

17.6 External links

- noble metal - chemistry Encyclopædia Britannica, online edition

- To see which bands cross the Fermi level, the Fermi surfaces of almost all the metals can be found at the Fermi Surface Database

- The following article might also clarify the correlation between *band structure* and the term *noble metal*: Hüger, E.; Osuch, K. (2005). "Making a noble metal of Pd". *EPL (Europhysics Letters)* **71** (2): 276. Bibcode:2005EL.....71..276H. doi:10.1209/epl/i2005-10075-5.

Chapter 18

Densities of the elements (data page)

Main article: Density

18.1 Density, solid phase

18.2 Density, liquid phase

18.3 Density, gas phase

18.4 Notes

- The suggested values for solid densities refer to "near room temperature" by default.

- The suggested values for liquid densities refer to "at the melting point" by default.

18.5 References

18.5.1 WEL

As quoted at http://www.webelements.com/ from these sources:

- A.M. James and M.P. Lord in *Macmillan's Chemical and Physical Data*, Macmillan, London, UK, 1992.

- D.R. Lide, (ed.) in *Chemical Rubber Company handbook of chemistry and physics*, CRC Press, Boca Raton, Florida, USA, 77th edition, 1996.

- J.A. Dean (ed) in *Lange's Handbook of Chemistry*, McGraw-Hill, New York, USA, 14th edition, 1992.

- G.W.C. Kaye and T.H. Laby in *Tables of physical and chemical constants*, Longman, London, UK, 15th edition, 1993.

18.5.2 CRC

As quoted from various sources in an online version of:

- David R. Lide (ed), *CRC Handbook of Chemistry and Physics, 84th Edition.* CRC Press. Boca Raton, Florida, 2003; Section 4, Properties of the Elements and Inorganic Compounds; Physical Constants of Inorganic Compounds

18.5.3 CR2

- David R. Lide (ed), *CRC Handbook of Chemistry and Physics, 84th Edition.* CRC Press. Boca Raton, Florida, 2003; Section 4, Properties of the Elements and Inorganic Compounds; Density of Molten Elements and Representative Salts

18.5.4 LNG

As quoted from an online version of:

- J.A. Dean (ed), *Lange's Handbook of Chemistry* (15th Edition), McGraw-Hill, 1999; Section 3; Table 3.2 Physical Constants of Inorganic Compounds

18.5.5 VDW

The following molar volumes and densities for the majority of the gaseous elements were calculated from the van der Waals equation of state, using the quoted values of the van der Waals constants. The source for the van der Waals constants and for the literature densities was: R. C. Weast (Ed.), *Handbook of Chemistry and Physics (53rd Edn.)*, Cleveland:Chemical Rubber Co., 1972.

18.5.6   Other

- KCH: Kuchling, Horst, *Taschenbuch der Physik*, 13. Auflage, Verlag Harri Deutsch, Thun und Frankfurt/Main, German edition, 1991. ISBN 3-8171-1020-0

- ICT: Washburn, E.W. (ed.), *International Critical Tables of Numerical Data, Physics, Chemistry and Technology* (1926–1930; 2003 Knovel online version)

  (a) Gray and Ramsay, *Proceedings of the Royal Society* (London). A. Mathematical and Physical Sciences. **84**: 536; (1911)

18.6   See also

Chapter 19

List of chemical element name etymologies

This is the **list of etymologies for all chemical element names**.

19.1 Table

19.2 See also

- List of chemical elements naming controversies
- Naming of elements

19.3 References

[1] "Aluminum in Online Etymological Dictionary, accessed March 9, 2010". Etymonline.com. Retrieved 2011-01-02.

[2] "Antimony | Define Antimony at Dictionary.com". Dictionary.reference.com. Retrieved 2011-01-02.

[3] "Online Etymology Dictionary". Etymonline.com. Retrieved 2011-01-02.

[4] **Antimony,**

- LSJ, *s.v.*, vocalisation, spelling, and declension vary; Endlich; Celsus, 6.6.6 ff; Pliny *Natural History* 33.33; Lewis and Short: *Latin Dictionary. OED*, s. *antimony.*
- *stimmi* is used by the Attic tragic poets of the 5th century BC. Later Greeks also used στίβι *(stibi)*, which is written in Latin by Celsus and Pliny in the first century AD. Pliny also names *stimi* [*sic*], *larbaris, alabaster* (Greek: ἀλάβαστρον), "very common platyophthalmos (πλατυόφθαλμος)", "wide-eye" in Greek (the description refers to the effects of the cosmetic). In Egyptian hieroglyphics, *mśdmt*; the vowels are uncertain, but in Coptic and according to an Arabic tradition, it is pronounced *mesdemet* (Albright; Sarton, quotes Meyerhof, the translator). In Arabic, the word for powdered Stibnite is kuhl.

[5] "Online Etymology Dictionary". Etymonline.com. Retrieved 2011-01-02.

[6] **Astatine**, An earlier name for astatine was **alabamine (Ab)**

[7] "Online Etymology Dictionary". Etymonline.com. Retrieved 2011-01-02.

[8] "Berkeley – Wiktionary". En.wiktionary.org. 2010-04-01. Retrieved 2011-01-02.

[9] At one time, beryllium was called **glucinium**, which is from Greek γλυκύς (glykys), which means "sweet", due to the sweet taste of its salts.

[10] "The American Heritage Dictionary of the English Language: beryl". Houghton Mifflin Company. 2000. Retrieved 2008-09-18.

[11] "Beryl in Online Etymological Dictionary, accessed March 9, 2010". Etymonline.com. Retrieved 2011-01-02.

[12] *see* **Naming controversy** *below*

[13] "Online Etymology Dictionary". Etymonline.com. Retrieved 2011-01-02.

[14] Gemoll W, Vretska K (1997). *Griechisch-Deutsches Schul- und Handwörterbuch (reek–German dictionary)*, 9th ed. öbvhpt. *ISBN 3-209-00108-1.*

[15] "Online Etymology Dictionary". Etymonline.com. Retrieved 2011-01-02.

[16] "Online Etymology Dictionary". Etymonline.com. Retrieved 2011-01-02.

[17] "Online Etymology Dictionary". Etymonline.com. Retrieved 2011-01-02.

[18] Some elements (particularly ancient elements) were associated with Greek (or Roman or others) gods or people, on Greek mythology (or other mythology), and with planets (or others in solar system), such as Mercury (mythology) – Mercury (planet) – Mercury (element), etc.
Also, astrological symbols (for the planets) (particularly ancient elements) also often used same each ancient alchemical symbols (for the element or its metal).

[19] Mike Campbell. "Meaning, Origin and History of the Name Ceres". Behind the Name. Retrieved 2011-01-02.

[20] "Online Etymology Dictionary". Etymonline.com. Retrieved 2011-01-02.

[21] "Online Etymology Dictionary". Etymonline.com. Retrieved 2011-01-02.

[22] "Online Etymology Dictionary". Etymonline.com. Retrieved 2011-01-02.

[23] "Online Etymology Dictionary". Etymonline.com. Retrieved 2011-01-02.

[24] **Darmstadtium**, some humorous scientists suggested the name **Policium**, because 110 is the emergency telephone number for the German police.

[25] Previous to discovery of some unknown elements, Prof. Dmitri Mendeleev predicted and described most of them appropriately properties, and fill the gaps in the table, on the basis of them position in his Periodic table. The properties of 4 predicted elements, **Eka-boron (Eb)**, **Eka-aluminium (El)**, **Eka-manganese (Em)**, and **Eka-silicon (Es)**, proved to be good predictors of Scandium, Gallium, Technetium and Germanium, respectively. The prefix *eka-*, from the Sanskrit, means "one" (places down from the known element in table), and is sometimes used in discussions about undiscovered elements, such as, Untriennium was referred into **Eka-actinium**. *see also: Mendeleev's predicted elements*

[26] Derived from a Latin masculine genitive.

[27] **Gold** in Sanskrit is जुवल *jval*; in Greek, χρυσός *(khrusos)*; in Chinese, ⬚ *(jīn)*.

[28] "Online Etymology Dictionary". Etymonline.com. Retrieved 2011-01-02.

[29] "Online Etymology Dictionary". Etymonline.com. Retrieved 2011-01-02.

[30] "Online Etymology Dictionary". Etymonline.com. Retrieved 2011-01-02.

[31] **Iron**, Benvéniste 1969 cit. dep

[32] **Lead**, Lead was mentioned in the Book of Exodus. Alchemists believed lead was the oldest metal and associated the element with Saturn.

[33] **Mendelevium**, "Mendeleyev" commonly spelt as Mendeleev, Mendeléef, or Mendelejeff, and first name sometimes spelt as Dmitry or Dmitriy

[34] **Mercury** – The Indian alchemy called *Rassayana*, which means "the way of mercury".

[35] **Neodymium** is frequently misspelled as *neodynium*

[36] "Online Etymology Dictionary". Etymonline.com. Retrieved 2011-01-02.

[37] Nickel in Online Etymological Dictionary. Retrieved December 12, 2008.

[38] "Online Etymology Dictionary". Etymonline.com. Retrieved 2011-01-02.

[39] **Nitrogen**, Pure gas is inert enough that Antoine Lavoisier referred to it as "Azote", means "without life", so this term has become the French for **Nitrogen** and later spread out to many other languages.

[40] "Online Etymology Dictionary". Etymonline.com. Retrieved 2011-01-02.

[41] "Online Etymology Dictionary". Etymonline.com. Retrieved 2011-01-02.

[42] "Online Etymology Dictionary". Etymonline.com. Retrieved 2011-01-02.

[43] "Online Etymology Dictionary". Etymonline.com. Retrieved 2011-01-02.

[44] Woods, Ian (2004). *The Elements: Platinum*. The Elements. Benchmark Books. ISBN 978-0-7614-1550-3.

[45] "Online Etymology Dictionary". Etymonline.com. Retrieved 2011-01-02.

[46] "Online Etymology Dictionary". Etymonline.com. Retrieved 2011-01-02.

[47] "Online Etymology Dictionary". Etymonline.com. Retrieved 2011-01-02.

[48] Harper, Douglas. "Potassium". *Online Etymology Dictionary*.

[49] The ancient Greek derivation of *Prometheus* from the Greek πρό *pro* ("before") + μανθάνω *manthano* ("learn"), thus "forethought", which engendered a contrasting brother Epimetheus, was a folk etymology; it is succinctly expressed in Servius' commentary on Virgil, *Eclogue* 6.42: "*Prometheus vir prudentissimus fuit, unde etiam Prometheus dictus est ἀπὸ τῆς προμηθείας, id est a providentia.*" Modern scientific linguistics suggests that the name derived from the Proto-Indo-European root that also produces the Vedic *pra math*, "to steal," hence *pramathyu-s*, "thief", cognate with "Prometheus", the thief of fire. The Vedic myth of fire's theft by Mātariśvan is an analog to the Greek account. *Pramantha* was the tool used to create fire. See: Fortson, Benjamin W. (2004). *Indo-European Language and Culture: An Introduction*. Blackwell Publishing, p. 27.; Williamson (2004), *The Longing for Myth in Germany*, 214–15; Dougherty, Carol (2006). *Prometheus*. p. 4.

[50] **Protactinium**; In 1913, Kasimir Fajans and Otto H. Göhring identified and named element 91 **Brevium**, from Latin *brevis*, which means "brief, short". The name was changed to "Protactinium" in 1918 and shortened to **Protactinium** in 1949.

[51] "Online Etymology Dictionary". Etymonline.com. Retrieved 2011-01-02.

[52] "Online Etymology Dictionary". Etymonline.com. Retrieved 2011-01-02.

[53] "soda". *soda*. Online Etymology Dictionary.

[54] In medieval Europe, **Sodanum** is Latin name of "a compound of sodium".

[55] *The Concise Dictionary of English Etymology – Sulfur*. Retrieved 2011-01-02.

[56] *The Origin of Medical Terms – Sulphur*. 2007-02-17. Retrieved 2011-01-02.

[57] *Magill's Survey of Science – Sulfur*. 2007-06-07. Retrieved 2011-01-02.

[58] "Online Etymology Dictionary". Etymonline.com. Retrieved 2011-01-02.

[59] "Sulfur | Define Sulfur at Dictionary.com". Dictionary.reference.com. Retrieved 2011-01-02.

[60] Harper, Douglas. "Sulfur". *Online Etymology Dictionary*.

[61] "Online Etymology Dictionary". Etymonline.com. Retrieved 2011-01-02.

[62] "Thule in Wordnik, accessed March 9, 2010". Wordnik.com. Retrieved 2011-01-02.

[63] **Tin** – The American Heritage Dictionary

[64] Česky (2010-07-09). "Zink – Wiktionary". En.wiktionary.org. Retrieved 2011-01-02.

[65] Pearse, Roger (2002-09-16). "Syriac Literature". Retrieved 2008-02-11.

[66] "Online Etymology Dictionary". Etymonline.com. Retrieved 2011-01-02.

19.4 Further reading

- Eric Scerri, *The Periodic System, Its Story and Its Significance*, Oxford University Press, New York, 2007.

Chapter 20

Abundance of the chemical elements

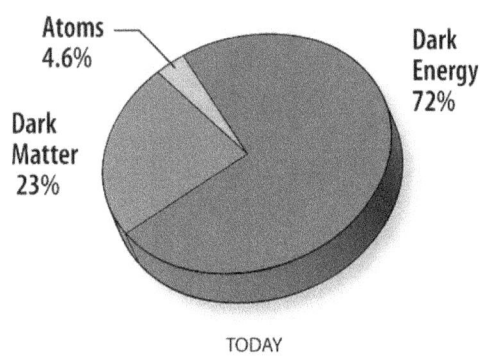

TODO: TODAY

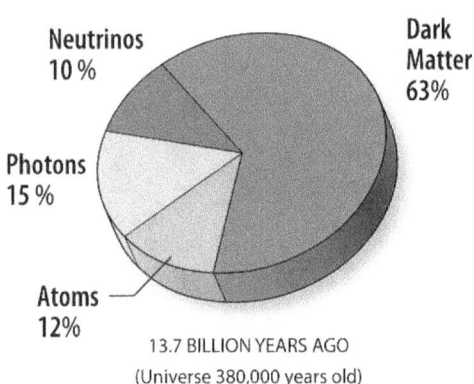

13.7 BILLION YEARS AGO
(Universe 380,000 years old)

Estimated proportions of matter, dark matter and dark energy in the universe. Only the fraction of the mass and energy in the universe labeled "atoms" is composed of chemical elements.

The **abundance** of a chemical element measures how common is the element relative to all other elements in a given environment. Abundance is measured in one of three ways: by the mass-fraction (the same as weight fraction); by the mole-fraction (fraction of atoms by numerical count, or sometimes fraction of molecules in gases); or by the volume-fraction. Volume-fraction is a common abundance measure in mixed gases such as planetary atmospheres, and is similar in value to molecular mole-fraction for gas mixtures at relatively low densities and pressures, and ideal gas mixtures. Most abundance values in this article are given as mass-fractions.

For example, the abundance of oxygen in pure water can be measured in two ways: the *mass fraction* is about 89%, because that is the fraction of water's mass which is oxygen. However, the *mole-fraction* is 33.3333...% because only 1 atom of 3 in water, H_2O, is oxygen.

As another example, looking at the *mass-fraction* abundance of hydrogen and helium in both the Universe as a whole and in the atmospheres of gas-giant planets such as Jupiter, it is 74% for hydrogen and 23-25% for helium; while the *(atomic) mole-fraction* for hydrogen is 92%, and for helium is 8%, in these environments. Changing the given environment to Jupiter's outer atmosphere, where hydrogen is diatomic while helium is not, changes the *molecular* mole-fraction (fraction of total gas molecules), as well as the fraction of atmosphere by volume, of hydrogen to about 86%, and of helium to 13%.[Note 1]

20.1 Abundance of elements in the Universe

See also: Stellar population, Cosmochemistry and Astrochemistry

The elements – that is, ordinary (baryonic) matter made of protons, neutrons, and electrons, are only a small part of the content of the Universe. Cosmological observations suggest that only 4.6% of the universe's energy (including the mass contributed by energy, $E = mc^2 \leftrightarrow m = E/c^2$) comprises the visible baryonic matter that constitutes stars, planets, and living beings. The rest is made up of dark energy (72%) and dark matter (23%).[2] These are forms of matter and energy believed to exist on the basis of scientific theory and observational deductions, but they have not been directly observed and their nature is not well understood.

Most standard (baryonic) matter is found in stars and interstellar clouds, in the form of atoms or ions (plasma), although it can be found in degenerate forms in extreme astrophysical settings, such as the high densities inside white dwarfs and neutron stars.

Hydrogen is the most abundant element in the Universe; helium is second. However, after this, the rank of abundance does not continue to correspond to the atomic number; oxygen has abundance rank 3, but atomic number 8. All others are substantially less common.

The abundance of the lightest elements is well predicted by the standard cosmological model, since they were mostly produced shortly (i.e., within a few hundred seconds) after the Big Bang, in a process known as Big Bang nucleosynthesis. Heavier elements were mostly produced much later, inside of stars.

Hydrogen and helium are estimated to make up roughly 74% and 24% of all baryonic matter in the universe respectively. Despite comprising only a very small fraction of the universe, the remaining "heavy elements" can greatly influence astronomical phenomena. Only about 2% (by mass) of the Milky Way galaxy's disk is composed of heavy elements.

These other elements are generated by stellar processes.[3][4][5] In astronomy, a "metal" is any element other than hydrogen or helium. This distinction is significant because hydrogen and helium are the only elements that were produced in significant quantities in the Big Bang. Thus, the metallicity of a galaxy or other object is an indication of stellar activity, after the Big Bang.

The following graph (note log scale) shows abundance of elements in our solar system. The table shows the twelve most common elements in our galaxy (estimated spectroscopically), as measured in parts per million, by mass.[1] Nearby galaxies that have evolved along similar lines have a corresponding enrichment of elements heavier than hydrogen and helium. The more distant galaxies are being viewed as they appeared in the past, so their abundances of elements appear closer to the primordial mixture. Since physical laws and processes are uniform throughout the universe, however, it is expected that these galaxies will likewise have evolved similar abundances of elements.

The abundance of elements in the Solar System (see graph) is in keeping with their origin from the Big Bang and nucleosynthesis in a number of progenitor supernova stars. Very abundant hydrogen and helium are products of the Big Bang, while the next three elements are rare since they had little time to form in the Big Bang and are not made in stars (they are, however, produced in small quantities by breakup of heavier elements in interstellar dust, as a result of impact by cosmic rays).

Beginning with carbon, elements have been produced in stars by buildup from alpha particles (helium nuclei), re-

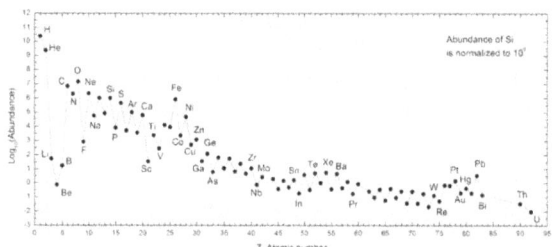

*Estimated abundances of the chemical **elements in the Solar system**. Hydrogen and helium are most common, from the Big Bang. The next three elements (Li, Be, B) are rare because they are poorly synthesized in the Big Bang and also in stars. The two general trends in the remaining stellar-produced elements are: (1) an alternation of abundance in elements as they have even or odd atomic numbers (the Oddo-Harkins rule), and (2) a general decrease in abundance, as elements become heavier. Iron is especially common because it represents the minimum energy nuclide that can be made by fusion of helium in supernovae.*

sulting in an alternatingly larger abundance of elements with even atomic numbers (these are also more stable). The effect of odd-numbered chemical elements generally being more rare in the universe was empirically noticed in 1914, and is known as the Oddo-Harkins rule.

Periodic table showing the cosmogenic origin of each element

Cosmogenesis: In general, such elements up to iron are made in large stars in the process of becoming supernovae. Iron-56 is particularly common, since it is the most stable element that can easily be made from alpha particles (being a product of decay of radioactive nickel-56, ultimately made from 14 helium nuclei). Elements heavier than iron are made in energy-absorbing processes in large stars, and their abundance in the universe (and on Earth) generally decreases with increasing atomic number.

20.1.1 Elemental abundance and nuclear binding energy

Loose correlations have been observed between estimated elemental abundances in the universe and the nuclear binding energy curve. Roughly speaking, the relative stability of various atomic isotopes has exerted a strong influence on the relative abundance of elements formed in the Big Bang, and during the development of the universe thereafter. [7] See the article about nucleosynthesis for the explanation on how certain nuclear fusion processes in stars (such as carbon burning, etc.) create the elements heavier than hydrogen and helium.

A further observed peculiarity is the jagged alternation between relative abundance and scarcity of adjacent atomic numbers in the elemental abundance curve, and a similar pattern of energy levels in the nuclear binding energy curve. This alternation is caused by the higher relative binding energy (corresponding to relative stability) of even atomic numbers compared to odd atomic numbers, and is explained by the Pauli Exclusion Principle.[8] The semi-empirical mass formula (SEMF), also called **Weizsäcker's formula** or the **Bethe-Weizsäcker mass formula**, gives a theoretical explanation of the overall shape of the curve of nuclear binding energy.[9]

20.2 Abundance of elements in the Earth

See also: Earth § Chemical composition

The Earth formed from the same cloud of matter that formed the Sun, but the planets acquired different compositions during the formation and evolution of the solar system. In turn, the natural history of the Earth caused parts of this planet to have differing concentrations of the elements.

The mass of the Earth is approximately 5.98×10^{24} kg. In bulk, by mass, it is composed mostly of iron (32.1%), oxygen (30.1%), silicon (15.1%), magnesium (13.9%), sulfur (2.9%), nickel (1.8%), calcium (1.5%), and aluminium (1.4%); with the remaining 1.2% consisting of trace amounts of other elements.[10]

The bulk composition of the Earth by elemental-mass is roughly similar to the gross composition of the solar system, with the major differences being that Earth is missing a great deal of the volatile elements hydrogen, helium, neon, and nitrogen, as well as carbon which has been lost as volatile hydrocarbons. The remaining elemental composition is roughly typical of the "rocky" inner planets, which formed in the thermal zone where solar heat drove volatile compounds into space. The Earth retains oxygen as the second-largest component of its mass (and largest atomic-fraction), mainly from this element being retained in silicate minerals which have a very high melting point and low vapor pressure.

20.2.1 Earth's detailed bulk (total) elemental abundance in table form

Click "show" at right, to show more numerical values in a full table. Note that these are ordered by atom-fraction abundance (right-most column), not mass-abundance.

An estimate[11] of the elemental abundances in the total mass of the Earth. Note that numbers are estimates, and they will vary depending on source and method of estimation. Order of magnitude of data can roughly be relied upon. ppb (atoms) is parts per billion, meaning that is the number of atoms of a given element in every billion atoms in the Earth.

20.2.2 Earth's crustal elemental abundance

Main article: Abundance of elements in Earth's crust
The mass-abundance of the nine most abundant elements in

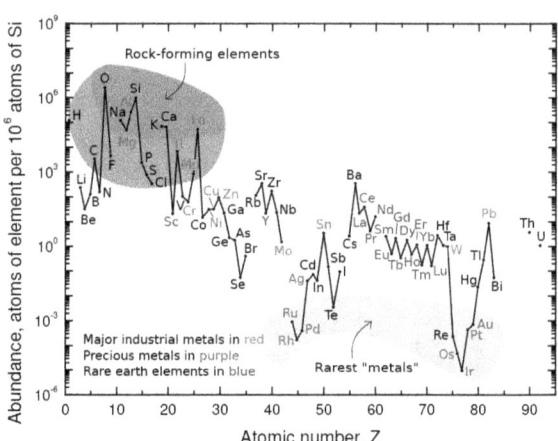

Abundance (atom fraction) of the chemical elements in Earth's upper continental crust as a function of atomic number. The rarest elements in the crust (shown in yellow) are the most dense. They were further rarefied in the crust by being siderophile (iron-loving) elements, in the Goldschmidt classification of elements. Siderophiles were depleted by being relocated into the Earth's core. Their abundance in meteoroid materials is relatively higher. Additionally, tellurium and selenium have been depleted from the crust due to formation of volatile hydrides.

the Earth's crust (see main article above) is approximately: oxygen 46%, silicon 28%, aluminum 8.2%, iron 5.6%, calcium 4.2%, sodium 2.5%, magnesium 2.4%, potassium 2.0%, and titanium 0.61%. Other elements occur at less than 0.15%.

The graph at left illustrates the relative atomic-abundance of the chemical elements in Earth's upper continental crust, which is relatively accessible for measurements and estimation. Many of the elements shown in the graph are classified into (partially overlapping) categories:

1. rock-forming elements (major elements in green field, and minor elements in light green field);

2. rare earth elements (lanthanides, La-Lu, and Y; labeled in blue);

3. major industrial metals (global production $> \sim 3 \times 10^7$ kg/year; labeled in red);

4. precious metals (labeled in purple);

5. the nine rarest "metals" — the six platinum group elements plus Au, Re, and Te (a metalloid) — in the yellow field.

Note that there are two breaks where the unstable elements technetium (atomic number: 43) and promethium (atomic number: 61) would be. These are both extremely rare, since on Earth they are only produced through the spontaneous fission of very heavy radioactive elements (for example, uranium, thorium, or the trace amounts of plutonium that exist in uranium ores), or by the interaction of certain other elements with cosmic rays. Both of the first two of these elements have been identified spectroscopically in the atmospheres of stars, where they are produced by ongoing nucleosynthetic processes. There are also breaks where the six noble gases would be, since they are not chemically bound in the Earth's crust, and they are only generated by decay chains from radioactive elements and are therefore extremely rare there. The twelve naturally occurring very rare, highly radioactive elements (polonium, astatine, francium, radium, actinium, protactinium, neptunium, plutonium, americium, curium, berkelium, and californium) are not included, since any of these elements that were present at the formation of the Earth have decayed away eons ago, and their quantity today is negligible and is only produced from the radioactive decay of uranium and thorium.

Oxygen and silicon are notably quite common elements in the crust. They have frequently combined with each other to form common silicate minerals.

Crustal rare-earth elemental abundance

"Rare" earth elements is a historical misnomer. The persistence of the term reflects unfamiliarity rather than true rarity. The more abundant rare earth elements are each similar in crustal concentration to commonplace industrial metals such as chromium, nickel, copper, zinc, molybdenum, tin, tungsten, or lead. The two least abundant rare earth elements (thulium and lutetium) are nearly 200 times more common than gold. However, in contrast to the ordinary base and precious metals, rare earth elements have very little tendency to become concentrated in exploitable ore deposits. Consequently, most of the world's supply of rare earth elements comes from only a handful of sources. Furthermore, the rare earth metals are all quite chemically similar to each other, and they are thus quite difficult to separate into quantities of the pure elements.

Differences in abundances of individual rare earth elements in the upper continental crust of the Earth represent the superposition of two effects, one nuclear and one geochemical. First, the rare earth elements with even atomic numbers ($_{58}$Ce, $_{60}$Nd, ...) have greater cosmic and terrestrial abundances than the adjacent rare earth elements with odd atomic numbers ($_{57}$La, $_{59}$Pr, ...). Second, the lighter rare earth elements are more incompatible (because they have larger ionic radii) and therefore more strongly concentrated in the continental crust than the heavier rare earth elements. In most rare earth ore deposits, the first four rare earth elements – lanthanum, cerium, praseodymium, and neodymium – constitute 80% to 99% of the total amount of rare earth metal that can be found in the ore.

20.2.3   Earth's mantle elemental abundance

Main article: Mantle (geology)

The mass-abundance of the eight most abundant elements in the Earth's crust (see main article above) is approximately: oxygen 45%, magnesium 23%, silicon 22%, iron 5.8%, calcium 2.3%, aluminum 2.2%, sodium 0.3%, potassium 0.3%.

The mantle differs in elemental composition from the crust in having a great deal more magnesium and significantly more iron, while having much less aluminum and sodium.

20.2.4   Earth's core elemental abundance

Due to mass segregation, the core of the Earth is believed to be primarily composed of iron (88.8%), with smaller amounts of nickel (5.8%), sulfur (4.5%), and less than 1% trace elements.[10]

20.2.5   Oceanic elemental abundance

For a complete list of the abundance of elements in the ocean, see Abundances of the elements (data page)#Sea water.

20.2.6   Atmospheric elemental abundance

The order of elements by volume-fraction (which is approximately molecular mole-fraction) in the atmosphere is nitrogen (78.1%), oxygen (20.9%),[12] argon (0.96%), followed by (in uncertain order) carbon and hydrogen because water vapor and carbon dioxide, which represent most of these two elements in the air, are variable components. Sulfur, phosphorus, and all other elements are present in significantly lower proportions.

According to the abundance curve graph (above right), argon, a significant if not major component of the atmosphere, does not appear in the crust at all. This is because the atmosphere has a far smaller mass than the crust, so argon remaining in the crust contributes little to mass-fraction there, while at the same time buildup of argon in the atmosphere has become large enough to be significant.

20.2.7   Abundances of elements in urban soils

For a complete list of the abundance of elements in urban soils, see Abundances of the elements (data page)#Urban soils.

Reasons for establishing

In the time of life existence, or at least in the time of the existence of human beings, the abundances of chemical elements within the Earth's crust have not been changed dramatically due to migration and concentration processes except the radioactive elements and their decay products and also noble gases. However, significant changes took place in the distribution of chemical elements. But within the biosphere not only the distribution, but also the abundances of elements have changed during the last centuries.

The rate of a number of geochemical changes taking place during the last decades in the biosphere has become catastrophically high. Such changes are often connected with human activities. To study these changes and to make better informed decisions on diminishing their adverse impact on living organisms, and especially on people, it is necessary to estimate the contemporary abundances of chemical elements in geochemical systems susceptible to the highest anthropogenic impact and having a significant effect on the development and existence of living organisms. One of such systems is the soil of urban landscapes. Settlements occupy less than 10% of the land area, but virtually the entire population of the planet lives within them. The main deposing medium in cities is soil, which ecological and geochemical conditions largely determine the life safety of citizens. So that, one of the priority tasks of the environmen-

tal geochemistry is to establish the average contents (abundances) of chemical elements in the soils of settlements.

Methods and results

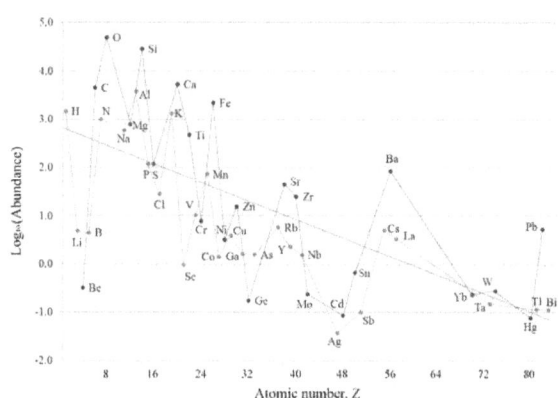

The half-logarithm graph of the abundances of chemical elements in urban soils. (Alekseenko and Alekseenko, 2014) Chemical elements are distributed extremely irregularly in urban soils, what is also typical for the Earth's crust. Nine elements (O, Si, Ca, C, Al, Fe, H, K, N) make the 97.68% of the considering geochemical system (urban soils). These elements and also Zn, Sr, Zr, Ba, and Pb essentially prevail over the trend line. Part of them could be considered as "inherited" from the concentrations in the Earth's crust; another part is explained as a result of intensive technogenic activity in the cities.

The geochemical properties of urban soils from more than 300 cities in Europe, Asia, Africa, Australia, and America were evaluated.[13] In each settlement samples were collected uniformly throughout the territory, covering residential, industrial, recreational and other urban areas. The sampling was carried out directly from the soil surface and specifically traversed pits, ditches and wells from the upper soil horizon. The number of samples in each locality ranged from 30 to 1000. The published data and the materials kindly provided by a number of geochemists were also incorporated into the research. Considering the great importance of the defined contents, quantitative and quantitative emission spectral, gravimetric, X-ray fluorescence, and partly neutron activation analyses were carried out in parallel approximately in the samples. In a volume of 3–5% of the total number of samples, sampling and analyses of the inner and external controls were conducted. Calculation of random errors and systematic errors allowed to consider the sampling and analytical laboratory work as good.

For every city the average concentrations of elements in soils were determined. To avoid the errors related to unequal number of samples, each city was then represented by only one "averaged" sample. The statistical processing of this data allowed to calculate the average concentrations, which can be considered as the abundances of chemical elements in urban soils.

This graph illustrates the relative abundance of the chemical elements in urban soils, irregularly decreasing in proportion with the increasing atomic masses. Therefore, the evolution of organisms in this system occurs in the conditions of light elements' prevalence. It corresponds to the conditions of the evolutional development of the living matter on the Earth. The irregularity of element decreasing may be somewhat connected, as stated above, with the technogenic influence. The Oddo-Harkins rule, which holds that elements with an even atomic number are more common than elements with an odd atomic number, is saved in the urban soils but with some technogenic complications. Among the considered abundances the even-atomic elements make 91.48% of the urban soils mass. As it is in the Earth's crust, elements with the 4-divisible atomic masses of leading isotope (oxygen — 16, silicon — 28, calcium — 40, carbon — 12, iron — 56) are sharply prevailing in urban soils.

In spite of significant differences between abundances of several elements in urban soils and those values calculated for the Earth's crust, the general patterns of element abundances in urban soils repeat those in the Earth's crust in a great measure. The established abundances of chemical elements in urban soils can be considered as their geochemical (ecological and geochemical) characteristic, reflecting the combined impact of technogenic and natural processes occurring during certain time period (the end of the 20th century–beginning of the 21st century). With the development of science and technology the abundances may gradually change. The rate of these changes is still poorly predictable. The abundances of chemical elements may be used during various ecological and geochemical studies.

20.3 Human body elemental abundance

Main article: Chemical makeup of the human body

By mass, human cells consist of 65–90% water (H_2O), and a significant portion of the remainder is composed of carbon-containing organic molecules. Oxygen therefore contributes a majority of a human body's mass, followed by carbon. Almost 99% of the mass of the human body is made up of six elements: oxygen, carbon, hydrogen, nitrogen, calcium, and phosphorus. The next 0.75% is made up of the next five elements: potassium, sulfur, chlorine, sodium, and magnesium. Only 17 elements are known for certain to be necessary to human life, with one additional element (fluorine) thought to be helpful for tooth enamel strength. A few more trace elements may play some role

in the health of mammals. Boron and silicon are notably necessary for plants but have uncertain roles in animals. The elements aluminium and silicon, although very common in the earth's crust, are conspicuously rare in the human body.[14]

Periodic table highlighting nutritional elements[15]

Periodic table highlighting dietary elements

20.4 See also

- Abundances of the elements (data page)

- Natural abundance (isotopic abundance)

- Primordial nuclide

20.5 References

20.5.1 Footnotes

[1] Croswell, Ken (February 1996). *Alchemy of the Heavens*. Anchor. ISBN 0-385-47214-5.

[2] WMAP- Content of the Universe

[3] Suess, Hans; Urey, Harold (1956). "Abundances of the Elements". *Reviews of Modern Physics* **28**: 53. Bibcode:1956RvMP...28...53S. doi:10.1103/RevModPhys.28.53.

[4] Cameron, A.G.W. (1973). "Abundances of the elements in the solar system". *Space Science Reviews* **15**: 121. Bibcode:1973SSRv...15..121C. doi:10.1007/BF00172440.

[5] Anders, E; Ebihara, M (1982). "Solar-system abundances of the elements". *Geochimica et Cosmochimica Acta* **46** (11): 2363. Bibcode:1982GeCoA..46.2363A. doi:10.1016/0016-7037(82)90208-3.

[6] Arnett, David (1996). *Supernovae and Nucleosynthesis* (First ed.). Princeton, New Jersey: Princeton University Press. ISBN 0-691-01147-8. OCLC 33162440.

[7] Bell, Jerry A.; GenChem Editorial/Writing Team (2005). "Chapter 3: Origin of Atoms". *Chemistry: a project of the American Chemical Society*. New York [u.a.]: Freeman. pp. 191–193. ISBN 978-0-7167-3126-9. Correlations between abundance and nuclear binding energy [Subsection title]

[8] Bell, Jerry A.; GenChem Editorial/Writing Team (2005). "Chapter 3: Origin of Atoms". *Chemistry: a project of the American Chemical Society*. New York [u.a.]: Freeman. p. 192. ISBN 978-0-7167-3126-9. The higher abundance of elements with even atomic numbers [Subsection title]

[9] Bailey, David. "Semi-empirical Nuclear Mass Formula". *PHY357: Strings & Binding Energy.* University of Toronto. Retrieved 2011-03-31.

[10] Morgan, J. W.; Anders, E. (1980). "Chemical composition of Earth, Venus, and Mercury". *Proceedings of the National Academy of Sciences* **77** (12): 6973–6977. Bibcode:1980PNAS...77.6973M. doi:10.1073/pnas.77.12.6973. PMC 350422. PMID 16592930.

[11] William F McDonough The composition of the Earth. quake.mit.edu

[12] Zimmer, Carl (3 October 2013). "Earth's Oxygen: A Mystery Easy to Take for Granted". *New York Times.* Retrieved 3 October 2013.

[13] Vladimir Alekseenko; Alexey Alekseenko (2014). "The abundances of chemical elements in urban soils". *Journal of Geochemical Exploration* (Elsevier B.V.) **147**: 245–249. doi:10.1016/j.gexplo.2014.08.003. ISSN 0375-6742.

[14] Table data from Chang, Raymond (2007). *Chemistry, Ninth Edition.* McGraw-Hill. p. 52. ISBN 0-07-110595-6.

[15] Ultratrace minerals. Authors: Nielsen, Forrest H. USDA, ARS Source: Modern nutrition in health and disease / editors, Maurice E. Shils ... et al.. Baltimore : Williams & Wilkins, c1999., p. 283-303. Issue Date: 1999 URI:

20.5.2 Notes

[1] Below Jupiter's outer atmosphere, volume fractions are significantly different from mole fractions due to high temperatures (ionization and disproportionation) and high density where the Ideal Gas Law is inapplicable.

20.5.3 Notations

- http://geopubs.wr.usgs.gov/fact-sheet/fs087-02/

- http://imagine.gsfc.nasa.gov/docs/dict_ei.html

20.6   External links

- List of elements in order of abundance in the Earth's crust (only correct for the twenty most common elements)

- Cosmic abundance of the elements and nucleosynthesis

- webelements.com Lists of elemental abundances for the Universe, Sun, meteorites, Earth, ocean, streamwater

Chapter 21

Timeline of chemical element discoveries

The discovery of the elements known to exist today is presented here in chronological order. The elements are listed generally in the order in which each was first defined as the pure element, as the exact date of discovery of most elements cannot be accurately defined.

Given is each element's name, atomic number, year of first report, name of the discoverer, and some notes related to the discovery.

21.1 Table

21.2 Unrecorded discoveries

21.3 Recorded discoveries

21.4 Unconfirmed discoveries

21.5 Graphics

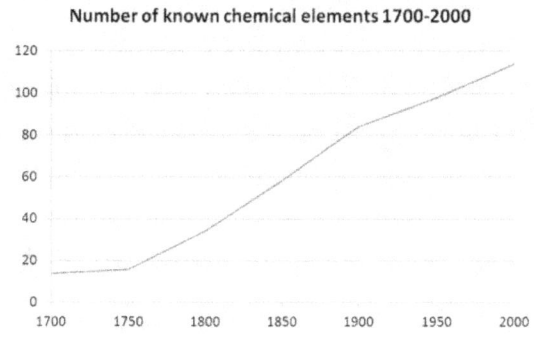

Development in discovery

21.6 See also

- History of the periodic table
- Periodic table
- The Mystery of Matter: Search for the Elements (2015 PBS film)

21.7 References

[1] "Copper History". Rameria.com. Retrieved 2008-09-12.

[2] CSA – Discovery Guides, A Brief History of Copper

[3] "The History of Lead – Part 3". Lead.org.au. Retrieved 2008-09-12.

[4] 47 Silver

[5] "Silver Facts – Periodic Table of the Elements". Chemistry.about.com. Retrieved 2008-09-12.

[6] "26 Iron". Elements.vanderkrogt.net. Retrieved 2008-09-12.

[7] Weeks, Mary Elvira; Leichester, Henry M. (1968). "Elements Known to the Ancients". *Discovery of the Elements*. Easton, PA: Journal of Chemical Education. pp. 29–40. ISBN 0-7661-3872-0. LCCCN 68-15217.

[8] "Notes on the Significance of the First Persian Empire in World History". Courses.wcupa.edu. Retrieved 2008-09-12.

[9] "History of Carbon and Carbon Materials – Center for Applied Energy Research – University of Kentucky". Caer.uky.edu. Retrieved 2008-09-12.

[10] "Chinese made first use of diamond". BBC News. 17 May 2005. Retrieved 2007-03-21.

[11] Ferchault de Réaumur, R-A (1722). *L'art de convertir le fer forgé en acier, et l'art d'adoucir le fer fondu, ou de faire des ouvrages de fer fondu aussi finis que le fer forgé (English translation from 1956)*. Paris, Chicago.

[12] Senese, Fred (September 9, 2009). "Who discovered carbon?". Frostburg State University. Retrieved 2007-11-24.

[13] "50 Tin". Elements.vanderkrogt.net. Retrieved 2008-09-12.

[14] "History of Metals". Neon.mems.cmu.edu. Retrieved 2008-09-12.

[15] "Sulfur History". Georgiagulfsulfur.com. Retrieved 2008-09-12.

[16] "Mercury and the environment — Basic facts". *Environment Canada, Federal Government of Canada.* 2004. Retrieved 2008-03-27.

[17] Craddock, P. T. et al. (1983), "Zinc production in medieval India", *World Archaeology* **15** (2), Industrial Archaeology, p. 13

[18] "30 Zinc". Elements.vanderkrogt.net. Retrieved 2008-09-12.

[19] Weeks, Mary Elvira (1933). "III. Some Eighteenth-Century Metals". *The Discovery of the Elements.* Easton, PA: Journal of Chemical Education. p. 21. ISBN 0-7661-3872-0.

[20] "Arsenic". Los Alamos National Laboratory. Retrieved 3 March 2013.

[21] SHORTLAND, A. J. (2006-11-01). "APPLICATION OF LEAD ISOTOPE ANALYSIS TO A WIDE RANGE OF LATE BRONZE AGE EGYPTIAN MATERIALS". *Archaeometry* **48** (4): 657–669. doi:10.1111/j.1475-4754.2006.00279.x.

[22] "15 Phosphorus". Elements.vanderkrogt.net. Retrieved 2008-09-12.

[23] "27 Cobalt". Elements.vanderkrogt.net. Retrieved 2008-09-12.

[24] "78 Platinum". Elements.vanderkrogt.net. Retrieved 2008-09-12.

[25] "28 Nickel". Elements.vanderkrogt.net. Retrieved 2008-09-12.

[26] "Bismuth". Los Alamos National Laboratory. Retrieved 3 March 2013.

[27] "12 Magnesium". Elements.vanderkrogt.net. Retrieved 2008-09-12.

[28] "01 Hydrogen". Elements.vanderkrogt.net. Retrieved 2008-09-12.

[29] Andrews, A. C. (1968). "Oxygen". In Clifford A. Hampel. *The Encyclopedia of the Chemical Elements.* New York: Reinhold Book Corporation. p. 272. LCCN 68-29938.

[30] "08 Oxygen". Elements.vanderkrogt.net. Retrieved 2008-09-12.

[31] Cook, Gerhard A.; Lauer, Carol M. (1968). "Oxygen". In Clifford A. Hampel. *The Encyclopedia of the Chemical Elements.* New York: Reinhold Book Corporation. pp. 499–500. LCCN 68-29938.

[32] Roza, Greg (2010). *The Nitrogen Elements: Nitrogen, Phosphorus, Arsenic, Antimony, Bismuth.* p. 7. ISBN 9781435853355.

[33] "07 Nitrogen". Elements.vanderkrogt.net. Retrieved 2008-09-12.

[34] "17 Chlorine". Elements.vanderkrogt.net. Retrieved 2008-09-12.

[35] "25 Manganese". Elements.vanderkrogt.net. Retrieved 2008-09-12.

[36] "56 Barium". Elements.vanderkrogt.net. Retrieved 2008-09-12.

[37] "42 Molybdenum". Elements.vanderkrogt.net. Retrieved 2008-09-12.

[38] "52 Tellurium". Elements.vanderkrogt.net. Retrieved 2008-09-12.

[39] IUPAC. "74 Tungsten". Elements.vanderkrogt.net. Retrieved 2008-09-12.

[40] "38 Strontium". Elements.vanderkrogt.net. Retrieved 2008-09-12.

[41] "Lavoisier". Homepage.mac.com. Retrieved 2008-09-12.

[42] "Chronology – Elementymology". Elements.vanderkrogt.net. Retrieved 2008-09-12.

[43] Lide, David R., ed. (2007–2008). "CRC Handbook of Chemistry and Physics" **4**. New York: CRC Press. p. 42. 978-0-8493-0488-0. |chapter= ignored (help)

[44] M. H. Klaproth (1789). "Chemische Untersuchung des Uranits, einer neuentdeckten metallischen Substanz". *Chemische Annalen* **2**: 387–403.

[45] E.-M. Péligot (1842). "Recherches Sur L'Uranium". *Annales de chimie et de physique* **5** (5): 5–47.

[46] "Titanium". Los Alamos National Laboratory. 2004. Retrieved 2006-12-29.

[47] Barksdale, Jelks (1968). The Encyclopedia of the Chemical Elements. Skokie, Illinois: Reinhold Book Corporation. pp. 732–38 "Titanium". LCCCN 68-29938.

[48] Browning, Philip Embury (1917). "Introduction to the Rarer Elements". *Kongl. Vet. Acad. Handl.* **XV**: 137.

[49] *Crell Anal.* **I**: 313. 1796. Missing or empty |title= (help)

[50] Vauquelin, Louis Nicolas (1798). "Memoir on a New Metallic Acid which exists in the Red Lead of Sibiria". *Journal of Natural Philosophy, Chemistry, and the Art* **3**: 146.

[51] "04 Beryllium". Elements.vanderkrogt.net. Retrieved 2008-09-12.

[52] "23 Vanadium". Elements.vanderkrogt.net. Retrieved 2008-09-12.

[53] "41 Niobium". Elements.vanderkrogt.net. Retrieved 2008-09-12.

[54] "73 Tantalum". Elements.vanderkrogt.net. Retrieved 2008-09-12.

[55] "46 Palladium". Elements.vanderkrogt.net. Retrieved 2008-09-12.

[56] "58 Cerium". Elements.vanderkrogt.net. Retrieved 2008-09-12.

[57] "76 Osmium". Elements.vanderkrogt.net. Retrieved 2008-09-12.

[58] "77 Iridium". Elements.vanderkrogt.net. Retrieved 2008-09-12.

[59] "45 Rhodium". Elements.vanderkrogt.net. Retrieved 2008-09-12.

[60] "19 Potassium". Elements.vanderkrogt.net. Retrieved 2008-09-12.

[61] "11 Sodium". Elements.vanderkrogt.net. Retrieved 2008-09-12.

[62] "05 Boron". Elements.vanderkrogt.net. Retrieved 2008-09-12.

[63] "09 Fluorine". Elements.vanderkrogt.net. Retrieved 2008-09-12.

[64] "53 Iodine". Elements.vanderkrogt.net. Retrieved 2008-09-12.

[65] "03 Lithium". Elements.vanderkrogt.net. Retrieved 2008-09-12.

[66] "48 Cadmium". Elements.vanderkrogt.net. Retrieved 2008-09-12.

[67] "34 Selenium". Elements.vanderkrogt.net. Retrieved 2008-09-12.

[68] "14 Silicon". Elements.vanderkrogt.net. Retrieved 2008-09-12.

[69] "13 Aluminium". Elements.vanderkrogt.net. Retrieved 2008-09-12.

[70] "35 Bromine". Elements.vanderkrogt.net. Retrieved 2008-09-12.

[71] "90 Thorium". Elements.vanderkrogt.net. Retrieved 2008-09-12.

[72] "57 Lanthanum". Elements.vanderkrogt.net. Retrieved 2008-09-12.

[73] "68 Erbium". Elements.vanderkrogt.net. Retrieved 2008-09-12.

[74] "65 Terbium". Elements.vanderkrogt.net. Retrieved 2008-09-12.

[75] "44 Ruthenium". Elements.vanderkrogt.net. Retrieved 2008-09-12.

[76] "55 Caesium". Elements.vanderkrogt.net. Retrieved 2008-09-12.

[77] Caesium

[78] "37 Rubidium". Elements.vanderkrogt.net. Retrieved 2008-09-12.

[79] "81 Thallium". Elements.vanderkrogt.net. Retrieved 2008-09-12.

[80] "49 Indium". Elements.vanderkrogt.net. Retrieved 2008-09-12.

[81] "02 Helium". Elements.vanderkrogt.net. Retrieved 2008-09-12.

[82] "31 Gallium". Elements.vanderkrogt.net. Retrieved 2008-09-12.

[83] "70 Ytterbium". Elements.vanderkrogt.net. Retrieved 2008-09-12.

[84] "67 Holmium". Elements.vanderkrogt.net. Retrieved 2008-09-12.

[85] "69 Thulium". Elements.vanderkrogt.net. Retrieved 2008-09-12.

[86] "21 Scandium". Elements.vanderkrogt.net. Retrieved 2008-09-12.

[87] "62 Samarium". Elements.vanderkrogt.net. Retrieved 2008-09-12.

[88] "64 Gadolinium". Elements.vanderkrogt.net. Retrieved 2008-09-12.

[89] "59 Praseodymium". Elements.vanderkrogt.net. Retrieved 2008-09-12.

[90] "60 Neodymium". Elements.vanderkrogt.net. Retrieved 2008-09-12.

[91] "32 Germanium". Elements.vanderkrogt.net. Retrieved 2008-09-12.

[92] "18 Argon". Elements.vanderkrogt.net. Retrieved 2008-09-12.

[93] "10 Neon". Elements.vanderkrogt.net. Retrieved 2008-09-12.

[94] "54 Xenon". Elements.vanderkrogt.net. Retrieved 2008-09-12.

[95] "84 Polonium". Elements.vanderkrogt.net. Retrieved 2008-09-12.

[96] "88 Radium". Elements.vanderkrogt.net. Retrieved 2008-09-12.

[97] Partington, J. R. (May 1957). "Discovery of Radon". *Nature* **179** (4566): 912. Bibcode:1957Natur.179..912P. doi:10.1038/179912a0.

[98] Ramsay, W.; Gray, R. W. (1910). "La densité de l'emanation du radium". *Comptes rendus hebdomadaires des séances de l'Académie des sciences* **151**: 126–128.

[99] "89 Actinium". Elements.vanderkrogt.net. Retrieved 2008-09-12.

[100] "63 Europium". Elements.vanderkrogt.net. Retrieved 2008-09-12.

[101] "71 Lutetium". Elements.vanderkrogt.net. Retrieved 2008-09-12.

[102] http://www.maik.ru/abstract/radchem/0/radchem0535_abstract.pdf

[103] "72 Hafnium". Elements.vanderkrogt.net. Retrieved 2008-09-12.

[104] Noddack, W.; Tacke, I.; Berg, O (1925). "Die Ekamangane". *Naturwissenschaften* **13** (26): 567. Bibcode:1925NW.....13..567.. doi:10.1007/BF01558746.

[105] "91 Protactinium". Elements.vanderkrogt.net. Retrieved 2008-09-12.

[106] Emsley, John (2001). *Nature's Building Blocks* ((Hardcover, First Edition) ed.). Oxford University Press. p. 347. ISBN 0-19-850340-7.

[107] "43 Technetium". Elements.vanderkrogt.net. Retrieved 2008-09-12.

[108] *History of the Origin of the Chemical Elements and Their Discoverers*, Individual Element Names and History, "Technetium"

[109] "87 Francium". Elements.vanderkrogt.net. Retrieved 2008-09-12.

[110] Adloff, Jean-Pierre; Kaufman, George B. (2005-09-25). Francium (Atomic Number 87), the Last Discovered Natural Element. *The Chemical Educator* **10** (5). [2007-03-26]

[111] "85 Astatine". Elements.vanderkrogt.net. Retrieved 2008-09-12.

[112] Close, Frank E. (2004). *Particle Physics: A Very Short Introduction*. Oxford University Press. p. 2. ISBN 978-0-19-280434-1.

[113] "93 Neptunium". Elements.vanderkrogt.net. Retrieved 2008-09-12.

[114] "94 Plutonium". Elements.vanderkrogt.net. Retrieved 2008-09-12.

[115] "95 Americium". Elements.vanderkrogt.net. Retrieved 2008-09-12.

[116] "96 Curium". Elements.vanderkrogt.net. Retrieved 2008-09-12.

[117] "97 Berkelium". Elements.vanderkrogt.net. Retrieved 2008-09-12.

[118] "98 Californium". Elements.vanderkrogt.net. Retrieved 2008-09-12.

[119] "99 Einsteinium". Elements.vanderkrogt.net. Retrieved 2008-09-12.

[120] "100 Fermium". Elements.vanderkrogt.net. Retrieved 2008-09-12.

[121] "101 Mendelevium". Elements.vanderkrogt.net. Retrieved 2008-09-12.

[122] "102 Nobelium". Elements.vanderkrogt.net. Retrieved 2008-09-12.

[123] "103 Lawrencium". Elements.vanderkrogt.net. Retrieved 2008-09-12.

[124] "104 Rutherfordium". Elements.vanderkrogt.net. Retrieved 2008-09-12.

[125] "105 Dubnium". Elements.vanderkrogt.net. Retrieved 2008-09-12.

[126] "106 Seaborgium". Elements.vanderkrogt.net. Retrieved 2008-09-12.

[127] "107 Bohrium". Elements.vanderkrogt.net. Retrieved 2008-09-12.

[128] "109 Meitnerium". Elements.vanderkrogt.net. Retrieved 2008-09-12.

[129] "108 Hassium". Elements.vanderkrogt.net. Retrieved 2008-09-12.

[130] "110 Darmstadtium". Elements.vanderkrogt.net. Retrieved 2008-09-12.

[131] "111 Roentgenium". Elements.vanderkrogt.net. Retrieved 2008-09-12.

[132] "112 Copernicium". Elements.vanderkrogt.net. Retrieved 2009-07-17.

[133] "Discovery of the Element with Atomic Number 112". www.iupac.org. 2009-06-26. Retrieved 2009-07-17.

[134] Oganessian, Yu. Ts.; Utyonkov, V. K.; Lobanov, Yu. V.; Abdullin, F. Sh.; Polyakov, A. N.; Shirokovsky, I. V.; Tsyganov, Yu. S.; Gulbekian, G. G.; Bogomolov, S. L.; Gikal, B.; Mezentsev, A.; Iliev, S.; Subbotin, V.; Sukhov, A.; Buklanov, G.; Subotic, K.; Itkis, M.; Moody, K.; Wild, J.; Stoyer, N.; Stoyer, M.; Lougheed, R. (October 1999). "Synthesis of Superheavy Nuclei in the ^{48}Ca + ^{244}Pu Reaction". *Physical Review Letters* **83** (16): 3154. Bibcode:1999PhRvL..83.3154O. doi:10.1103/PhysRevLett.83.3154.

[135] Oganessian, Yu. Ts.; Utyonkov, V. K.; Lobanov, Yu. V.; Abdullin, F. Sh.; Polyakov, A. N.; Shirokovsky, I. V.; Tsyganov, Yu. S.; Gulbekian, G. G.; Bogomolov, S. L.; Gikal, B.; Mezentsev, A.; Iliev, S.; Subbotin, V.; Sukhov, A.; Ivanov, O.; Buklanov, G.; Subotic, K.; Itkis, M.; Moody, K.; Wild, J.; Stoyer, N.; Stoyer, M.; Lougheed, R.; Laue, C.; Karelin, Ye.; Tatarinov, A. (2000). "Observation of the decay of ^{292}116". *Physical Review C* **63**: 011301. Bibcode:2001PhRvC..63a1301O. doi:10.1103/PhysRevC.63.011301.

[136] Oganessian, Yu. Ts.; Utyonkov, V. K.; Lobanov, Yu. V.; Abdullin, F. Sh.; Polyakov, A. N.; Sagaidak, R. N.; Shirokovsky, I. V.; Tsyganov, Yu. S.; Voinov, A. A.; Gulbekian, G.; Bogomolov, S.; Gikal, B.; Mezentsev, A.; Iliev, S.; Subbotin, V.; Sukhov, A.; Subotic, K.; Zagrebaev, V.; Vostokin, G.; Itkis, M.; Moody, K.; Patin, J.; Shaughnessy, D.; Stoyer, M.; Stoyer, N.; Wilk, P.; Kenneally, J.; Landrum, J.; Wild, J.; Lougheed, R. (2006). "Synthesis of the isotopes of elements 118 and 116 in the ^{249}Cf and ^{245}Cm+^{48}Ca fusion reactions". *Physical Review C* **74** (4): 044602. Bibcode:2006PhRvC..74d4602O. doi:10.1103/PhysRevC.74.044602.

[137] Oganessian, Yu. Ts.; Utyonkov, V. K.; Dmitriev, S. N.; Lobanov, Yu. V.; Itkis, M. G.; Polyakov, A. N.; Tsyganov, Yu. S.; Mezentsev, A. N.; Yeremin, A. V.; Voinov, A.; Sokol, E.; Gulbekian, G.; Bogomolov, S.; Iliev, S.; Subbotin, V.; Sukhov, A.; Buklanov, G.; Shishkin, S.; Chepygin, V.; Vostokin, G.; Aksenov, N.; Hussonnois, M.; Subotic, K.; Zagrebaev, V.; Moody, K.; Patin, J.; Wild, J.; Stoyer, M.; Stoyer, N. et al. (2005). "Synthesis of elements 115 and 113 in the reaction ^{243}Am + ^{48}Ca". *Physical Review C* **72** (3): 034611. Bibcode:2005PhRvC..72c4611O. doi:10.1103/PhysRevC.72.034611.

[138] Oganessian, Yu. Ts.; Abdullin, F. Sh.; Bailey, P. D.; Benker, D. E.; Bennett, M. E.; Dmitriev, S. N.; Ezold, J. G.; Hamilton, J. H.; Henderson, R. A.; Itkis, M. G.; Lobanov, Yu. V.; Mezentsev, A. N.; Moody, K. J.; Nelson, S. L.; Polyakov, A. N.; Porter, C. E.; Ramayya, A. V.; Riley, F. D.; Roberto, J. B.; Ryabinin, M. A.; Rykaczewski, K. P.; Sagaidak, R. N.; Shaughnessy, D. A.; Shirokovsky, I. V.; Stoyer, M. A.; Subbotin, V. G.; Sudowe, R.; Sukhov, A. M.; Tsyganov, Yu. S. et al. (April 2010). "Synthesis of a New Element with Atomic Number Z=117". *Physical Review Letters* **104** (14): 142502. Bibcode:2010PhRvL.104n2502O. doi:10.1103/PhysRevLett.104.142502. PMID 20481935.

21.8 External links

- History of the Origin of the Chemical Elements and Their Discoverers Last updated by Boris Pritychenko on March 30, 2004

- History of Elements of the Periodic Table

- Timeline of Element Discoveries

- Discovery of the Elements - The Movie - YouTube (1:18)

- The History Of Metals Timeline. A timeline showing the discovery of metals and the development of metallurgy.

21.9   Text and image sources, contributors, and licenses

21.9.1   Text

- **Chemical element** *Source:* https://en.wikipedia.org/wiki/Chemical_element?oldid=681724445 *Contributors:* AxelBoldt, Vicki Rosenzweig, Mav, Bryan Derksen, The Anome, Tarquin, Css, Andre Engels, Youssefsan, Rmhermen, William Avery, Heron, Mercury610, Stevertigo, Patrick, RTC, Menchi, Ixfd64, Eric119, Minesweeper, Kosebamse, Ahoerstemeier, Suisui, Александър, Glenn, Andres, Evercat, Mxn, Schneelocke, Mulad, Reddi, Stone, Jay, Taxman, Thue, Bevo, Xevi~enwiki, Geraki, Archivist~enwiki, Donarreiskoffer, Gentgeen, Robbot, Fredrik, Bitwise-Man, Yelyos, Nurg, Romanm, Arkuat, Merovingian, Pingveno, Academic Challenger, PxT, Rursus, Texture, Roscoe x, Caknuck, Wikibot, Ebeisher, Jimduck, Hexii, Centrx, Giftlite, DocWatson42, Christopher Parham, Tom harrison, HangingCurve, Xerxes314, Everyking, No Guru, Alison, Bensaccount, Rpyle731, Mboverload, R. fiend, Blankfaze, Antandrus, Beland, Karol Langner, Thincat, Icairns, Trevor MacInnis, Mike Rosoft, Sdrawkcab, HedgeHog, EugeneZelenko, Discospinster, Rich Farmbrough, FT2, Cacycle, FiP, Vsmith, Xezbeth, Nvj, Mani1, Blade Hirato~enwiki, SpookyMulder, Bender235, Andrejj, RJHall, MisterSheik, Kwamikagami, Mwanner, Laurascudder, Shanes, RoyBoy, Bookofjude, Danshil, Femto, CDN99, Bobo192, Smalljim, Cmdrjameson, SpeedyGonsales, Man vyi, Jojit fb, PeterisP, Obradovic Goran, Nsaa, Ranveig, Jumbuck, Kuratowski's Ghost, Alansohn, Hi ruwen, Loa, Paleorthid, Riana, Walkerma, Jaw959, Bantman, Sobolewski, Wtmitchell, Maxkirk1, TenOfAllTrades, LFaraone, H2g2bob, Bsadowski1, Skatebiker, Gene Nygaard, Kay Dekker, Benoni, Thryduulf, Firsfron, Alvis, SNPP, OwenX, ScottDavis, Benbest, Polyparadigm, Pol098, WadeSimMiser, Dozenist, Terence, Sengkang, CharlesC, Wayward, 陳鵬宇帥哥, Mandarax, Tslocum, RichardWeiss, Graham87, Ryoung122, Chun-hian, Kushboy, DePiep, Effeietsanders, Drbogdan, Saperaud~enwiki, Rjwilmsi, Kinu, Strait, VogonFord, Tangotango, Crazynas, ScottJ, Brighterorange, ThePoorGuy, Yamamoto Ichiro, FlaBot, RobertG, Nivix, Ayla, Glenn L, Physchim62, Snailwalker, Imnotminkus, King of Hearts, Chobot, GangofOne, DVdm, Korg, NSR, Gwernol, Banaticus, Wavelength, Hawaiian717, Alchemy pete, Phantomsteve, RussBot, Ismaeelah, Limulus, Quintusdecimus, Wimt, RadioKirk, NawlinWiki, Grafen, Brythain, Peter Delmonte, E rulez, Raven4x4x, Nick C, Zwobot, Werdna, SamuelRiv, 21655, Cynicism addict, Orbis 3, NielsenGW, Peter, Johnpseudo, Katieh5584, Kungfuadam, JDspeeder1, Luk, ChemGardener, Itub, Winick88, Mexistache, Yakudza, SmackBot, Dreamer.redeemer, CarbonCopy, Hydrogen Iodide, Melchoir, C.Fred, Bomac, Jagged 85, Jrockley, Delldot, Edgar181, HalfShadow, Alsandro, Yamaguchi先生, Gilliam, Aaron of Mpls, Skizzik, Chris the speller, Kurykh, Persian Poet Gal, MK8, Miquonranger03, MalafayaBot, CherryT~enwiki, SchfiftyThree, Bonaparte, Deli nk, Aclwon, DHNbot~enwiki, Sbharris, Darth Panda, MaxSem, Modest Genius, Can't sleep, clown will eat me, MyNameIsVlad, Kristbg, Yidisheryid, Booshank, EvelinaB, Rrburke, Andy120290, RedHillian, SundarBot, Nakon, Lordshaun, Het, Sadi Carnot, Curly Turkey, Khazar, John, Euchiasmus, Kipala, Sir Nicholas de Mimsy-Porpington, Tony Corsini, Anoop.m, Madris, IronGargoyle, Hvn0413, Beetstra, Maksim L., Funnybunny, Iridescent, Joseph Solis in Australia, Shoeofdeath, J Di, Cbrown1023, Gil Gamesh, Civil Engineer III, Courcelles, Tawkerbot2, Lahiru k, JForget, Svlad Jelly, CmdrObot, Irwangatot, FunPika, Van helsing, NickW557, WeggeBot, Darren10000, Nmacu, Funnyfarmofdoom, Nilfanion, TJDay, Nick Y., Steel, Kaldosh, Rifleman 82, Gogo Dodo, Julian Mendez, Tawkerbot4, Carstensen, Christian75, DumbBOT, Obrian7, Lee, Matwilko, Mydoghasworms, Smeazel, Thijs!bot, Epbr123, Headbomb, Canada Jack, Marek69, Davidhorman, Kiran201193, ZeekyH.bomb, SusanLesch, Escarbot, Morgana The Argent, KrakatoaKatie, AntiVandalBot, Jj137, Farosdaughter, Gdo01, Istartfires, Nousakan, Res2216firestar, MER-C, Skomorokh, Plantsurfer, Hamsterlopithecus, Cynwolfe, GoodDamon, Acroterion, Moni3, Karlhahn, Bongwarrior, VoABot II, AuburnPilot, Edmund372, WhatamIdoing, Dirac66, 28421u2232nfenfcenc, Thibbs, TekNOSX, Glen, DerHexer, JaGa, T55648L, Pax:Vobiscum, TheRanger, Gwern, Rickterp, Nietzscheanlie, MartinBot, ChemNerd, Rettetast, R'n'B, AlexiusHoratius, PrestonH, Smokizzy, Tgeairn, J.delanoy, Pharaoh of the Wizards, Trusilver, Bogey97, Rhinestone K, Maurice Carbonaro, Metrax, Rod57, Chaveyd, Warut, Pcfjr9, Mufka, Tanaats, Ionescuac, Lilwik, KylieTastic, Lordaraq, Rpr117, Jamesofur, DorganBot, Treisijs, Useight, Fusion Power, Xiahou, CardinalDan, Daz643, Xenonice, Sumo su, 28bytes, VolkovBot, ABF, Jeff G., JoeDeRose, FutharkRed, Philip Trueman, TXiKiBoT, The Original Wildbear, Gary Levell, Daydreambeliever15, Billiards, Qxz, Lradrama, Martin451, LeaveSleaves, Karlengblom, Inx272, Michelle192837, Brainmuncher, Sarc37, Wolfrock, Synthebot, Burntsauce, Riversong, 555zozo555, Onceonthisisland, AlleborgoBot, EmxBot, Glennklockwood, Demmy100, SieBot, Stever Augustus, Buccaneerande, Sonicology, PlanetStar, Winchelsea, Gerakibot, Dawn Bard, Viskonsas, Caltas, Calabraxthis, JerrySteal, Keilana, Toddst1, Tiptoety, Qst, Oda Mari, Wombatcat, Oxymoron83, Jdaloner, Lightmouse, Mjkhfg, Tombomp, Sjn28, Maelgwnbot, Wuhwuzdat, Tesi1700, Maralia, Dolphin51, Zbisasimone, Nergaal, Escape Orbit, Into The Fray, Romit3, SallyForth123, Twinsday, Elassint, ClueBot, PleasantPheasant, The Thing That Should Not Be, Rodhullandemu, VQuakr, J8079s, Boing! said Zebedee, Xenon54, Daracul, Firzen the Great, Neander7hal, Excirial, Alexbot, PrincealiG, Eeekster, Clutchmetal, ParisianBlade, Helenginn, LarryMorseDCOhio, Muro Bot, La Pianista, Calor, Thingg, Pzoxicuvybtnrm, Teleomatic, Johnuniq, SoxBot III, DumZiBoT, RMFan1, Drjezza, PseudoOne, Pgallert, Klemox, Vianello, ZooFari, Bir el Arweh, MystBot, Buckeyenaaashun, Good Olfactory, ElMeBot, Pj13, Addbot, ERK, Proofreader77, Trygve94, Roentgenium111, Some jerk on the Internet, DOI bot, Element16, Guoguo12, 1266asdsdjapg, Sedsa1, Hda3ku, Friginator, Fieldday-sunday, KorinoChikara, Shirtwaist, Cst17, Googleguy1234, Challengedgenius, CarsracBot, Azim5498, Greasel5, TStein, Quercus solaris, Numbo3-bot, Ehrenkater, Tide rolls, Bfigura's puppy, Lightbot, Jan eissfeldt, Gail, Mr.pomo, Fryed-peach, Luckas-bot, Yobot, Dor Cohen, Moda yahia, Les boys, Legobot II, Newportm, Ajh16, THEN WHO WAS PHONE?, 13lade94, Synchronism, Davelo99, DiverDave, AnomieBOT, Grahamching, 1exec1, NoPity2, IRP, JackieBot, Piano non troppo, Bhujerban, Kingpin13, Rathla, Flewis, Materialscientist, Pepo13, Susiban227, Citation bot, Adsfk1236io54lklk, Neurolysis, Chemeditor, Huklkl, LilHelpa, Xqbot, Phazvmk, RJav, Roftltime8, Timir2, Sionus, Intelati, Capricorn42, 4twenty42o, Grim23, Nakumi, Scream1013, Michele600, Dominicp2007, GrouchoBot, Jermentr, Jhbdel, Omnipaedista, RibotBOT, Otnick, Doulos Christos, Eugene-elgato, Franman3024, Erik9, A.amitkumar, FrescoBot, Remotelysensed, Tobby72, Goodbye Galaxy, Wik1ped1a is meant 2 be vanda1ised, Machine Elf 1735, Commit charge, Citation bot 1, Nirmos, ElmentPT, Pinethicket, I dream of horses, RedBot, Mikespedia, Leodescal, FoxBot, Double sharp, یقلع فشاک, Ticklewickleukulele, Hostel369, Dinamik-bot, Vrenator, Nemesis of Reason, Aoidh, Diannaa, Jamietw, Mitchellpuss, Lala2034, McPoopyPee, Sandman888, DARTH SIDIOUS 2, Whisky drinker, Mean as custard, Shanoor212, The Utahraptor, RjwilmsiBot, TjBot, TomT0m, EmausBot, WikitanvirBot, Gfoley4, ScottyBerg, Syncategoremata, RA0808, XinaNicole, Emperorcheston, Wikipelli, Prowster, JSquish, Cogiati, Fæ, StringTheory11, Bryce Carmony, A930913, Wayne Slam, Frigotoni, Ocaasi, Mpb4lol, Tolly4bolly, L1A1 FAL, Resprinter123, Ericlaermans, L Kensington, Puffin, YOSF0113, Negovori, Moocow121, ChuispastonBot, Matthewrbowker, DASHBotAV, Socialservice, ResearchRave, Petrb, ClueBot NG, Kieran0397, LeastCommonAncestor, Jack Greenmaven, Satellizer, Chester Markel, Erichusk, Lanthanum-138, Delusion23, Cntras, O.Koslowski, Auchansa, Metaknowledge, Sehgalamit, Theopolisme, नगरज मंथा, Diyar se, Helpful Pixie Bot, Ronaldo47, RobertGustafson, JohnSRoberts99, ಉಮೆ ಬಾಲಿಗಾ, Sergeyshar, Javon12javon, Gob Lofa, Bibcode Bot, Markovnikov, BG19bot, Sandbh, Wiki13, AvocatoBot, CA Mendeleev, The Whispering Wind, AdventurousSquirrel, Jenjenniferjenni, Soerfm, Jamhol1234, Seanpkenny, Doctor Lipschitz, Aisteco, Justincheng12345-bot, Zarospwnz, Jimw338, Althealster, Mechknight117, Soulbust, EnzaiBot, Hans5620, BrightStarSky, Dexbot, Mogism,

Lugia2453, Megachemistrygenius, Bulba2036, Frosty, Jnargus, Gaby263612, Unliquidating, Rainbow Shifter, Unholyherget, Kslinker5493, Genius567812, Jerry123456, Astredita, Rbpltr, Filedelinkerbot, SharpQuillPen, PaulZapata, Smartguy1234567891040420, Redsamuraii, IiKkEe, Forbidden User, TerryAlex, Rezve59, The Clipper, E757, Ur moms a nice person, Orduin, Kalemh24, MKZombie, Dsrawlings, KasparBot and Anonymous: 989

- **Chemical substance** *Source:* https://en.wikipedia.org/wiki/Chemical_substance?oldid=682428759 *Contributors:* Marj Tiefert, Bryan Derksen, Tarquin, LA2, Enchanter, Montrealais, Patrick, Michael Hardy, Tompagenet, Mkweise, Ahoerstemeier, Notheruser, AugPi, Andres, Media lib, Ham sandwich, Tpbradbury, Geraki, Robbot, Academic Challenger, 88888888, Hadal, UtherSRG, Centrx, Giftlite, Sampo, Iridium77, Everyking, Bensaccount, Unconcerned, Eequor, Kandar, Andycjp, Alexf, Antandrus, DragonflySixtyseven, Grunners, Bluemask, Discospinster, Cacycle, Vsmith, CDN99, Bobo192, Guiltyspark, Nsaa, Alansohn, Gary, Interiot, Lectonar, Walkerma, Hohum, Snowolf, RainbowOfLight, Wimvandorst, Vuo, Gene Nygaard, Richwales, Commander Keane, Damicatz, Macaddct1984, Kralizec!, MrSomeone, Graham87, V8rik, Jake Wartenberg, SMC, Vegaswikian, Durin, DoubleBlue, Yamamoto Ichiro, Ewlyahoocom, Gurch, Mark J, Physchim62, DVdm, Gwernol, Yurik-Bot, Wavelength, Sceptre, Flameviper, Reo On, SpuriousQ, Gaius Cornelius, Pseudomonas, NawlinWiki, Syrthiss, DeadEyeArrow, Wknight94, Silverchemist, Light current, Nikkimaria, Closedmouth, Reyk, Jolt76, DaltinWentsworth, CIreland, Kf4bdy, Itub, SmackBot, Slashme, Pavlovič, Kimon, AnonUser, Bomac, Edgar181, Gilliam, Skizzik, Chris the speller, Kurykh, Persian Poet Gal, Bduke, Fuzzform, EncMstr, SchftftyThree, RayAYang, Achristl, FordPrefect42, Sbharris, Colonies Chris, Hallenrm, Brinerustle, Can't sleep, clown will eat me, Frap, Onorem, Yidisheryid, Addshore, PsychoCola, Zophar1, COMPFUNK2, Springnuts, SashatoBot, Accurizer, Minna Sora no Shita, IronGargoyle, Ckatz, Beetstra, Meco, Mets501, Iridescent, Wjejskenewr, IvanLanin, DavidOaks, Pudeo, Ale jrb, BoH, Fried Gold, Harej bot, Neelix, Pro bug catcher, JJC1138, Corpx, Christian75, Garik, Calvero JP, Epbr123, Headbomb, Marek69, FreeKresge, Dawnseeker2000, CTZMSC3, AntiVandalBot, Widefox, Seaphoto, Prolog, Blu3d, Myanw, JAnDbot, Plantsurfer, Kerotan, Penubag, Parsecboy, VoABot II, JamesBWatson, Latifshaikh20, Daarznieks, DerHexer, Wdflake, Seba5618, SquidSK, Mithras6, Phoogenb, Hdt83, MartinBot, Bbi5291, ChemNerd, CommonsDelinker, Paranomia, J.delanoy, Pharaoh of the Wizards, Ginsengbomb, McSly, HiLo48, NewEnglandYankee, EyebrowOnVacation, KylieTastic, Joshua Issac, Fragbase, WJBscribe, Treisijs, FanCon, Neod4000, CardinalDan, Idioma-bot, Lights, VolkovBot, CWii, Indubitably, JohnBlackburne, AlnoktaBOT, Omar737, Crohnie, Anna Lincoln, Qyt, Seraphim, Leafyplant, Goodlilstevie, Synthebot, Dozer199, Ratear, Enviroboy, Cvf-ps, Kbrose, SieBot, Scarian, WereSpielChequers, Caltas, Yintan, Maha101, Keilana, Bob98133, Oxymoron83, Nuttycoconut, Jdaloner, Alex.muller, LonelyMarble, Paulinho28, Tony Webster, JL-Bot, Laburke, Atif.t2, Tanvir Ahmmed, ClueBot, Binksternet, Fyyer, Trevor242, Uncle Milty, SuperHamster, CounterVandalismBot, Graysen98, Soccerfreakbry, Excirial, Jusdafax, PixelBot, Abrech, Ykhwong, Coinmanj, NuclearWarfare, Arjayay, Hullcrushdepth, Kemo1993, Thingg, Horselover Frost, Bobhasissues, Saenzc, Vanished user uih38riiw4hjlsd, DumZiBoT, KingsOfHearts, Dexter siu, Zhile, Ismx24, Mkcas, Skarebo, SilvonenBot, ElMeBot, Osarius, Addbot, Metagraph, Fieldday-sunday, GD 6041, Leszek Jańczuk, LaaknorBot, CarsracBot, Glane23, AndersBot, Quercus solaris, 5 albert square, Jasoneric, Nnedass, Tide rolls, WikiDreamer Bot, SIgcat, Alfie66, Jack who built the house, Legobot II, Fryderyk, عالم, محبوب, AnomieBOT, Shootbamboo, Jim1138, IRP, Dwayne, Odlaw, Materialscientist, Spy5295, The High Fin Sperm Whale, Virionspiral, Auther175, Jameson812, Rehansaccount, LilHelpa, Princesstwilighter, Xqbot, Cureden, Capricorn42, Frink99887, NFD9001, Ilyilyily, C+C, Rsmn, Lo28, Earlypsychosis, RibotBOT, حامد میرزاحسینی, Karlluo, Doulos Christos, Sophus Bie, Lordvisucius, JayJay, Pepper, Orhanghazi, Finalius, Ιωάννης Καραμήτρος, Drew R. Smith, Krish Dulal, 闪闪的红星, Pinethicket, I dream of horses, A8UDI, V.narsikar, RedBot, Eagle-0, Mnjh, Lrobb95, ActivExpression, ConsumerEducation, Vrenator, Zvn, Seahorseruler, Merlinsorca, Diannaa, Suffusion of Yellow, Wikiman1222, Tbhotch, DARTH SIDIOUS 2, MMS2013, Galloping Moses, Wintonian, EmausBot, Randy madman on roids, Pavlo Chemist, RenamedUser01302013, Tommy2010, ZéroBot, Fæ, Deeas, Jande417, Chemicalinterest, Jdwwilson, Wayne Slam, Ocaasi, Wagino 20100516, Aleksander Sestak, Donner60, Orange Suede Sofa, Theislikerice, Krazie808, Neil P. Quinn, Rocketrod1960, Socialservice, ClueBot NG, AluminumFear, MelbourneStar, Pika32141, Widr, Antiqueight, Wegothim, BG19bot, Kwells1989, Swagggurl2, MusikAnimal, Frze, AvocatoBot, Mark Arsten, IraChesterfield, ZRRSZR, Snow Blizzard, Bleep234, Shubh12, GENIUS28, IsraphelMac, Webclient101, Reatlas, Leprof 7272, Riddler23tron, PlanetEditor, Www.winner.com, Uberaccount, Gayu3, Haminoon, Whitelawkirk, Susan.grayeff, WZ-121, Suelru, Ananagram, Wiki3457, Littlemonkey14, BaconMuncher57, TheQ Editor, Dfreshlut, Jimmyblocks, Last Man Never, Yakuza jackman, Argi 12345, Dog Poopie, KasparBot, Adarshjchandran, Eballard1214, LandedEagle and Anonymous: 534

- **Synthetic element** *Source:* https://en.wikipedia.org/wiki/Synthetic_element?oldid=678348218 *Contributors:* The Anome, D, Tim Starling, DopefishJustin, Eric119, Alfio, Stone, Zemat~enwiki, Raul654, LMT1, Mintleaf~enwiki, Garth 187, Icairns, JamesTeterenko, Reflex Reaction, Brianhe, Vsmith, HasharBot~enwiki, Kuratowski's Ghost, Sl, Walkerma, Gene Nygaard, Blaxthos, Forteblast, Benbest, Prashanthns, Ryoung122, DePiep, Vegaswikian, Kolbasz, Chobot, Mordicai, Bgwhite, YurikBot, Borgx, MSJapan, Elizabeyth, Jpeob, Wknight94, LeonardoRob0t, Cf-frost, Itub, SmackBot, Incnis Mrsi, Provelt, Edgar181, Kurykh, Chlewbot, Kingdon, Qmwne235, Lambiam, A. Parrot, CumbiaDude, A876, Dharma6662000, Plantsurfer, Tzittnan, Warut, 7yu6, ChaosCon343, VolkovBot, Soliloquial, Broadbot, Phe-bot, GlassCobra, Bentogoa, Nergaal, Denisarona, ClueBot, Wild quinine, SchreiberBike, XLinkBot, HexaChord, Getsnoopy, Addbot, Roentgenium111, Mitchtn14, Jasper Deng, Numbo3-bot, Peridon, Zoolicious, Luckas-bot, Yobot, THEN WHO WAS PHONE?, IRP, Jhjk, Double sharp, Ticklewickleukulele, Dinamik-bot, Horsebrutality, DexDor, Mandolinface, Hoejriis, Quantumor, Hang Li Po, Xanchester, ClueBot NG, This lousy T-shirt, Lanthanum-138, RobertGustafson, AnandVivekSatpathi, Jeffreyahn99, Sparkie82, Doltybolty, Glacialfox, Ashutoshpandey123, The Illusive Man, ChrisGualtieri, Arcandam, Ranze, LegionMammal978, Taborgen, DJTY5 and Anonymous: 90

- **Isotope** *Source:* https://en.wikipedia.org/wiki/Isotope?oldid=682628957 *Contributors:* Trelvis, Marj Tiefert, Mav, Bryan Derksen, Tarquin, AstroNomer~enwiki, Malcolm Farmer, Stokerm, William Avery, Peterlin~enwiki, Spiff~enwiki, Patrick, Alan Peakall, Shyamal, Shellreef, Liftarn, Ixfd64, Tango, Minesweeper, Ahoerstemeier, Snoyes, Suisui, BigFatBuddha, Julesd, Glenn, Andres, Kaihsu, Hectorthebat, Rl, Smack, Dysprosia, Taxman, Jusjih, Palefire, Shantavira, Robbot, Merovingian, Sunray, Bkell, Hadal, Alan Liefting, Giftlite, Mikez, Lethe, Xerxes314, Bensaccount, Guanaco, Bovlb, Prosfilaes, Christopherlin, Wmahan, Antandrus, Kaldari, Icairns, Urhixidur, Adashiel, Bluemask, DanielCD, KNewman, Discospinster, C12H22O11, Mani1, MarkS, ESkog, A purple wikiuser, Walden, Femto, CDN99, Bobo192, Viriditas, Mytildebang, I9Q79oL78KiL0QTFHgyc, Deryck Chan, Obradovic Goran, Jumbuck, Alansohn, Gary, GRider, Cjthellama, Hu, Malo, Ayeroxor, Cburnett, Mikeo, Zoohouse, DV8 2XL, Adrian.benko, Stemonitis, Gmaxwell, Thryduulf, OwenX, Camw, Tripodics, Kurzon, Bratsche, Tylerni7, Clemmy, GregorB, SCEhardt, Sin-man, Graham87, Magister Mathematicae, Kbdank71, DePiep, Jclemens, Sjö, Rjwilmsi, Astronaut, Strait, Quiddity, Feydey, Watcharakorn, FlaBot, Gurch, Physchim62, King of Hearts, Chobot, Sharkface217, GangofOne, Bgwhite, YurikBot, Wavelength, RobotE, JWB, Jimp, Wolfmankurd, Pip2andahalf, Phantomsteve, Petiatil, Stephenb, Cryptic, Wimt, Anomalocaris, NawlinWiki, Wiki alf, Grafen, Chick Bowen, Welsh, Długosz, Dooky, Semperf, Kkmurray, Black Falcon, Silverchemist, Citynoise, Closedmouth, E Wing, KGasso, JuJube, Roberto DR, JoanneB, CWenger, Junglecat, Mjroots, AssistantX, GrinBot~enwiki, Cookiedog, SkerHawx, Serendipodous, 闪闪的 robot,

Luk, Itub, MacsBug, SmackBot, FocalPoint, Jclerman, Incnis Mrsi, CoderDennis, Shoy, Unyoyega, Pgk, C.Fred, Davewild, Edgar181, Half-Shadow, Gilliam, Skizzik, Chris the speller, Master Jay, Father McKenzie, Master of Puppets, Miquonranger03, MalafayaBot, DHN-bot~enwiki, Cassivs, Sbharris, Vladislav, Gurps npc, Rrburke, TKD, Rainmonger, Addshore, Mr.Z-man, SundarBot, Megamix, Radagast83, Khukri, Nibuod, Nakon, G716, Drphilharmonic, DMacks, Clicketyclack, Will Beback, SashatoBot, ArglebargleIV, Silvem, John, Ckatz, Smith609, Stwalkerster, FadieZ, Ryulong, MTSbot~enwiki, SmokeyJoe, Cadaeib, KJS77, Iridescent, Theone00, J Di, Cbrown1023, Sam Li, Witchyrose, Courcelles, Heliomance, Tawkerbot2, JForget, CmdrObot, Tanthalas39, Rambam rashi, RedRollerskate, FlyingToaster, Myrddin1977, Stephen Luce, Prakharbirla, Myasuda, Nmacu, Jlking3, HPaul, Christian75, MagnusGallant, Theadder, Daniel Olsen, Gimmetrow, Satori Son, JamesAM, Thijs!bot, Epbr123, Looskuh, Mojo Hand, Headbomb, Pjvpjv, Marek69, John254, A3RO, Jbwst, Tellyaddict, Jklumker, D.H, Tocharianne, CTZMSC3, Escarbot, Mentifisto, Hmrox, AntiVandalBot, Yuanchosaan, Yonatan, Luna Santin, Seaphoto, Prolog, Cheif Captain, Chuchunezumi, LibLord, Spencer, Charles Clark, Myanw, JAnDbot, Leuko, MER-C, Nthep, Fetchcomms, Andonic, Dcooper, Dricherby, WRonG, Magioladitis, Bongwarrior, VoABot II, Transcendence, Jespinos, Quantockgoblin, Dougz1, Catgut, Indon, Animum, Dirac66, Allstarecho, StuFifeScotland, DerHexer, Hans Moravec, MartinBot, Ryanrulz 11, BetBot~enwiki, Anaxial, R'n'B, AlexiusHoratius, Pbroks13, Interwal, J.delanoy, Trusilver, EscapingLife, Rhinestone K, Uncle Dick, Yonidebot, -jmac-, Keesiewonder, Shotime900, Rod57, BillyZane, BaseballDetective, HOUZI, Pyrospirit, AntiSpamBot, Richard D. LeCour, NewEnglandYankee, Ohms law, Nehakhat, Cmichael, Greenpuppy333, Hakkahakkabazoom, BrianScanlan, Davecrosby uk, Deor, 28bytes, VolkovBot, DagnyB, Jmocenigo, Arnd Klotz, Philip Trueman, Martinevans123, SamMichaels, TXiKiBoT, Technopat, Quilbert, Rei-bot, Z.E.R.O., Charlesdrakew, Qxz, Liko81, Martin451, Slysplace, Unvandalizor, Wenli, Billinghurst, Hey jude, don't let me down, Sw607813, Cosmo737, !dea4u, WatermelonPotion, Ceranthor, K10wnsta, Angel crystal88, AlleborgoBot, Logan, Oba.coskun, Viridium, SaltyBoatr, Wsycng, Juanmantoya, SieBot, Dusti, Sonicology, PlanetStar, Tiddly Tom, Moonriddengirl, Ray23713111, Caltas, Cwkmail, Yintan, Kaypoh, Flyer22, Oda Mari, Oxymoron83, Nuttycoconut, SH84, Ks0stm, Jruderman, AcroX, Kdebens, Alpine McRiper, Sean.hoyland, Manipulator, Pinkadelica, Nergaal, Denisarona, Jons63, Pehsfo, Sidhu ghanta, Atif.t2, Sfan00 IMG, ClueBot, WilliamRoper, Snigbrook, The Thing That Should Not Be, Meisterkoch, Themole12, Hsheller, Hplommer, Nuclearmedzors, DragonBot, Jusdafax, Hello Control, Rhododendrites, Sun Creator, Radiogenic, PhySusie, Promethean, Kanxkawii, 7, Gryphn, Versus22, DumZiBoT, Crazy Boris with a red beard, Tarheel95, Fastily, Spitfire, Nellyb1993, Avoided, Badgernet, Noctibus, Thatguyflint, Sami Lab, Addbot, Proofreader77, Jonny.sinclair, Willking1979, AVand, DOI bot, Fyrael, Icycomputer, Ronhjones, Fieldday-sunday, CanadianLinuxUser, Leszek Jańczuk, The birch tree and the dandelion, WFPM, Glane23, AndersBot, Chzz, Colinho22, Nanzilla, AgadaUrbanit, Ehrenkater, VASANTH S.N., Tide rolls, Verazzano, Zorrobot, Luckas-bot, Makeachange10, Yobot, Senator Palpatine, Newportm, Mirandamir, KamikazeBot, IW.HG, AnomieBOT, DemocraticLuntz, Jim1138, IRP, 9258fahsflkh917fas, Godwotan, Shoopmawhoop, AdjustShift, Ulric1313, Materialscientist, Greatspacegibbon, Citation bot, Vuerqex, St00j, GB fan, ArthurBot, Haidata, Xqbot, J G Campbell, Capricorn42, Skippydogue, Tad Lincoln, Ansonchen88, BLP-outrageous move logs, Jiffe, Apbiologyrocks, Shadowjams, Erik9, BoomerAB, Dcrunner, Prari, FrescoBot, LucienBOT, Wikipe-tan, Saehrimnir, Icorrectu, Xhaoz, Saiarcot895, Citation bot 1, Thwait, Pshent, Pinethicket, Elockid, Edderso, Tanweer Morshed, Kazasik.3, MJ94, Gralco8, Minivip, Fumitol, Shanmugamp7, Merlion444, December21st2012Freak, Veneventura, Tim1357, Double sharp, TobeBot, Mercy11, Twinckletoes911, Throwaway85, Callanecc, Vrenator, Alex2009258, 777sms, Afirtree, DARTH SIDIOUS 2, DexDor, NerdyScienceDude, WildBot, Skamecrazy123, DASHBot, EmausBot, Kourosch44, Gbyers72, Howy9814, RA0808, Ashiel7, XinaNicole, Tommy2010, Wikipelli, Hhhippo, Akhil 0950, JSquish, ZéroBot, Fæ, Monterey Bay, SCStrikwerda, Wayne Slam, IGeMiNix, L Kensington, Tomásdearg92, Donner60, RockMagnetist, Shashank artemis fowl, TheRadicalPi, -revi, Xrayburst1, Xonqnopp, ClueBot NG, PegLegTuna, Accelerometer, Ulflund, Sledhead22, Millermk, Omaro2000, 123Hedgehog456, Widr, Reify-tech, Minecamph, Mr. Credible, Electriccatfish2, Bibcode Bot, Gauravjuvekar, Priyansh verma, Hallows AG, Wiki13, MusikAnimal, Metricopolus, Mark Arsten, Editerjhon, Tp GATE, Pikachu Bros., Kodi55, Anbu121, Sinemet25-250, LHcheM, EuroCarGT, BrightStarSky, Rehmanshahid, Timeweaver, Hmainsbot1, Webclient101, Mogism, Kingaustin42, 2010ipo, Burzuchius, Sfgiants1995, Telfordbuck, Reatlas, Epicgenius, Cavisson, Hubbard96, Elephantsandbacon, Wongchufeng, Meemz05, Rbarhoush12, J.meija, Jwratner1, Quenhitran, Paul2520, Apopert, DSIM123456789, Abitslow, 7Sidz, Dannyzhaofb, Lachlan Newland, Prathamesh Rajput, Dr.Shawn7, Isamiatehreem, 2d6d399, Wyeh, Random user-hobo, KasparBot, JJMC89, Georgehat, Qwerty13245lolmlg and Anonymous: 922

- **Periodic table** *Source:* https://en.wikipedia.org/wiki/Periodic_table?oldid=682400345 *Contributors:* AxelBoldt, Dreamyshade, Chuck Smith, Lee Daniel Crocker, Mav, Bryan Derksen, Timo Honkasalo, The Anome, Tarquin, DanKeshet, Rjstott, Andre Engels, XJaM, Christian List, PierreAbbat, Heron, Fonzy, Youandme, Olivier, Someone else, Bob Jonkman, Patrick, Infrogmation, Michael Hardy, Erik Zachte, TMC, Kwertii, Dan Koehl, Shellreef, Taras, Wapcaplet, Ixfd64, Dcljr, Tomi, Eric119, Kosebamse, Egil, Mdebets, Ahoerstemeier, Stan Shebs, Ronz, Jpatokal, Theresa knott, Snoyes, Suisui, Den fjättrade ankan~enwiki, Kragen, Salsa Shark, Cyan, Stefan-S, Poor Yorick, Kwekubo, Jiang, Eirik (usurped), Mxn, BRG, Smack, Schneelocke, Jengod, Okome~enwiki, Emperorbma, EL Willy, Eszett, Adam Bishop, Reddi, Stone, Piolinfax, Dtgm, Selket, Tpbradbury, Rarb, Maximus Rex, Nv8200pa, Tempshill, Bevo, Traroth, Shizhao, Stormie, Dpbsmith, Bcorr, Secretlondon, Jusjih, Just another user 2, Darthchaos, Jeffq, Lumos3, Denelson83, Jni, Nofutureuk, Gromlakh, Gentgeen, Robbot, Phisite, Juve82, Fredrik, Chris 73, WormRunner, Altenmann, Romanm, Naddy, Lowellian, WebElements, Yosri, Rfc1394, Texture, Hippietrail, Caknuck, Bkell, David Edgar, Borislav, Eliashedberg, Radagast, David Gerard, Giftlite, DocWatson42, Haeleth, Ævar Arnfjörð Bjarmason, Tom harrison, Lupin, Everyking, Bkonrad, No Guru, NeoJustin, Bensaccount, Zaphod Beeblebrox, AJim, Avsa, Yekrats, Dmmaus, Archenzo, Brockert, Darrien, SWAdair, Bobblewik, Deus Ex, Edcolins, Lucky 6.9, Peter Ellis, Gadfium, Zed0, Ran, Antandrus, Ctachme, PDH, Jossi, Exigentsky, Kesac, Vbs, Icairns, Sam Hocevar, Clemwang, Karl Dickman, Adashiel, Iwilcox, EagleOne, Mike Rosoft, Alkivar, D6, Andrew11, Poccil, Zarxos, EugeneZelenko, Felix Wan, A-giau, Noisy, Discospinster, Rich Farmbrough, KarlaQat, Cacycle, Inkypaws, Vsmith, Samboy, Joeclark, SpookyMulder, Bender235, TerraFrost, Sunborn, Klenje, RJHall, El C, Kwamikagami, Shanes, Briséis~enwiki, RoyBoy, Femto, Semper discens, Grick, Bobo192, AlHalawi, Whosyourjudas, Nyenyec, Reinyday, Clawson, Cwolfsheep, Dbchip, Giraffedata, SpeedyGonsales, Jojit fb, Nk, Eddideigel, Conget~enwiki, Jhd, Conny, Stephen G. Brown, Danski14, Honeycake, Orzetto, Alansohn, Mo0, Atlant, Keenan Pepper, Plumbago, Sl, Damnreds, AzaToth, Mac Davis, Caesura, Blobglob, Wtmitchell, ClockworkSoul, Unconventional, Helixblue, Stephan Leeds, Harej, RJFJR, Skatebiker, Computerjoe, GabrielF, Ghirlandajo, HGB, Feline1, Weyes, Lucent, Philthecow, Cimex, TigerShark, Benbest, Mpatel, Schzmo, U10ajf, Bluemoose, CharlesC, Waldir, SeventyThree, EarthmatriX, MarcoTolo, Cataclysm, V8rik, Qwertyus, Kbdank71, FreplySpang, DePiep, Dwaipayanc, Canderson7, Drbogdan, Saperaud~enwiki, Angusmclellan, Joe Decker, Koavf, Oblivious, SeanMack, Shalmanese, Sango123, Ptdecker, Yamamoto Ichiro, RobertG, Pumeleon, Nivix, Pathoschild, RexNL, Gurch, Kolbasz, Brendan Moody, Scerri, Alphachimp, Kri, Dalta~enwiki, Glenn L, Physchim62, Imnotminkus, Chobot, Visor, Jared Preston, DVdm, Bgwhite, Gwernol, EamonnPKeane, Roboto de Ajvol, Mercury McKinnon, YurikBot, Wavelength, Hairy Dude, Deeptrivia, Phantomsteve, RussBot, Vlad4599, Fabartus, SpuriousQ, IanManka, Stephenb, Rintrah, Alvinrune, Schoen, Rsrikanth05, Bovineone, Wimt, Stassats, Anomalocaris, EngineerScotty, NawlinWiki, Wiki alf, E123, Test-tools~enwiki, Jaxl, Terfili, Yahya Abdal-Aziz, Mkouklis, Nick, Ragesoss, Dhollm, Cholmes75, Dmoss, Matticus78, RUL3R, AdiJapan, Ryanminier, Juanpdp, Hv,

- **Period (periodic table)** *Source:* https://en.wikipedia.org/wiki/Period_(periodic_table)?oldid=677103549 *Contributors:* Mav, The Anome, Dcljr, Ahoerstemeier, Suisui, Александър, Xnybre, Jusjih, Gentgeen, Robbot, Yekrats, Kandar, Ctachme, 1297, Icairns, LiSrt, Mike Rosoft, Kwamikagami, Fremsley, Jumbuck, Alansohn, Wtmitchell, Velella, Feline1, Blaxthos, Woohookitty, Benbest, Tckma, M412k, DePiep, PHenry, MZMcBride, Jared999, Glenn L, Chobot, Bgwhite, Roboto de Ajvol, YurikBot, RobotE, Grafen, Ms2ger, Adilch, Josh3580, Allens, Kungfuadam, GrinBot~enwiki, SmackBot, Melchoir, Eskimbot, Skizzik, DHN-bot~enwiki, Shunpiker, VMS Mosaic, SashatoBot, Lambiam, Rigadoun, Stwalkerster, Jimbuckar00, N2e, Rifleman 82, Thijs!bot, Epbr123, Headbomb, Nick Number, Northumbrian, Tharkon, Jayron32, Legolost, .anacondabot, Easchiff, DerHexer, Kayau, MartinBot, R'n'B, Tgeairn, RockMFR, Reedy Bot, ThsTorturedSoul, VolkovBot, AlnoktaBOT, Soliloquial, TXiKiBoT, Rei-bot, Qxz, Pimemorizer, Luuva, Pstuart, Insanity Incarnate, SieBot, Rdx-77, Keilana, Plainandsimple, ClueBot, Wikijens, ChandlerMapBot, BOTarate, WikHead, SilvonenBot, HexaChord, Addbot, Ronhjones, Laurinavicius, LinkFA-Bot, Tide rolls, Legobot, आशीष भटनागर, Luckas-bot, Yobot, AnomieBOT, DemocraticLuntz, IRP, JackieBot, Brane.Blokar, MauritsBot, Xqbot, YBG, Jakwra, GrouchoBot, Kieryh, Amaury, Erik9bot, LucienBOT, Pinethicket, Rushbugled13, RedBot, Double sharp, RjwilmsiBot, TjBot, EmausBot, Noobatron30, Golfandme, ZéroBot, StringTheory11, Bethastrong, Isarra, Palosirkka, ClueBot NG, Lanthanum-138, Marechal Ney, Widr, Shovan Luessi, Helpful Pixie Bot, Strike Eagle, Titodutta, Sandbh, Vivek7de, ChrisGualtieri, JYBot, Tripodno1, Suradnik50, Vanamonde93, Eyesnore, George8211, Monkbot, Rider ranger47, Zschrambo and Anonymous: 141

- **Main sequence** *Source:* https://en.wikipedia.org/wiki/Main_sequence?oldid=680764262 *Contributors:* Mav, Bryan Derksen, Zundark, Malcolm Farmer, XJaM, William Avery, Roadrunner, Patrick, Michael Hardy, Alfio, Looxix~enwiki, Rboatright, Angela, Mark Foskey, Susurrus, Samw, Pizza Puzzle, WhisperToMe, Robbot, Babbage, Rursus, Timrollpickering, Tobycat, Modeha, Aetheling, Pmcray, David Gerard, Stirling Newberry, Lestatdelc, Kaldari, Thincat, Icairns, Moverton, Rich Farmbrough, Vsmith, Mani1, Wadewitz, RJHall, Kwamikagami, Art LaPella, Jon the Geek, Grick, Bobo192, Wayfarer, Eric Kvaalen, ATG, Stillnotelf, GeorgeStepanek, Jun-Dai, Blaxthos, WilliamKF, Nuno Tavares, Pol098, Arrkhal, Bebenko, RichardWeiss, Lawrence King, Graham87, BD2412, Rjwilmsi, Raddick, Mike s, Brighterorange, Maxim Razin, SiriusB, Chobot, Sharkface217, Gwernol, Kdehl, The Rambling Man, YurikBot, Wavelength, Spacepotato, Sceptre, Hydrargyrum, Ksyrie, Kvn8907, Ragesoss, Bozoid, Gadget850, Perry Middlemiss, Chaos syndrome, Junglecat, Hal peridol, SmackBot, NickyMcLean, Moeron, Tom Lougheed, Unyoyega, C.Fred, Saros136, Hibernian, Adun12, Jerome Charles Potts, Modest Genius, Tlusťa, ConMan, Salamurai, Erimus, JorisvS, Like tears in rain, Dan Gluck, Dekaels~enwiki, Røed, Mssgill, JForget, David s graff, Ruslik0, Funnyfarmofdoom, Mato, Jayen466, Michael C Price, Thijs!bot, Epbr123, Oerjan, Marek69, Ialsoagree, AntiVandalBot, Orionus, Spartaz, JAnDbot, Plantsurfer, East718, Acroterion, Penubag, WolfmanSF, Secret Squïrrel, Bongwarrior, VoABot II, Ronstew, Kokin, Catgut, DerHexer, Stargazer360, Geboy, Leyo, Fellwalker57, El0i, J.delanoy, Pharaoh of the Wizards, DrKiernan, Rrostrom, All Is One, Gzkn, Dividing, Nwbeeson, Bcp67, Idioma-bot, X!, Larryisgood, Philip Trueman, GimmeBot, BotKung, Jmac1962, James McBride, FKmailliW, Junkinbomb, Timb66, Rfts, Gerakibot, Snideology, Radzewicz, Techman224, Jruderman, Ethan1701, Freewayguy, Loren.wilton, ClueBot, Ukabia, DanielDeibler, Boing! said Zebedee, Blanchardb, Rotational, Piledhigheranddeeper, Auntof6, DragonBot, Estirabot, Maradona01, Thingg, The Sock of Maelgwn, Stefanole, Spitfire, LoneStar77, Addbot, DOI bot, Landon1980, CanadianLinuxUser, Jaydec, Lightbot, CountryBot, Legobot, Luckas-bot, Aldebaran66, Rrokkedd, KamikazeBot, Kristen Eriksen, Galoubet, Citation bot, Eumolpo, Ulysse2000, LovesMacs, The Fiddly Leprechaun, DSisyphBot, Thatfeel, Lithopsian, TopHyatt, Moxy, Matteobachetti, FrescoBot, Pepper, 28club, Millslogle, Tom.Reding, IVAN3MAN, RjwilmsiBot, Limequat, Bhawani Gautam, Regancy42, TGCP, EmausBot, John of Reading, Orphan Wiki, Mhdkandil, Dewritech, RenamedUser01302013, Jmencisom, StringTheory11, Davido44, H3llBot, Tolly4bolly, Michaelandsandy, Whoop whoop pull up, ClueBot NG, CocuBot, A520, Mgribov, Alex Nico, Jj1236, Kevin Gorman, Ryan Vesey, MerlIwBot, Bibcode Bot, Solomon7968, Snow Blizzard, Ggreybeard, Oznitecki, StarryGrandma, Cyberbot II, Rhlozier, Frosty, Malerooster, Acetotyce, Melonkelon, Nouvelle Planète, Dillon128, Kogge, Monkbot, Yangch17, Tetra quark, PlaQu, Sdgedfegw and Anonymous: 207

- **Stellar nucleosynthesis** *Source:* https://en.wikipedia.org/wiki/Stellar_nucleosynthesis?oldid=677634121 *Contributors:* Bryan Derksen, Clintp, Xavic69, Looxix~enwiki, Ahoerstemeier, Aarchiba, Med, Charles Matthews, Stismail, BitwiseMan, Rursus, Giftlite, Sendhil, Mboverload, Karol Langner, Frau Holle, ELApro, Rich Farmbrough, Pjacobi, Vsmith, Eric Forste, RJHall, Art LaPella, Rajah, Wrs1864, Gene Nygaard, Richard-Weiss, Rjwilmsi, Tim!, Ground Zero, Alvin-cs, Chobot, Whosasking, Kjlewis, 8bitW, Jimp, Witan, Anomalocaris, Uber nemo, Zzuuzz, Ninly, Modify, Allens, Otto ter Haar, Cmglee, MacsBug, Unyoyega, Jrockley, Mak17f, Aksi great, Pfhreak, Joseph Solis in Australia, FatBastardInk, Van helsing, WeggeBot, Sckirklan, Epbr123, Headbomb, Nyme, Magioladitis, WolfmanSF, Cgingold, Dirac66, MartinBot, Maurice Carbonaro, JohnFarnham1, Rominandreu, Liveste, DadaNeem, DAID, Sheliak, Camrn86, Seattle Skier, UnitedStatesian, Hvgap2, SieBot, Timb66, Skylark42, Ioverka, Pionade, ClueBot, GorillaWarfare, Scog, ברוקולי, Addbot, DOI bot, LatitudeBot, Lightbot, Margin1522, Luckas-bot, Yobot, Citation bot, Marshallsumter, NOrbeck, GrouchoBot, FrescoBot, Tom.Reding, عباد مجاهد ديراني, CaptRik, ZéroBot, StringTheory11, Arbnos, ClueBot NG, Frietjes, Helpful Pixie Bot, Calabe1992, Bibcode Bot, Danzanfran, Ghostsarememories, Writ Keeper, Zedshort, Garuda0001, Reatlas, Monkbot, WAFred, Boazkat and Anonymous: 38

- **Big Bang nucleosynthesis** *Source:* https://en.wikipedia.org/wiki/Big_Bang_nucleosynthesis?oldid=681754212 *Contributors:* Vicki Rosenzweig, AstroNomer~enwiki, Roadrunner, Space Cadet, PaulDSP, Bueller 007, LittleDan, Schneelocke, Reddi, Phil Boswell, Korath, Sanders muc, Peak, Rursus, Harp, Art Carlson, Herbee, Anville, Dmmaus, Eroica, JoJan, Karol Langner, Deglr6328, Pjacobi, Vsmith, SpookyMulder, Brian0918, RJHall, Pilatus, Art LaPella, Army1987, I9Q79oL78KiL0QTFHgyc, GeorgeStepanek, Jheald, Oleg Alexandrov, Camw, BlaiseFEgan, Wdanwatts, Joke137, Grundle, Qwertyus, Rjwilmsi, Oo64eva, Mishuletz, Goudzovski, Phoenix2~enwiki, Chobot, Amaurea, Rmbyoung, YurikBot, Sir48, Fobos~enwiki, Uber nemo, Enormousdude, Modify, Ilmari Karonen, Cmglee, SmackBot, Dauto, Bluebot, Kashami, Silly rabbit, Sbharris, Colonies Chris, Ligulembot, GodBlessTheNet, JorisvS, Stevebritgimp, Getjonas, Mssgill, George100, Vyznev Xnebara, Jsd, Gregbard, Thijs!bot, Markus Pössel, Headbomb, John254, Peter Gulutzan, DPdH, Uruiamme, Orionus, Nipisiquit, VoABot II, ThoHug, LorenzoB, DerHexer, Geboy, MartinBot, Pagw, Peter Chastain, Eliz81, BobEnyart, Vegasprof, Wesino, Biglovinb, Juliancolton, Sheliak, VolkovBot, TXiKiBoT, Calwiki, Thrawn562, OlavN, Broadbot, UnitedStatesian, BotKung, Pamputt, SwordSmurf, Newsaholic, Gdude95, Ashdabash, SieBot, Jdaloner, Escape Artist Swyer, ClueBot, CLCalver, ChandlerMapBot, NuclearWarfare, DumZiBoT, TimothyRias, Chanakal, Addbot, Shiba6, Uruk2008, DOI bot, Njaelkies Lea, Yobot, AnomieBOT, Tad Lincoln, Rainald62, Physdragon, Citation bot 1, Gil987, Dogaru Florin, Pinethicket, I dream of horses, Edderso, Jonesey95, Tom.Reding, Pmokeefe, Footwarrior, Double sharp, RobertMfromLI, RjwilmsiBot, DASH-Bot, XinaNicole, GenyAncalagon, Ad3l, Rcsprinter123, Zueignung, ClueBot NG, Kevin pirotto, Bibcode Bot, Krastanov, AvocatoBot, Zedshort, Mrt3366, Garuda0001, Wjs64, James floodhall, Frosty, Rsenk326, Jwratner1, HamiltonFromAbove, Anrnusna, Monkbot, Sofia Koutsouveli, ComicsAreJustAllRight, Tetra quark, Phseek, TychosElk and Anonymous: 105

- **Supernova nucleosynthesis** *Source:* https://en.wikipedia.org/wiki/Supernova_nucleosynthesis?oldid=673226455 *Contributors:* Tedernst, Ken Arromdee, Rursus, Crimson30, Mateuszica, Karol Langner, Mrtrey99, Vsmith, RJHall, CDN99, Bradkittenbrink, Tycho, AlexTiefling, Scott-

Davis, Drbogdan, Rjwilmsi, Strait, MarSch, Srleffler, Chobot, Beanyk, Uber nemo, Cadillac, Georgewilliamherbert, Modify, Cmglee, MacsBug, SmackBot, Eskimbot, Bluebot, Darth Panda, Rogermw, Voyajer, Mystman666, Jmnbatista, Flyguy649, Friendly Neighbour, CmdrObot, Fokion, King Hildebrand, James E B, Underpants, Headbomb, Orionus, Seddon, Magioladitis, Rhadamante, DinoBot, DAID, Sheliak, Claydonald, UnitedStatesian, BotKung, Wasted Sapience, Dpeinador~enwiki, Peterckw, Nergaal, Masterblooregard, Roberto Mura, Addbot, Roentgenium111, DOI bot, Lightbot, Ibmua, Tom.Reding, Jvandonsel, Jusses2, Arbnos, ClueBot NG, Helpful Pixie Bot, Calabe1992, Bibcode Bot, Kristaoz, Zedshort, Андрей Бондарь, Mfuerte, TheJJJunk, Coconutporkpie and Anonymous: 42

- **Nucleosynthesis** *Source:* https://en.wikipedia.org/wiki/Nucleosynthesis?oldid=682288599 *Contributors:* Vicki Rosenzweig, Mav, Roadrunner, Artsygeek, Andres, Samw, Cherkash, Epo~enwiki, Dcoetzee, Reddi, Stone, Stormie, Cmbant, Korath, Arkuat, Rursus, Xanzzibar, GreatWhiteNortherner, Giftlite, Harp, Herbee, Curps, Gzornenplatz, Sidar, Karol Langner, Tdent, D6, Perey, Pjacobi, Vsmith, Eric Forste, TaintedMustard, Voxadam, Oliphaunt, Benbest, Rjwilmsi, Strait, Goudzovski, Fogelmatrix, Chobot, Spacepotato, Sir48, Beanyk, BeastRHIT, Uber nemo, Light current, Modify, GrinBot~enwiki, Cmglee, Nekura, SmackBot, KnowledgeOfSelf, Chris the speller, Sbharris, Colonies Chris, Siffler~enwiki, Krich, Nakon, OhioFred, JorisvS, Iridescent, Zaphody3k, Kurtan~enwiki, Colonel Marksman, Van helsing, MrFizyx, Lokal Profil, Myasuda, Hga, James E B, Agony, Thijs!bot, Epbr123, Headbomb, Gierszep, Orionus, IanOsgood, Dosbears, Jingxin, Jessicapierce, Marhault, J.delanoy, Xarqi, Drake Dun, Rex07, DorganBot, Idioma-bot, Aucitypops, 28bytes, ABF, Szymanda, Claydonald, JhsBot, UnitedStatesian, BotKung, Samuelih, Michael Frind, Scarian, Paradoctor, Breakyunit, LeoBC, Auntof6, DragonBot, Taxa, Shinkolobwe, HexaChord, Addbot, DOI bot, Climbingfool, Loupeter, Yobot, Reindra, Azcolvin429, AnomieBOT, Piano non troppo, Materialscientist, Citation bot, LilHelpa, Xqbot, Tucsoncasey, Mnmngb, Thehelpfulbot, FrescoBot, Citation bot 1, Tom.Reding, Double sharp, Extra999, Nucleosynthesis, Kaiomai, Androstachys, Egjensen, Hovgiv, Virtual Loïc, RockMagnetist, Starbuster39, Terraflorin, Llightex, ClueBot NG, Law of Entropy, Helpful Pixie Bot, BG19bot, M Behnia, Zedshort, BattyBot, Khazar2, Martiantenor, Mogism, James floodhall, Vanamonde93, UnTrueOrUnSimplified, Linuxjava, Trackteur, WAFred, Tetra quark, Boazkat, Larry8000 and Anonymous: 97

- **Chronology of the universe** *Source:* https://en.wikipedia.org/wiki/Chronology_of_the_universe?oldid=682686009 *Contributors:* AxelBoldt, Bryan Derksen, The Anome, Roadrunner, Heron, Modemac, Edward, Patrick, Boud, PhilipMW, Earth, Looxix~enwiki, Mkweise, Ahoerstemeier, Jebba, Darkwind, EdH, Timwi, A1r, Reddi, Doradus, Zoicon5, Maximus Rex, Topbanana, Lord Emsworth, TravelingDude, Chrisjj, Flockmeal, BenRG, Robbot, Astronautics~enwiki, Jotomicron, Gandalf61, Raeky, Tobias Bergemann, Ancheta Wis, Giftlite, Lethe, Bkonrad, BrendanRyan, Eequor, Edcolins, OldakQuill, Coldacid, Opera hat, Piotrus, CSTAR, Tothebarricades.tk, Icairns, JDoolin, Lumidek, Darksun, Njh@bandsman.co.uk, D6, FT2, Pjacobi, ArnoldReinhold, Dave souza, Dbachmann, El C, Art LaPella, Drhex, CDN99, Bobo192, Army1987, Evolauxia, Maurreen, I9Q79oL78KiL0QTFHgyc, La goutte de pluie, Holdek, Pearle, Demi, *Kat*, RJFJR, Geraldshields11, Vuo, Gene Nygaard, Mindmatrix, FeanorStar7, Jeff3000, Duncan.france, Colin Watson, Joke137, Alan Canon, Christopher Thomas, RichardWeiss, Surnólë, Zalasur, Drbogdan, Rjwilmsi, Zbxgscqf, Captain Disdain, Mike Peel, SeanMack, Krash, Yamamoto Ichiro, Nihiltres, Harmil, Who, SouthernNights, Pathoschild, Mathrick, Masnevets, Wjfox2005, YurikBot, Spacepotato, Mushin, Vuvar1, Jimp, TheDoober, Ericorbit, Stephenb, NawlinWiki, Siddiqui, Chrisbrl88, Raven4x4x, Pudist, Dbfirs, Bota47, Superiority, Leptictidium, Lenis, Enormousdude, Aremisasling, Arthur Rubin, Tevildo, GraemeL, Peyna, CWenger, ScoutNZ, Kevin, Geoffrey.landis, Albert ip, RG2, Hirudo, Paul Erik, GrinBot~enwiki, Serendipodous, KnightRider~enwiki, SmackBot, RDBury, MrDemeanour, Mehranwahid, Ashill, WildElf, AnOddName, David G Brault, Gilliam, BirdValiant, Saros136, JRSP, Bluebot, Stevenwagner, MrDrBob, Mergatroidal, Colonies Chris, Scwlong, Vladis1av, Hve, Chlewbot, Andy120290, Jgoulden, Huon, Emre D., John D. Croft, Pwjb, Yevgeny Kats, CIS, SashatoBot, Lambiam, Swatjester, Writtenonsand, Count Caspian, Ckatz, Ex nihil, JHunterJ, Stwalkerster, SQGibbon, Hypnosifl, JeffW, Vanished user, Pegasus1138, Chetvorno, Fairyhairycarpetfluff, CWY2190, MaxEnt, Ramitmahajan, LouisBB, Gogo Dodo, Huysman, Michael C Price, Abtract, Thijs!bot, Fournax, Quarkchord, Sopranosmob781, Headbomb, Najro, Marek69, James086, Second Quantization, Peter Gulutzan, AntiVandalBot, Obiwankenobi, Prolog, DarkAudit, Jdfekete, Shlomi Hillel, JAnDbot, MER-C, Xeno, Rothorpe, Ophion, Dward Fardhard, VoABot II, Coredumperror, Catgut, Cgingold, Tenacious221, DinoBot, MartinBot, Anaxial, CommonsDelinker, Leyo, J.delanoy, Love Krittaya, ARTE, Clemm, Madhava 1947, STBotD, Fuenfundachtzig, GrahamHardy, Idioma-bot, Sheliak, Black Kite, VolkovBot, TreasuryTag, Jmrowland, Cosmic Latte, Andrius.v, Clarince63, Seraphim, UnitedStatesian, Madhero88, Hanjabba, Richwil, James McBride, Hibou21, MCTales, Alcmaeonid, NHRHS2010, Thw1309, Weezymagic, Llehctimj, Momo san, Oxymoron83, Globaleducator, Steven Crossin, Sean.hoyland, Namiluko, Denisarona, Escape Orbit, Martarius, ClueBot, Snigbrook, Taffboyz, Gawaxay, Wanderer57, Wysprgr2005, Cp111, Nondescripts, TheOldJacobite, CounterVandalismBot, Thomas Kist, ChandlerMapBot, Rprpr, DragonBot, Excirial, Gnome de plume, CrazyChemGuy, Eeekster, Scog, Soccerguy7735, Panos84, Subash.chandran007, DumZiBoT, XLinkBot, Tarlneustaedter, Scottman162, Jmacwiki, Aunt Entropy, CosmologyProfessor, Maldek, Addbot, Some jerk on the Internet, Medich1985, Spiderbo13, Ersik, E.pajer, Markkujn, 84user, F Notebook, Tide rolls, 豫, Whatismetric, Yinweichen, Yobot, TaBOTzerem, THEN WHO WAS PHONE?, Robert Treat, AnomieBOT, Materialscientist, Citation bot, LilHelpa, Obersachsebot, HenryCorp, Wikikone, ProtectionTaggingBot, Omnipaedista, The Wiki Octopus, Smallman12q, Kikuyu3, Sae1962, Oashi, Citation bot 1, Dogaru Florin, Pinethicket, Tom.Reding, Naturehead, Seryo93, Saayiit, Extra999, Canuck100, JLincoln, Orca54, DeltaAce, Michaelbernal, Carlborden, DexDor, Tesseract2, EmausBot, GoingBatty, Rowan casey, Solomonfromfinland, Hhhippo, John Cline, Cogiati, Susfele, Medeis, Thedropsoffire, Mikeyfresh1556, Makecat, Wagino 20100516, Kirothereaper, RockMagnetist, Pandeist, AUN4, ClueBot NG, Astrocog, Jack Greenmaven, Gilderien, Piast93, Bped1985, James childs, Snotbot, Frietjes, Tinyplanetmusic, Cpluhar, Finchy0, Apopp8333, Bibcode Bot, Lowercase sigmabot, Rarelight, Cadiomals, Harizotoh9, Andi2011, Duxwing, Amphibio, ShizlGzngar, Justincheng12345-bot, Soulbust, Avneref, Wjs64, Mixert, 069952497a, SomeFreakOnTheInternet, AlbertoEspancho, IvanderClarent, Jwratner1, Kogge, SpeedEvil, Concord hioz, Monkbot, Sofia Koutsouveli, Wikipedian 2, Officer Sid X, Julietdeltalima, OmoiEgaite, Tetra quark, Xytreyum, Isambard Kingdom, Therewasnoothernameguys1, KSFT, Esadri21 and Anonymous: 347

- **Chemical bond** *Source:* https://en.wikipedia.org/wiki/Chemical_bond?oldid=682498967 *Contributors:* AxelBoldt, Sodium, Tarquin, Rgamble, William Avery, Peterlin~enwiki, Dwmyers, Patrick, Michael Hardy, Dan Koehl, Ixfd64, Kricxjo, WeißNix, Darkwind, Александър, Ghewgill, Smack, Raven in Orbit, Hashar, Stephenw32768, Timwi, Lfh, Gentgeen, Robbot, DHN, Quadalpha, Tobias Bergemann, Marc Venot, Centrx, Giftlite, No Guru, Eequor, Falcon Kirtaran, Pcarbonn, Quadell, Karol Langner, NoPetrol, Frangibility, Discospinster, Rich Farmbrough, Cacycle, Vsmith, Mani1, Djordjes, Sten, Shanes, Bobo192, Michaf~enwiki, La goutte de pluie, Jojit fb, Obradovic Goran, Haham hanuka, Pearle, Yougi, Passw0rd, Jumbuck, Alansohn, Mo0, Velella, RainbowOfLight, Dirac1933, Shoefly, Redvers, Zntrip, Nuno Tavares, OwenX, Mindmatrix, Simon Shek, Rparson, StradivariusTV, WadeSimMiser, JeremyA, MONGO, Tabletop, Kmg90, Lol~enwiki, Ghostofgauss, Palica, RuM, Kesla, V8rik, Jobnikon, Rjwilmsi, Yamamoto Ichiro, FlaBot, Gseryakov, RexNL, Gurch, Takometer, Physchim62, Imnotminkus, Korg, YurikBot, Wavelength, Ashleyisachild, RobotE, Phantomsteve, RussBot, Shawn81, Ironist, Rsrikanth05, Alynna Kasmira, NawlinWiki, Welsh, Irishguy, Speedevil, Tony1, Bucketsofg, Zythe, Kortoso, User27091, Enormousdude, Closedmouth, Reyk, Anclation~enwiki, SalM, Katieh5584,

RG2, Ásgeir IV.~enwiki, GrinBot~enwiki, Itub, SmackBot, FocalPoint, Bobet, Reedy, Bomac, KocjoBot~enwiki, Jrockley, Delldot, Onebrave-monkey, Edgar181, Xaosflux, Ohnoitsjamie, Andy M. Wang, Bduke, Dreg743, Bethling, SchfiftyThree, DHN-bot~enwiki, Sbharris, Colonies Chris, Quaque, Can't sleep, clown will eat me, Celarnor, SundarBot, Krich, Nakon, BWDuncan, Richard001, Rebooted, G N Frykman, Smoke-foot, DMacks, SteveLower, Sadi Carnot, Vanished user 9i39j3, TempyIncursion, Scientizzle, Solon.KR, Gobonobo, Rundquist, Peterlewis, Aleenf1, IronGargoyle, Munita Prasad, Beetstra, Mathlizard, Amitch, HisSpaceResearch, BananaFiend, JoeBot, InvisibleK, CmdrObot, Ale jrb, GargoyleMT, WeggeBot, Cracker017, Linberry, UncleBubba, Rifleman 82, Shirulashem, Emmett5, EnglishEfternamn, JamesAM, Thijs!bot, Epbr123, 24fan24, Marek69, Speedyboy, Baron162, Tellyaddict, Mentifisto, AntiVandalBot, Isilanes, Gökhan, JAnDbot, Husond, Kruckenberg.1, PhilKnight, Beaumont, Kerotan, Connormah, VoABot II, JNW, Albmont, WODUP, Brusegadi, Catgut, Zhanghia, Daarznieks, Eldumpo, Dirac66, 28421u2232nfenfcenc, User A1, Cpl Syx, DerHexer, Excesses, MartinBot, ChemNerd, Rettetast, Leyo, Mayrell, Manticore, J.delanoy, Pharaoh of the Wizards, Trusilver, Crase, P.wormer, Petersec, Jcwf, NewEnglandYankee, KChiu7, Cmichael, Elenseel, Useight, Vatic7, Lights, Rasengan 13, 28bytes, VolkovBot, Csellersatmicdsdotorg, Stefan Kruithof, VasilievVV, Ryan032, Philip Trueman, TXiKiBoT, The Original Wildbear, Lradrama, LeaveSleaves, Shanata, Lerdthenerd, Anna512, NHRHS2010, Manvi ac, EJF, Minestrone Soup, SieBot, OMCV, Tiddly Tom, BotMultichill, Krawi, Winchelsea, WRK, Flyer22, Xkfusionxk, Oda Mari, Arbor to SJ, Oxymoron83, Techman224, KathrynLybarger, Macy, Mygerardromance, WikiLaurent, Dabomb87, Superbeecat, Nergaal, Efe, Explicit, ImageRemovalBot, WikipedianMarlith, WikiBotas, ClueBot, GorillaWarfare, The Thing That Should Not Be, DesertAngel, Drmies, Meganismaria, Estevoaei, Harland1, Officer781, Auntof6, Say2anniyan, 718 Bot, Excirial, Jusdafax, Eeekster, NuclearWarfare, Jotterbot, DrHood4, Morel, Razorflame, Versus22, Pzoxicuvybtnrm, SoxBot III, Vanished User 1004, DumZiBoT, Darkicebot, XLinkBot, Dark Mage, Duncan, Biffa1990, Finooccidere, Thatguyflint, Addbot, DOI bot, Wsvlqc, Fieldday-sunday, SpillingBot, LaaknorBot, EconoPhysicist, Scott1221, AndersBot, Johntrin, LinkFA-Bot, West.andrew.g, Tyw7, Kisbesbot, Species8473, Tide rolls, Pietrow, Teles, Gail, Zorrobot, Ettrig, Bssquirrel, Legobot, Luckas-bot, Yobot, Sirsparksalot, Smartyashas, 잃어버린세계, Eric-Wester, AnomieBOT, Ciphers, Jim1138, Walrus heart, 9258fahsflkh917fas, RandomAct, Flewis, Materialscientist, Citation bot, Vulcan Hephaestus, Xqbot, Jeffy386, Intelati, Capricorn42, Firefox767, Ruy Pugliesi, GrouchoBot, Pavitb, Wizardist, Backpackadam, RibotBOT, Shadowjams, Thehelpfulbot, FrescoBot, Muhammad barma, JMS Old Al, SenaiERI, Citation bot 1, Pinethicket, Piandcompany, Thecurran91, Nafile, John Christian D. Cañada, Gamewizard71, Noatakzak, Vrenator, Bluefist, Obikwan13, Jfmantis, 808daman, EmausBot, Thecreator09, John of Reading, RenamedUser01302013, Wikipelli, Hhhippo, Daonguyen95, Fæ, Akerans, AManWithNoPlan, L Kensington, Donner60, Sailsbystars, Niro061196, Zac, DASHBotAV, Whoop whoop pull up, Petrb, Xanchester, 49greene1915, ClueBot NG, Gareth Griffith-Jones, Satellizer, Chester Markel, Sloberyleaf, Widr, Helpful Pixie Bot, HMSSolent, Lowercase sigmabot, Vibhor1997, Mynameis-noted, PhnomPencil, Wiki13, Mark Arsten, IraChesterfield, Rm1271, Hassan540, Samtez, Srishty singh, Shawn Worthington Laser Plasma, Glacialfox, Anbu121, Madprofessional, Mdann52, Total wiki guy, Blaksabath1, Minkevich, Mathar.hasan, Timothy Gu, Chemya, Dexbot, Webclient101, Lugia2453, Leprof 7272, Epicgenius, 23sheena23, Ananthkamath1995, DavidLeighEllis, Ugog Nizdast, Khenta37, Mdann52(alt), John Doppler, Mahusha, Monkbot, Silberbuschc17, SantiLak, Gabewalker5464, BethNaught, ZBarnes2271, Alango1998, Y-S.Ko, Blueworld-Speccie, Chickenmaster251, MacPoli1, IndrasisSD, GeneralizationsAreBad, KasparBot, RJ raghava, CyberWarfare and Anonymous: 718

- **Chemical compound** *Source:* https://en.wikipedia.org/wiki/Chemical_compound?oldid=682607776 *Contributors:* AxelBoldt, Tarquin, AstroNomer~enwiki, Malcolm Farmer, Tim Starling, Lexor, Kku, Menchi, Ixfd64, Ahoerstemeier, Ronz, Snoyes, Angela, AugPi, Mxn, Raven in Orbit, Schneelocke, Saltine, Taxman, Geraki, Jni, Donarreiskoffer, Gentgeen, Robbot, Pigsonthewing, Fredrik, Altenmann, Romanm, Catbar, Hadal, JesseW, Wikibot, Mushroom, Marc Venot, Centrx, Giftlite, Aphaia, Everyking, Eequor, Utcursch, Pgan002, OverlordQ, Kraton, Mzajac, Zfr, Tsemii, Syvanen, Grunt, Corti, Mike Rosoft, Spiffy sperry, Discospinster, Cacycle, Vsmith, Xezbeth, Paul August, Brian0918, Aranel, Roy-Boy, ~K, Bobo192, Robotje, Elipongo, Guiltyspark, Knucmo2, Alansohn, Wirawan0, ABCD, Walkerma, Snowolf, Velella, Super-Magician, Cburnett, Yuckfoo, Tm1000, Simon Shek, Kurzon, Hurricane Angel, Plrk, Palica, V8rik, Cuchullain, Rjwilmsi, DeadlyAssassin, Tawker, Bhadani, Titoxd, FlaBot, ACrush, Nihiltres, Gurch, Chobot, Sbrools, Banaticus, DerrickOswald, YurikBot, Wavelength, Chaser, Rsrikanth05, Thane, NawlinWiki, Wiki alf, Robertvan1, DarthVader, DeadEyeArrow, Private Butcher, Elkman, Nlu, Dv82matt, Silverchemist, FF2010, Closedmouth, Јованбб, KGasso, LeonardoRob0t, Willtron, Spliffy, Katieh5584, Kungfuadam, Monk of the highest order, Moomoomoo, Paul Erik, GrinBot~enwiki, DVD R W, Itub, Prodego, KnowledgeOfSelf, Bggoldie~enwiki, Bomac, Delldot, SmartGuy Old, Gilliam, Rmosler2100, Cowman109, Quinsareth, Roscelese, SchfiftyThree, Bonaparte, CyberSach, Octahedron80, DHN-bot~enwiki, Sbharris, Xchbla423, Can't sleep, clown will eat me, Chlewbot, Malindi man, JonHarder, Rrburke, VMS Mosaic, SundarBot, Khoikhoi, Nakon, Blake-, Smokefoot, Inositol, Sashato-Bot, Dak, Thopper, Microchip08, UberCryxic, Scientizzle, Kipala, TheNeon, Yanwen, Rawmustard, Beetstra, Noah Salzman, Mr Stephen, Dbo789, Optakeover, Nageeb, Mapleleafedge, Zapvet, Wizard191, Civil Engineer III, Redtitan, Tawkerbot2, JForget, Betaeleven, CmdrObot, Ale jrb, Harej bot, Benwildeboer, RobertLovesPi, Mato, EdiOnjales, Palffy, Pascal.Tesson, Christian75, DumbBOT, Alaibot, SpK, Omicron-persei8, Nol888, RickDC, Thijs!bot, Epbr123, Kablammo, Stanislav87, Marek69, John254, AgentPeppermint, Escarbot, Band geek13, Anti-VandalBot, Michael phan, Seaphoto, Gökhan, Res2216firestar, JAnDbot, Leuko, MER-C, Ericoides, Instinct, Imoeng, Meeples, Bongwarrior, VoABot II, Dekimasu, ToaRabin, Catgut, WhatamIdoing, Animum, Robotman1974, Dirac66, Vssun, TehBrandon, Talon Artaine, DerHexer, Khalid Mahmood, Pax:Vobiscum, Nietzscheanlie, MartinBot, Poeloq, Anaxial, Burnedthru, CommonsDelinker, Leyo, Tgeairn, AnimaniacRu, Trusilver, Rhinestone K, Dwarf King, Davidprior, Jess4909, DarkFalls, McSly, Gurchzilla, Pyrospirit, RoboMaxCyberSem, NewEnglandYankee, Cometstyles, Burzmali, Treisijs, Skeetypeety, Doctoroxenbriery, Useight, Rexparry sydney, Idioma-bot, Idarin, X!, VolkovBot, CWii, Indubitably, VasilievVV, Philip Trueman, TXiKiBoT, Vipinhari, Lradrama, Martin451, Leafyplant, ^demonBot2, Psyche825, Ekor, Madhero88, Mwilso24, AJmon13, Synthebot, LiLy.xXx, IvaB~enwiki, Brianga, Spamer123, Nagy, Logan, Struway, Gregelectro13, AngChenrui, SieBot, Sonicology, AS, BotMultichill, Jauerback, Fabullus, Gerakibot, Caltas, Keilana, RucasHost, Flyer22, Radon210, Captain Yankee, Rollosm, Oxymoron83, Faradayplank, Baseball Bugs, Kosack, Fkobylecki, Iain99, Alex.muller, Sunrise, Web107, Jacob.jose, ML-Est, Superbeecat, Denisarona, Llywelyn2000, ClueBot, GorillaWarfare, The Thing That Should Not Be, Jazzman831, VsBot, Amoeba91, Niceguyedc, LonelyBeacon, I am a violinist, Puchiko, JustinClarkCasey, DragonBot, Mkativerata, Sosurmum, Tyler, Jotterbot, Saebjorn, Friedlibend und tapfer, Bald Zebra, Tired time, Thingg, Aitias, Iamsabs, Versus22, Qwfp, Party, DumZiBoT, Darkicebot, Mjharrison, Pichpich, Rawr123321, Mifter, JinJian, Addbot, Some jerk on the Internet, Tcncv, Captain-tucker, Ronhjones, Vishnava, CanadianLinuxUser, Leszek Jańczuk, Download, CarsracBot, Quercus solaris, Xie Hao Zhe, Tide rolls, OlEnglish, Luckas-bot, 2D, Newportm, II MusLiM HyBRiD II, Amirobot, Tempodivalse, Orion11M87, AnomieBOT, Shootbamboo, DemocraticLuntz, 1exec1, Daniele Pugliesi, Floozybackloves, Infinituslacuna, AdjustShift, Kingpin13, Sz-iwbot, Law, Mikey1183, Materialscientist, ImperatorExercitus, The High Fin Sperm Whale, Citation bot, ArthurBot, Xqbot, Civjaty, RJav, महाकाल, Addihockey10, Bihco, 4twenty42o, Nasnema, Grim23, GrouchoBot, Frosted14, حامد میرزاحسینی, Bellerophon, Amaury, Boudiicca, Schnoebi, Sesu Prime, R8R Gtrs, VS6507, HJ Mitchell, DivineAlpha, Citation bot 1, Pinethicket, I dream of horses, KAPITALIST88, Serols, WriteRight1stTime, Jauhienij, Yoiloper, Diannaa, Sirkablaam, Reach Out to the Truth, Brambleclawx, DARTH SIDIOUS 2, RjwilmsiBot, Bento00, Simon Jäkle, NerdyScienceDude, Slon02, Skamecrazy123, DASHBot, EmausBot, Orphan Wiki, Gfoley4, JawsBrody, Astridtate9, Super48paul,

Awesomeness16and, Bibcode Bot, Kevin451~enwiki, Cor234, Lowercase sigmabot, Krenair, Coasthill, MusikAnimal, Planetary Chaos Redux, BPositive, Mark Arsten, CarloMartinelli, Contian2001, GayBlade111111, CitationCleanerBot, Benavidezt, Applesarereallycool100, SuchA-Cute, Loriendrew, ChrisGualtieri, EuroCarGT, Dexbot, Webclient101, TwoTwoHello, Frosty, Haroldmedina, Little green rosetta, CarlesMillan, Cindy sabellina, Telfordbuck, Stanlyblake, Xingq, Ugog Nizdast, Jaredzimmerman (WMF), Jwratner1, Ginsuloft, Hkmaher01, Arthur.nnf, Disgurl, Jinnyskeans, JaconaFrere, Alexxxxxxla, Monkbot, Vcnmblgvhb,jb, Lilythechanger123, Derfla1234, Arthurfoy30, Maths314, 2002bp, Iiicceeeyy, Mjj4life, MacPoli1, KasparBot, Mckburton and Anonymous: 1175

- **Noble gas** *Source:* https://en.wikipedia.org/wiki/Noble_gas?oldid=679000009 *Contributors:* AxelBoldt, Carey Evans, Derek Ross, LC~enwiki, Mav, Tarquin, Andre Engels, Christian List, DrBob, Fonzy, Hephaestos, Olivier, Edward, Shellreef, Ahoerstemeier, Jimfbleak, Rlandmann, Salsa Shark, Nikai, GCarty, Smack, Lenaic, Stone, Jake Nelson, Tero~enwiki, Paul-L~enwiki, Taoster, Betterworld, Fvw, Shantavira, Donarreiskoffer, Gentgeen, Robbot, Sander123, Jakohn, Romanm, Merovingian, Rfc1394, Flauto Dolce, Meelar, Mervyn, Hadal, JackofOz, Robinh, Lupo, Dina, Giftlite, DocWatson42, Herbee, Monedula, Karn, Everyking, Slyguy, Kandar, Quackor, Andycjp, Antandrus, 1297, Icairns, Sam Hocevar, Gscshoyru, Deglr6328, Adashiel, Mike Rosoft, DanielCD, Discospinster, Rich Farmbrough, Vsmith, Ponder, Paul August, NeilTarrant, Geoking66, RJHall, El C, Dnwq, Shanes, Remember, Sietse Snel, Art LaPella, Femto, Marco Polo, Shenme, Viriditas, SpeedyGonsales, Severious, Obradovic Goran, Haham hanuka, Nsaa, Eddideigel, Orangemarlin, Ranveig, Jumbuck, Alansohn, Gary, Jared81, Keenan Pepper, Plumbago, Cjthellama, InShaneee, Suruena, Vuo, Gene Nygaard, Feline1, Kay Dekker, Boothy443, Woohookitty, Cimex, TarmoK, LOL, Pol098, WadeSimMiser, Mpatel, Wayward, Shanedidona, Palica, Stevey7788, Paxsimius, Graham87, David Levy, DePiep, Canderson7, Rjwilmsi, Mfwills, Vary, HappyCamper, Matjlav, Vuong Ngan Ha, RobertG, Latka, Nihiltres, Strangnet, RexNL, Mjp797, DevastatorIIC, Goudzovski, Scerri, Kri, King of Hearts, Chobot, DVdm, Bgwhite, EamonnPKeane, YurikBot, Chaser, Ollie holton, Yyy, NawlinWiki, Bachrach44, Jaxl, Adamn, Semperf, Zirland, Bota47, T-rex, Thetoaster3, Wknight94, Tetracube, Leptictidium, Phgao, Zzuuzz, Adilch, Scoutersig, Keepiru, HereToHelp, Katieh5584, NeilN, GrinBot~enwiki, DVD R W, Tom Morris, That Guy, From That Show!, Itub, Anthony Duff, SmackBot, FocalPoint, Tarret, KnowledgeOfSelf, Shoy, Kilo-Lima, Edgar181, Gaff, Aksi great, Gilliam, Isaac Dupree, Pslawinski, Durova, Bluebot, Bduke, SchfiftyThree, Moshe Constantine Hassan Al-Silverburg, CSWarren, Robth, DHN-bot~enwiki, Darth Panda, Gyrobo, Tsca.bot, Eric Olson, MJCdetroit, Rrburke, Aldaron, Nakon, Eganev, Pwjb, Smokefoot, DMacks, Wizardman, The undertow, SashatoBot, Lambiam, Krashlandon, Titus III, John, Zaphraud, Jaganath, Breno, Anoop.m, IronGargoyle, JHunterJ, Slakr, Beetstra, Noah Salzman, Optimale, Kpengboy, SandyGeorgia, AdultSwim, MTSbot~enwiki, BranStark, Iridescent, StephenBuxton, Jaksmata, Tawkerbot2, Swampgas, JForget, Irwangatot, Ruslik0, CompRhetoric, Dorothybaez, Bentleymrk, Gogo Dodo, A Softer Answer, Hibou8, SteveMcCluskey, Mattisse, Thijs!bot, Epbr123, Kablammo, Headbomb, JaimeAnnaMoore, Straussian, Werdnanoslen, Dantheman531, Mentifisto, AntiVandalBot, Luna Santin, Opelio, Cinnamon42, Scepia, LibLord, Xnuiem, Gökhan, IanOsgood, Nicholas Tan, Easchiff, Animaly2k2, Magioladitis, Bongwarrior, VoABot II, Hasek is the best, Ling.Nut, LorenzoB, Thibbs, Vssun, DerHexer, JaGa, Awolnetdiva, Mattinbgn, Hdt83, ChemNerd, Polartsang, CommonsDelinker, AlexiusHoratius, Leyo, J.delanoy, Pharaoh of the Wizards, Phillip.northfield, Hans Dunkelberg, Dhruv17singhal, Uncle Dick, Jeri Aulurtve, Extransit, Cpiral, Bombhead, Wandering Ghost, Shay Guy, Coppertwig, Plasticup, Andraaide, Belovedfreak, Acey365, NewEnglandYankee, Najlepszy, Numerjeden, Matthardingu, KChiu7, WinterSpw, Brvman, Wilhelm meis, Squids and Chips, Idioma-bot, Wikieditor06, Deor, VolkovBot, Eakka, JGHowes, Jeff G., Chris Dybala, AlnoktaBOT, Nousernamesleft, Philip Trueman, Photonikonman, TXiKiBoT, Tavix, GimmeBot, Muro de Aguas, Rei-bot, Slysplace, Enigmaman, Davidmwhite, CephasE, Sylent, TinribsAndy, Owainbut, AlleborgoBot, Surfrat60793, SieBot, Calliopejen1, OTAVIO1981, Graham Beards, BotMultichill, ToePeu.bot, Jauerback, Nathan, Triwbe, Agesworth, Keilana, Flyer22, The Evil Spartan, Arbor to SJ, Sohelpme, Scorpion451, Enok Walker, Lightmouse, OKBot, Nielg, Nimbusania, Nergaal, Escape Orbit, PerpetualSX, Runtishpaladin, UKe-CH, Martarius, ClueBot, Artichoker, The Thing That Should Not Be, Cygnis insignis, Manbearpig4, Franamax, Blanchardb, Piledhigheranddeeper, ChandlerMapBot, Puchiko, DragonBot, Excirial, Sidias300, GngstrMNKY, Jusdafax, Finch-HIMself, Estirabot, Poigol5043, Cenarium, Zomno, Jotterbot, Bellax22, Chaser (away), Werson, Boatcolour, SeanFarris, Thingg, Aitias, Dank, Versus22, RexxS, Boleyn, Neuralwarp, Feinoha, Little Mountain 5, Skarebo, Frood, Freestyle-69, CalumH93, Addbot, Mr0t1633, Roentgenium111, DOI bot, Theleftorium, Popopee, Ronhjones, Jncraton, Moosehadley, CanadianLinuxUser, AnnaFrance, Jasper Deng, Alchemist-hp, Numbo3-bot, Tide rolls, Zorrobot, Angrysockhop, Arimareiji, Legobot, Seresin.public, Luckas-bot, ZX81, Yobot, IsFari, TaBOT-zerem, Rsquire3, Bloody Mary (folklore), KamikazeBot, Widey, Synchronism, Andme2, AnomieBOT, Lolcopter666, Jcsdude, Navneethmohan, Jim1138, IRP, Law, Materialscientist, Citation bot, E2eamon, Maxis ftw, ArthurBot, Xqbot, Capricorn42, Nickkid5, Tad Lincoln, Turk oğlan, NocturneNoir, Lop242438, Pmlineditor, GrouchoBot, Doulos Christos, Antonjad, Jilkmarine, Smot94, Robo37, OgreBot, Citation bot 1, AstaBOTh15, Pinethicket, HRoestBot, Calmer Waters, I own in the bed, Marine79, Double sharp, TobeBot, Yopure, 777sms, Navy101, Reach Out to the Truth, Minimac, DARTH SIDIOUS 2, AXRL, Mean as custard, RjwilmsiBot, Japheth the Warlock. Ripchip Bot, Salvio giuliano, Deagle AP, EmausBot, WikitanvirBot, RA0808, Jordan776, Wikipelli, P. S. F. Freitas, AvicBot, ZéroBot, Fingerginger1, Maxviwe, StringTheory11, H3llBot, Makecat, Wagino 20100516, L Kensington, Donner60, Whoop whoop pull up, JohnMCrain, Mjbmrbot, Special Cases, Washington Irving Esquire, ClueBot NG, Rich Smith, Jack Greenmaven, Hon-3s-T, Skoot13, Ethanpiot, Lanthanum-138, Widr, Lolm8, Bibcode Bot, Swamphlosion, Lowercase sigmabot, Gluonman, TCN7JM, Iankhou, Sandbh, MusikAnimal, Altaïr, WikisucksKNOBlegasses, VictorParker, Jimbo2440, Tycho Magnetic Anomaly-1, Softballbaby984, ThomasRules, BattyBot, Justincheng12345-bot, Abilanin, ChrisGualtieri, EuroCarGT, Dexbot, Webclient101, TwoTwoHello, King jakob c, RandomLittleHelper, Reatlas, Cteung, Ugog Nizdast, Ginsuloft, Noyster, DudeWithAFeud, Skr15081997, Matthewweber12, HotHabenero, Hotta stuffu, Sony Vark XIII, Monkbot, HiYahhFriend, ZYjacklin, Narky Blert, Jodihe93, Selimozd20, Anbgsm07, UZawMoeNaing, SandKitty256, Supdiop, KasparBot, Sat cheat and Anonymous: 591

- **Noble metal** *Source:* https://en.wikipedia.org/wiki/Noble_metal?oldid=678876296 *Contributors:* Kpjas, The Anome, DopefishJustin, Ike9898, Furrykef, Robbot, Rursus, Enochlau, Everyking, Icairns, Abdull, Rich Farmbrough, Vsmith, Shad0, Art LaPella, Whosyourjudas, Viriditas, Keenan Pepper, Sligocki, Wtshymanski, Benbest, Polyparadigm, Knuckles, Bluemoose, Mandarax, Enz, Margosbot~enwiki, Rune.welsh, Common Man, Physchim62, Chobot, Roboto de Ajvol, YurikBot, Wavelength, Borgx, Jimp, Hede2000, Bhny, Voyevoda, Mccready, Rayc, Skittle, Itub, SmackBot, Edgar181, Bluebot, Cybercobra, RandomP, Smokefoot, JorisvS, Wizard191, Metre01, Patrickwooldridge, N2e, Rifleman 82, Bsdaemon, Alaibot, Satori Son, Thijs!bot, Muaddeeb, Escarbot, Gioto, Wayiran, Amberroom, PhilKnight, McDoobAU93, STBotD, Idioma-bot, Christophenstein, TXiKiBoT, JhsBot, UnitedStatesian, Bcharles, Cmjayakumar, SieBot, ToePeu.bot, KoshVorlon, Sanya3, Chem-awb, Fangjian, Twinsday, Michał Sobkowski, Piledhigheranddeeper, Tablemajorrt5, Scyldscefing, EgraS, DumZiBoT, BodhisattvaBot, Heeero60, Addbot, DOI bot, Numbo3-bot, Donfbreed, Smackeldorf, AnomieBOT, Magog the Ogre 2, Materialscientist, Citation bot, Xqbot, GrouchoBot, Asfarer, DrilBot, Pinethicket, Achim1999, Bgpaulus, Double sharp, LilyKitty, Mtz1010, WikitanvirBot, Rami radwan, Chemicalinterest, Scientific29, Yugo312, ClueBot NG, Wikiphysicsgr, Hari Eswar SM, Sasakubo1717, Bibcode Bot, BG19bot, Softballbaby984, Nedgreiner, Wieldthespade. T.J.S.1 and Anonymous: 65

21.9.2   Images

- **File:2006-02-13_Drop-impact.jpg** *Source:* https://upload.wikimedia.org/wikipedia/commons/f/f8/2006-02-13_Drop-impact.jpg *License:* CC-BY-SA-3.0 *Contributors:* Picture taken and uploaded by Roger McLassus. *Original artist:* Roger McLassus

- **File:Aegirine-233494.jpg** *Source:* https://upload.wikimedia.org/wikipedia/commons/5/54/Aegirine-233494.jpg *License:* CC BY-SA 3.0 *Contributors:* http://www.mindat.org/photo-233494.html *Original artist:* Rob Lavinsky / iRocks.com

- **File:Ambox_important.svg** *Source:* https://upload.wikimedia.org/wikipedia/commons/b/b4/Ambox_important.svg *License:* Public domain *Contributors:* Own work, based off of Image:Ambox scales.svg *Original artist:* Dsmurat (talk · contribs)

- **File:Améthystre_sceptre2.jpg** *Source:* https://upload.wikimedia.org/wikipedia/commons/a/a3/Am%C3%A9thystre_sceptre2.jpg *License:* CC BY-SA 3.0 *Contributors:* Own work *Original artist:* Didier Descouens

- **File:Andradite-172390.jpg** *Source:* https://upload.wikimedia.org/wikipedia/commons/d/d1/Andradite-172390.jpg *License:* CC BY-SA 3.0 *Contributors:* http://www.mindat.org/photo-172390.html *Original artist:* Rob Lavinsky / iRocks.com

- **File:ArTube.jpg** *Source:* https://upload.wikimedia.org/wikipedia/commons/2/2f/ArTube.jpg *License:* CC BY-SA 2.5 *Contributors:* user-made *Original artist:* User:Pslawinski

- **File:Argon-glow.jpg** *Source:* https://upload.wikimedia.org/wikipedia/commons/5/53/Argon-glow.jpg *License:* CC BY 3.0 *Contributors:* http://images-of-elements.com/argon.php *Original artist:* Jurii

- **File:Argon_Spectrum.png** *Source:* https://upload.wikimedia.org/wikipedia/commons/3/37/Argon_Spectrum.png *License:* CC BY-SA 3.0 *Contributors:* Own work http://goiphone5.com/ *Original artist:* Abilanin

- **File:Argon_discharge_tube.jpg** *Source:* https://upload.wikimedia.org/wikipedia/commons/8/87/Argon_discharge_tube.jpg *License:* GFDL 1.2 *Contributors:* Own work *Original artist:* Alchemist-hp (<a href='//commons.wikimedia.org/wiki/User_talk:Alchemist-hp' title='User talk: Alchemist-hp'>talk</a>) (www.pse-mendelejew.de)

- **File:Arthur_Stanley_Eddington.jpg** *Source:* https://upload.wikimedia.org/wikipedia/commons/2/24/Arthur_Stanley_Eddington.jpg *License:* Public domain *Contributors:* This image is available from the United States Library of Congress's Prints and Photographs division under the digital ID ggbain.38064.
  This tag does not indicate the copyright status of the attached work. A normal copyright tag is still required. See Commons:Licensing for more information. *Original artist:* George Grantham Bain Collection (Library of Congress)

- **File:Asbestos_with_muscovite.jpg** *Source:* https://upload.wikimedia.org/wikipedia/commons/c/cd/Asbestos_with_muscovite.jpg *License:* Public domain *Contributors:* Own work *Original artist:* Aram Dulyan (User:Aramgutang)

- **File:Asterisks_one.svg** *Source:* https://upload.wikimedia.org/wikipedia/commons/4/49/Asterisks_one.svg *License:* CC BY-SA 3.0 *Contributors:* Own work *Original artist:* DePiep

- **File:Asterisks_one_(right).svg** *Source:* https://upload.wikimedia.org/wikipedia/commons/1/1c/Asterisks_one_%28right%29.svg *License:* CC BY-SA 3.0 *Contributors:* Own work *Original artist:* DePiep

- **File:Asterisks_two.svg** *Source:* https://upload.wikimedia.org/wikipedia/commons/3/3f/Asterisks_two.svg *License:* CC BY-SA 3.0 *Contributors:* Own work *Original artist:* DePiep

- **File:Barium_unter_Argon_Schutzgas_Atmosphäre.jpg** *Source:* https://upload.wikimedia.org/wikipedia/commons/1/16/Barium_unter_Argon_Schutzgas_Atmosph%C3%A4re.jpg *License:* Public domain *Contributors:* yes *Original artist:* Matthias Zepper

- **File:Biotite-Orthoclase-229808.jpg** *Source:* https://upload.wikimedia.org/wikipedia/commons/d/d9/Biotite-Orthoclase-229808.jpg *License:* CC BY-SA 3.0 *Contributors:* http://www.mindat.org/photo-229808.html *Original artist:* Rob Lavinsky / iRocks.com

- **File:Bromine_vial_in_acrylic_cube.jpg** *Source:* https://upload.wikimedia.org/wikipedia/commons/3/35/Bromine_vial_in_acrylic_cube.jpg *License:* CC BY-SA 3.0 de *Contributors:* Own work *Original artist:* Alchemist-hp (pse-mendelejew.de)

- **File:CMB_Timeline300_no_WMAP.jpg** *Source:* https://upload.wikimedia.org/wikipedia/commons/6/6f/CMB_Timeline300_no_WMAP.jpg *License:* Public domain *Contributors:* Original version: NASA; modified by Ryan Kaldari *Original artist:* NASA/WMAP Science Team

- **File:CNO_Cycle.svg** *Source:* https://upload.wikimedia.org/wikipedia/commons/2/21/CNO_Cycle.svg *License:* CC BY-SA 3.0 *Contributors:* ? *Original artist:* ?

- **File:Calcite-Galena-elm56c.jpg** *Source:* https://upload.wikimedia.org/wikipedia/commons/9/92/Calcite-Galena-elm56c.jpg *License:* CC BY-SA 3.0 *Contributors:* Image: http://www.irocks.com/db_pics/pics/elm56c.jpg, Description: http://www.pink.com/ *Original artist:* Rob Lavinsky / iRocks.com

- **File:Carnotite-201050.jpg** *Source:* https://upload.wikimedia.org/wikipedia/commons/b/b8/Carnotite-201050.jpg *License:* CC BY-SA 3.0 *Contributors:* http://www.mindat.org/photo-201050.html *Original artist:* Rob Lavinsky / iRocks.com

- **File:Carson_Fall_Mt_Kinabalu.jpg** *Source:* https://upload.wikimedia.org/wikipedia/commons/5/57/Carson_Fall_Mt_Kinabalu.jpg *License:* CC BY-SA 3.0 *Contributors:* Own work *Original artist:* Sze Sze SOO

- **File:Cinnabar_on_Dolomite.jpg** *Source:* https://upload.wikimedia.org/wikipedia/commons/9/9e/Cinnabar_on_Dolomite.jpg *License:* CC BY-SA 3.0 *Contributors:* Own work *Original artist:* JJ Harrison (jjharrison89@facebook.com)

- **File:Coloured-transition-metal-solutions.jpg** *Source:* https://upload.wikimedia.org/wikipedia/commons/5/57/Coloured-transition-metal-solutions.jpg *License:* Public domain *Contributors:* ? *Original artist:* ?

- **File:Commons-logo.svg** *Source:* https://upload.wikimedia.org/wikipedia/en/4/4a/Commons-logo.svg *License:* ? *Contributors:* ? *Original artist:* ?

- **File:Copper.jpg** *Source:* https://upload.wikimedia.org/wikipedia/commons/f/fb/Copper.jpg *License:* CC BY 3.0 *Contributors:* http://images-of-elements.com/copper.php *Original artist:* Jurii

- **File:Cosmic_History_020622_b.jpg** *Source:* https://upload.wikimedia.org/wikipedia/commons/1/1e/Cosmic_History_020622_b.jpg *License:* Public domain *Contributors:* http://map.gsfc.nasa.gov/media/020622/index.html *Original artist:* NASA / WMAP Science Team

- **File:Crab_Nebula.jpg** *Source:* https://upload.wikimedia.org/wikipedia/commons/0/00/Crab_Nebula.jpg *License:* Public domain *Contributors:* HubbleSite: gallery, release. *Original artist:* NASA, ESA, J. Hester and A. Loll (Arizona State University)

- **File:DIMendeleevCab.jpg** *Source:* https://upload.wikimedia.org/wikipedia/commons/c/c8/DIMendeleevCab.jpg *License:* Public domain *Contributors:* Transferred from ru.wikipedia
  Original artist: —. Original uploader was Serge Lachinov at ru.wikipedia

- **File:Discovery_of_chemical_elements.svg** *Source:* https://upload.wikimedia.org/wikipedia/commons/3/3d/Discovery_of_chemical_elements.svg *License:* CC BY-SA 3.0 *Contributors:* Wikimedia Commons. *Original artist:* Sandbh

- **File:Discovery_of_neon_isotopes.JPG** *Source:* https://upload.wikimedia.org/wikipedia/commons/e/e6/Discovery_of_neon_isotopes.JPG *License:* Public domain *Contributors:*
  Original artist: ?

- **File:Edelmetalle.jpg** *Source:* https://upload.wikimedia.org/wikipedia/commons/9/90/Edelmetalle.jpg *License:* CC-BY-SA-3.0 *Contributors:* de:Image:Edelmetalle.jpg. *Original artist:* de:User:Tomihahndorf.

- **File:Elbaite-121353.jpg** *Source:* https://upload.wikimedia.org/wikipedia/commons/a/ae/Elbaite-121353.jpg *License:* CC BY-SA 3.0 *Contributors:* http://www.mindat.org/photo-121353.html *Original artist:* Rob Lavinsky / iRocks.com

- **File:Electron_affinity_of_the_elements.svg** *Source:* https://upload.wikimedia.org/wikipedia/commons/6/6c/Electron_affinity_of_the_elements.svg *License:* CC BY-SA 3.0 *Contributors:* Based on Electron affinities of the elements 2.png by Sandbh. *Original artist:* DePiep

- **File:Electron_dot.svg** *Source:* https://upload.wikimedia.org/wikipedia/commons/0/02/Electron_dot.svg *License:* CC BY-SA 2.5 *Contributors:* ? *Original artist:* ?

- **File:Electron_shell_010_Neon_-_no_label.svg** *Source:* https://upload.wikimedia.org/wikipedia/commons/3/3e/Electron_shell_010_Neon_-_no_label.svg *License:* CC BY-SA 2.0 uk *Contributors:* http://commons.wikimedia.org/wiki/Category:Electron_shell_diagrams (corresponding labeled version) *Original artist:* commons:User:Pumbaa (original work by commons:User:Greg Robson)

- **File:Elemental_abundances.svg** *Source:* https://upload.wikimedia.org/wikipedia/commons/0/09/Elemental_abundances.svg *License:* Public domain *Contributors:* http://pubs.usgs.gov/fs/2002/fs087-02/ *Original artist:* Gordon B. Haxel, Sara Boore, and Susan Mayfield from USGS; vectorized by User:michbich

- **File:Elementspiral_(polyatomic).svg** *Source:* https://upload.wikimedia.org/wikipedia/commons/c/ce/Elementspiral_%28polyatomic%29.svg *License:* CC BY-SA 3.0 *Contributors:* Own work *Original artist:* DePiep

- **File:Empirical_atomic_radius_trends.png** *Source:* https://upload.wikimedia.org/wikipedia/commons/b/bc/Empirical_atomic_radius_trends.png *License:* GFDL *Contributors:* Own work *Original artist:* StringTheory11

- **File:Endohedral_fullerene.png** *Source:* https://upload.wikimedia.org/wikipedia/commons/e/e1/Endohedral_fullerene.png *License:* GFDL *Contributors:* Own work *Original artist:* Hajv01

- **File:Epidote_Oisans.jpg** *Source:* https://upload.wikimedia.org/wikipedia/commons/b/bf/Epidote_Oisans.jpg *License:* CC BY 4.0 *Contributors:* Own work *Original artist:* Didier Descouens

- **File:Ethanol-2D-skeletal.png** *Source:* https://upload.wikimedia.org/wikipedia/commons/f/fc/Ethanol-2D-skeletal.png *License:* Public domain *Contributors:* ? *Original artist:* ?

- **File:First_Ionization_Energy.svg** *Source:* https://upload.wikimedia.org/wikipedia/commons/1/1d/First_Ionization_Energy.svg *License:* CC BY-SA 3.0 *Contributors:* http://commons.wikimedia.org/wiki/File:Erste_Ionisierungsenergie_PSE_color_coded.png *Original artist:* User: Sponk

- **File:Folder_Hexagonal_Icon.svg** *Source:* https://upload.wikimedia.org/wikipedia/en/4/48/Folder_Hexagonal_Icon.svg *License:* Cc-by-sa-3.0 *Contributors:* ? *Original artist:* ?

- **File:FusionintheSun.svg** *Source:* https://upload.wikimedia.org/wikipedia/commons/7/78/FusionintheSun.svg *License:* CC BY-SA 3.0 *Contributors:* Own work *Original artist:* Borb

- **File:Glenn_Seaborg_-_1964.jpg** *Source:* https://upload.wikimedia.org/wikipedia/commons/4/47/Glenn_Seaborg_-_1964.jpg *License:* Public domain *Contributors:* NAIL Control Number: NWDNS-326-COM-12
  NARA (enter "Glenn Seaborg" in search form under Digital Copies tab) *Original artist:* Atomic Energy Commission. (1946 - 01/19/1975)

- **File:Gnome-searchtool.svg** *Source:* https://upload.wikimedia.org/wikipedia/commons/1/1e/Gnome-searchtool.svg *License:* LGPL *Contributors:* http://ftp.gnome.org/pub/GNOME/sources/gnome-themes-extras/0.9/gnome-themes-extras-0.9.0.tar.gz *Original artist:* David Vignoni

- **File:Gold-mz4b.jpg** *Source:* https://upload.wikimedia.org/wikipedia/commons/8/8c/Gold-mz4b.jpg *License:* CC BY-SA 3.0 *Contributors:* Image: http://www.irocks.com/db_pics/pics/mz4b.jpg, Description: http://www.mindat.org/photo-37436.html *Original artist:* Rob Lavinsky / iRocks.com

- **File:Goodyear-blimp.jpg** *Source:* https://upload.wikimedia.org/wikipedia/commons/2/2a/Goodyear-blimp.jpg *License:* Public domain *Contributors:* user-made *Original artist:* Derek Jensen (Tysto)

- **File:Grossular-ww51a.jpg** *Source:* https://upload.wikimedia.org/wikipedia/commons/7/71/Grossular-ww51a.jpg *License:* CC BY-SA 3.0 *Contributors:* Image: http://www.irocks.com/db_pics/pics/ww51a.jpg, Description: http://www.mindat.org/photo-121044.html *Original artist:* Rob Lavinsky / iRocks.com

- **File:HEUraniumC.jpg** *Source:* https://upload.wikimedia.org/wikipedia/commons/d/d8/HEUraniumC.jpg *License:* Public domain *Contributors:* http://web.archive.org/web/20050829231403/http://web.em.doe.gov/takstock/phochp3a.html *Original artist:* ?

21.9.3   Content license

- Creative Commons Attribution-Share Alike 3.0